RATIONALITY IN EXTENSIVE FORM GAMES

THEORY AND DECISION LIBRARY

General Editors: W. Leinfellner (*Vienna*) and G. Eberlein (*Munich*)

Series A: Philosophy and Methodology of the Social Sciences

Series B: Mathematical and Statistical Methods

Series C: Game Theory, Mathematical Programming and Operations Research

Series D: System Theory, Knowledge Engineering an Problem Solving

SERIES C: GAME THEORY, MATHEMATICAL PROGRAMMING AND OPERATIONS RESEARCH

VOLUME 29

Scope: Particular attention is paid in this series to game theory and operations research, their formal aspects and their applications to economic, political and social sciences as well as to socio-biology. It will encourage high standards in the application of game-theoretical methods to individual and social decision making.

The titles published in this series are listed at the end of this volume.

RATIONALITY IN EXTENSIVE FORM GAMES

by

ANDRÉS PEREA
University of Maastricht, The Netherlands and
Universidad Carlos III de Madrid, Spain

KLUWER ACADEMIC PUBLISHERS
BOSTON / DORDRECHT / LONDON

A C.I.P. Catalogue record for this book is available from the Library of Congress.

ISBN 978-1-4419-4918-9

Published by Kluwer Academic Publishers,
P.O. Box 17, 3300 AA Dordrecht, The Netherlands.

Sold and distributed in North, Central and South America
by Kluwer Academic Publishers,
101 Philip Drive, Norwell, MA 02061, U.S.A.

In all other countries, sold and distributed
by Kluwer Academic Publishers,
P.O. Box 322, 3300 AH Dordrecht, The Netherlands.

Printed on acid-free paper

To Cati

Contents

Preface

I would like to use this preface to thank some persons and institutions which have been important during the various stages of writing this book. First of all, I am grateful to Kluwer Academic Publishers for giving me the opportunity to write this book. I started writing the book in 1998 while I was working at the Departament d'Economia i d'Història Econòmica at Universidad Autònoma de Barcelona, and continued the writing job from september 1998 to september 2000 at the Departamento de Economía at Universidad Carlos III de Madrid. The book has been completed while I was visiting the Department of Quantitative Economics at the University of Maastricht from october 2000 to august 2001. I wish to thank these three departments for their hospitality. The book has improved substantially by comments and critique from the following persons who have read parts of the manuscript: Geir Asheim, Eric van Damme, János Flesch, Mari-Angeles de Frutos, Diego Moreno, Hans Peters, Antonio Romero and Dries Vermeulen. I should also mention my discussions with Peter Wakker about the decision-theoretic foundations of noncooperative game theory, which have had an important impact on various parts of the book. Finally, I wish to express my warmest gratitude to my parents, my brother and my sister, and, last but not least, to Cati, to whom I dedicate this book.

Maastricht, July 2001

Chapter 1

Introduction

Noncooperative game theory is concerned with situations where several persons, with possibly different interests, reach decisions independently, and where the final consequence depends upon each of the decisions chosen. In addition, the outcome of the decision making process may be influenced by events that are beyond the decision makers' control, such as random events determined by nature. We refer to such environments as *noncooperative games*, and the decision makers are called *players*. The basic question underlying noncooperative game theory is: which decisions may be viewed "rational" in a given noncooperative game? As far as one-person decision making is concerned, "rationality" reflects the situation where the decision maker holds some preference relation over the decisions available to him, and chooses a most preferred decision. The same principle could be applied to noncooperative games by assuming that each player holds a preference relation over his decisions, and by imposing that players reach optimal decisions, given their preferences. There is, however, an important distinction between one-person decision problems and noncooperative games that complicates a purely decision theoretic approach to the latter. The problem is that every player in a noncooperative game, when evaluating his decisions, should take into account that his opponents will act in their own interest, given *their* preferences. As such, a theory of rational decision making in noncooperative games should not only require players to act optimally given their preferences, but should also impose that the players' preferences be compatible with optimal behavior by the opponents.

In order to illustrate this distinction, consider the following simple example. Suppose that a decision maker has to choose one of two urns, say urn 1 and urn 2. He knows that in one of the urns there is a black ball, and in the other urn there is a white ball, but he does not know in which urn the black ball is. We may assume that a referee distributes the balls before the decision maker chooses. If the decision maker chooses urn 1 and finds the black ball, he earns 10 euros. If he chooses urn 1 and finds the white ball, he should pay 10 euros. If he chooses urn 2 and finds the black ball, he should pay 9 euros. Finally, if he chooses urn 2 and finds the white ball, he earns 1 euro. This decision problem may be represented by the following table.

	(b, w)	(w, b)
urn 1	10	-10
urn 2	1	-9

Here, (b, w) represents the configuration where the black ball is in urn 1 and the white ball is in urn 2, whereas (w, b) represents the opposite configuration. The numbers in the table are the payoffs to the decision maker for each decision and each possible configuration of the balls. In this problem, it seems reasonable for the decision maker to prefer urn 1 over urn 2. Intuitively, such preferences arise if the decision maker "does not consider it too unlikely" that the black ball is in urn 1. In turn, such beliefs about the configuration of the balls seem perfectly legitimate since the referee may be assumed indifferent between the two configurations.

Now, assume that in the decision problem above, the referee is to *pay* the respective amount to the decision maker if the latter earns money, whereas the referee *receives* the respective amount in case the decision maker has to pay. The payoffs for both the decision maker and the referee are represented by table below.

	(b, w)	(w, b)
urn 1	$10, -10$	$-10, 10$
urn 2	$1, -1$	$-9, 9$

The first payoff corresponds to the decision maker, whereas the second payoff corresponds to the referee. As before, the referee selects the configuration of the balls before the decision maker chooses the urn. In this case, it does no longer seem reasonable for the decision maker to prefer urn 1. The decision maker, namely, should realize that it is always in the referee's interest to choose the configuration (w, b) since, irrespective of the choice of the urn, this configuration yields a higher payoff to the referee than (b, w). If the decision maker indeed believes that the referee chooses (w, b), then he should prefer urn 2 and not urn 1. In other words, the decision maker's preference relation in which he prefers urn 1 is not compatible with optimal behavior by the referee.

It is the aim of this monograph to present rationality criteria for noncooperative games that reflect the principle above, namely that players are to base their preferences upon the conjecture that each of their opponents will behave optimally. Within this context, we restrict our attention to so-called *finite extensive form games*. By the latter, we mean noncooperative games for which the rules of decision making may be represented by a directed, rooted tree with a finite number of nodes and edges (see Section 2.1), allowing for an explicit dynamic structure in the decision making process. We shall assume throughout that the players' preferences over their decisions may be represented by a *subjective probability distribution* on the opponents' decisions, and a *utility function* assigning some real number to each of the possible consequences. Informally speaking, the subjective probability distribution may be interpreted as some measure of likelihood which the player assigns to each of the opponents' decisions, whereas the utility of a consequence may be seen as some measure for the "desirability" of this particular consequence. We refer to player i's subjective probability distribution as player i's *conjecture about the opponents' behavior.*

Each of the rationality criteria discussed in this book takes the players' utility functions as given, and imposes restrictions solely on the players' subjective probability distributions on the opponents' decisions. A common feature of the different criteria is that they require players to only assign positive probability to "optimal decisions" by the opponents. The various criteria differ, however, in their formalization of this requirement. It may happen, for instance, that a player holds some conjecture about the opponents' behavior at the beginning of the game, but that this conjecture is contradicted during the course of the game. In this case, the player is asked to revise his conjecture about the opponents' behavior, as to make his conjecture compatible with the events observed so far. Some questions which arise are: (1) should the player's decision also be optimal with respect to his *revised* conjecture or not? and (2) should the player, when revising his conjecture, still assume that his opponents are acting optimally or not? We shall see that the different rationality concepts discussed here will provide different answers to these, and other, questions.

The outline of this monograph is as follows. In Chapter 2 we lay out some basic definitions that will be used throughout the book, such as the definition of an extensive form game, the players' conjectures about the opponents' behavior and a way to model the revision of such conjectures during the course of the game. In Chapters 3 to 6 it is assumed that the players are informed about the conjectures held by the opponents about the other players' behavior. Within this particular framework, we present various rationality criteria that put restrictions on the players' conjectures and revision of conjectures. There are three principles which have been highly influential for the development of these criteria: *backward induction, Nash equilibrium* and *forward induction.* Chapter 3 discusses rationality criteria which are based on the principles of backward induction and Nash equilibrium, or some combinations of these. In Chapter 4 we focus on the concept of *sequential rationality* which may be viewed as some generalization of backward induction to the general class of extensive form games. Chapter 5 analyzes a series of concepts which provide different formalizations of the forward induction principle. Chapter 6 is devoted to the question how different rationality concepts react to certain transformations of extensive form games. In Chapter 7 it is no longer assumed that players are informed about the conjectures held by the opponents about the other players' behavior. We explore the consequences of dropping this assumption by discussing some rationality concepts designed for this particular context, such as *rationalizability* and *extensive form rationalizability.*

Let me conclude the introduction with some remarks on the scope and limitations of this book. The monograph does by no means offer a comprehensive overview of rationality concepts in extensive form games, since the literature on this field is simply too large to be covered completely. Instead, I have concentrated on a rather subjective selection of topics, realizing that many interesting issues, such as communication, evolutionary stability, learning and bounded rationality, are left untouched. The material presented in this book, however, is mostly self-contained in the sense that complete proofs are given for almost all results. At the few places where no proof is given, the reader is referred to the paper or book where this proof can be found. The mathematical prerequisite for this book is fairly mild, since for Chapters 2 to

6 only some elementary knowledge of calculus, linear algebra and probability theory is required, whereas Chapter 7, in addition, uses some elementary concepts from measure theory.

Chapter 2

Extensive Form Games

In this chapter we present the model of extensive form games that will be used throughout this book. As a first step, Section 2.1 introduces a formal way to represent the "rules" of the game, which we refer to as the *extensive form structure* of a game. These rules specify the possible decisions which are available to the players at each point in time, the order in which decisions are to be taken, and the information available to the players whenever they are called upon to move. Before the game starts each player makes a decision plan, specifying what to do at every situation that can potentially occur during the course of the game. Such decision plans are called *strategies*. As in one-person decision problems, each player is assumed to hold some preference relation on his set of strategies, and *rational* players are expected to choose a most preferred strategy. Throughout the book, we make the assumption that the preferences of a player may be characterized by two components: (1) a subjective probability distribution held by this player on the opponents' strategy choices, and (2) a utility function for this player assigning to every possible outcome of the game some real number. We refer to (1) as the player's *conjecture* about the opponents' behavior, and to (2) as the *utility function* for this player. Section 2.2 provides the decision-theoretical tools necessary to formally represent the participants in an extensive form game as decision makers under uncertainty. In Section 2.3 these tools are applied to our specific context of extensive form games. In Section 2.4 we discuss an alternative way of representing the players' conjectures, namely by means of *local* subjective probability distributions. Section 2.5 is concerned with the question how players revise their conjectures about the opponents' behavior if their initial conjectures have been falsified by the course of the game. To this purpose, we apply the notion of local subjective probability distributions introduced in Section 2.4. Finally, in Section 2.6 we investigate different notions of "optimal strategies" and "optimal actions" against conjectures about the opponents' behavior.

2.1 Extensive Form Structures and Strategies

We start this chapter by laying out the rules of an extensive form game, gathered in a mathematical object called the *extensive form structure* of a game. Kuhn (1953) and von Neumann and Morgenstern (1947) were probably the first to provide a general model for the rules of an extensive form game, and this section is mainly based upon their seminal work. As in Kuhn (1953), we model the timing of the decisions by means of a *directed rooted tree*. The latter is an object (X, E) consisting of a finite set X of *nodes* and a finite set E of *directed edges* between nodes in X. Every directed edge e may be represented by an ordered pair (x, y) with $x, y \in X, x \neq y$, meaning that e is an edge which goes from x to y. We assume that if there is some edge from x to y, then there is no edge from y to x. We say that there is a *path* from node x to node y if there is a sequence of neighbouring edges $(x_0, x_1), (x_1, x_2), (x_2, x_3), ..., (x_{n-1}, x_n)$ in E with $x_0 = x$ and $x_n = y$. In this case, we say that y *follows* x. In a directed rooted tree, there is some node $x_0 \in X$ such that for every $x \in X$ there is a unique path from x_0 to x. It may be verified easily that in this case, there is a *unique* node x_0 with this property. This node x_0 is called the *root* of (X, E). In a directed rooted tree, we therefore not only have that each node is connected with the root, but moreover each node may be identified with the unique path that goes from the root to this node.

The game-theoretical interpretation of the directed rooted tree (X, E) is as follows. The root x_0 represents the beginning of the game. Every node that is followed by some other node is called a *non-terminal node*, whereas the remaining nodes are called *terminal nodes*. The set of terminal nodes is denoted by Z. Every non-terminal node represents a state in which either a player has to reach a decision or a *chance move* occurs. In the former case, the node is called a *decision node*, and in the latter case we speak about a *chance node*. By a chance move we mean an event whose outcome can not be influenced by the players, but which influences the final outcome of the game. One could think, for instance, of a referee tossing a coin. Chance moves are also called *moves of nature*.

If the node is a *decision node*, the set of edges leaving this node represents the set of possible decisions or *actions* available to the player at this state. In the case of a *chance node*, the outgoing edges are the possible outcomes of the chance move (heads or tails). Every action or outcome of a chance move thus moves the game from one node to another, until a terminal node is reached. In this case, the game ends here.

The information structure in the game is modeled by so-called *information sets*, which are collections of non-terminal nodes belonging to the same player. An information set should be interpreted as a collection of states among which the corresponding player can not distinguish a-priori. Or, put in another way, if one of the nodes in this information set is reached, the player knows that the game has reached one of the nodes in this information set, without knowing exactly which node (if the information set contains more than one node, of course). Since a player should know his set of available actions each time he is called upon to move, the set of possible actions should be the same for every node belonging to the same information set. Otherwise, the player could extract extra information about the true node from the set of possible actions, implying that the player knows more than the information set tells him, which would be a contradiction. If the information set contains only one node, the player knows that this node is reached and is therefore completely informed about everything that has occurred in the game so far. The formal definition of an extensive form structure is as follows.

Definition 2.1.1 *An extensive form structure is a tuple $\mathcal{S} = (T, P, \mathcal{H}, \mathcal{A}, \tau)$ where.*
(1) $T = (X, E)$ is a directed rooted tree with a finite set of nodes X, a finite set of directed edges E and root x_0. By $E(x)$ we denote the set of edges leaving the node x. By Z we denote the set of terminal nodes.
(2) I is the finite set of players and $P : X \backslash Z \to I \cup \{N\}$ is the player labelling function assigning to every non-terminal node $x \in X \backslash Z$ some label $P(x) \in I \cup \{N\}$. Here, $P(x) = N$ indicates that a chance move occurs at node x, whereas $P(x) \in I$ indicates the player that has to move at x. For every player i, let X_i be the set of nodes at which player i has to move.
(3) $\mathcal{H} = (H_i)_{i \in I}$ where for each i, H_i is a partition of the set of nodes X_i controlled by player i. The sets $h \in H_i$ are called information sets for player i. Every information set h should satisfy the following conditions: (a) every path in T intersects h at most once and (b) every node in h has the same number of outgoing edges.
(4) $\mathcal{A} = (A(h))_{h \in H}$, where for each information set h, $A(h)$ is a partition of the set of edges $\cup_{x \in h} E(x)$ leaving the information set h. Every set $a \in A(h)$ is called an action at h. The partition $A(h)$ should be such that $E(x) \cap a$ contains exactly one edge for every action $a \in A(h)$ and every $x \in h$. We assume that there are at least two actions at every information set.
(5) τ is a function assigning to every chance node a strictly positive probability distribution on the set of outgoing edges. Every outgoing edge c should be interpreted as a possible outcome of the chance move and $\tau(c)$ is the objective probability that nature chooses c.

Here, a *probability distribution* on a finite set Y is a function $p : Y \to [0, 1]$ with $\sum_{y \in Y} p(y) = 1$. We say that p is *strictly positive* if $p(y) > 0$ for all $y \in Y$. Condition (a) in (3) guarantees that a player can always distinguish states which occurred *before* he took some action from states that will occur *after* taking this action. Suppose, namely, that the game reaches an information set and the corresponding player chooses an action here. Since condition (a) prohibits the game tree to cross this information set twice, the player can distinguish all future nodes from the nodes in this information set. Condition (b) in (3) simply makes sure that the number of available actions is the same for all nodes belonging to the same information set. Part (4) states that an action at an information set with several nodes formally corresponds to a collection of outgoing edges: one edge for each node of the information set. By $H = \cup_{i \in I} H_i$ we denote the collection of all information sets, whereas $A = \cup_{h \in H} A(h)$ is the collection of all actions. In order to illustrate the definition of an extensive form structure, consider Figure 2.1. The tree should be read from the left to the right. The numbers at the nodes indicate the player that controls this node, whereas the letter N indicates that a chance move occurs at this node. The label at an edge leaving a decision node specifies the action to which this edge corresponds. If the node is a chance node, the probabilities for the outcomes of the chance move are written at the corresponding outgoing edges. If an information set contains more than one node, the nodes in

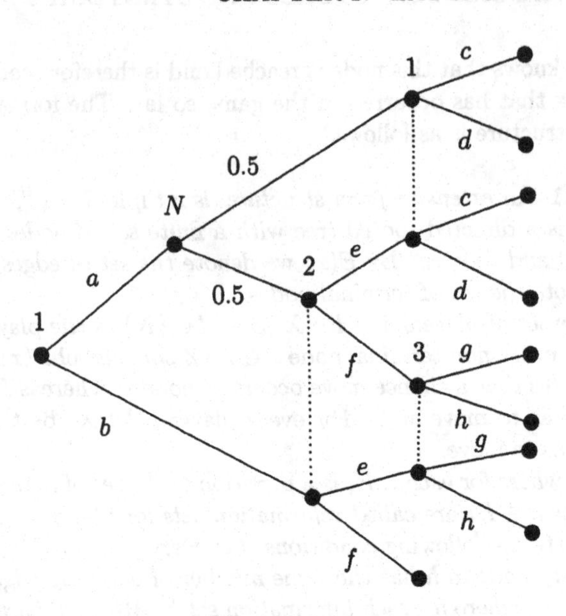

Figure 2.1

this information set are connected by a dashed line. In this case, all nodes in the information set are controlled by the same player and the player label is written only at the upper node. The game starts with player 1, who can choose among the actions a and b. If he chooses b, player 2's information set is reached. Since this information set contains two nodes, player 2 does not know whether the upper or the lower node has been reached. The upper node represents the event in which player 1 has chosen a and, subsequently, nature has chosen the lower outcome, whereas the lower node should be interpreted as the event in which player 1 has chosen b. Player 2, upon reaching his information set, therefore knows than one of both events has taken place, without knowing which one.

An additional assumption we impose upon the extensive form structure is that it satisfies *perfect recall,* as defined in Kuhn (1953). Intuitively, perfect recall states that a player, at each of his information sets, always remembers the actions he took at previous information sets. Since by definition actions available at different information sets are different, perfect recall implies that a player, at each of his information sets h, always remembers which of his previous information sets have been reached by the course of the game.

Definition 2.1.2 *An extensive form structure S is said to satisfy perfect recall for player i if for every information set $h \in H_i$ and every pair of nodes $x, y \in h$ the path from the root to x contains the same collection of player i actions as the path from the root to y. We say that S satisfies perfect recall if it satisfies perfect recall for all players i.*

It may be verified that the extensive form structure in Figure 2.1 satisfies perfect

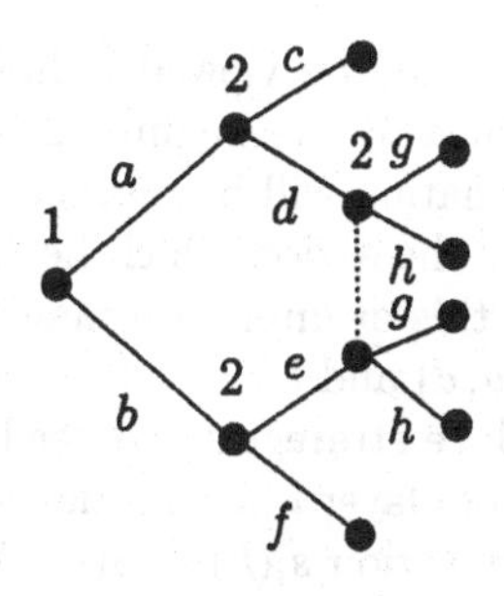

Figure 2.2

ecall. We need only check player 1 here, since players 2 and 3 control only one nformation set. Player 1's second information set contains two nodes, say x and . The path to x and the path to y contain the same set of actions for player 1, amely $\{a\}$. Therefore, perfect recall is satisfied. Figure 2.2 presents an extensive form tructure that does not satisfy perfect recall. Consider player 2's last information set. et the nodes in this information set be x and y, where x is the upper node. Then, the ath to x contains the set of player 2 actions $\{d\}$ whereas the path to y contains the et of player 2 actions $\{e\}$. Since both are different, perfect recall is violated. In this xample, player 2, when called upon to play at his last information set, has forgotten nformation about player 1's behavior that he previously possessed. Namely, if player chooses between a and b at the beginning of the game, player 2 observes the action hosen by player 1 since the decision nodes which immediately follow a and b are in ifferent player 2 information sets. However, at his last information set player 2 no onger remembers the action chosen by player 1. To see this, assume that player 2, t his final information set, would still know which action player 1 had chosen. In his case, player 2 could distinguish node x from node y since x is preceded by action , whereas node y is preceded by action b. But then, x and y should be in different nformation sets.

Now that we have clarified the rules of an extensive form game, we turn to the uestion how players plan their decisions in the game. Such decision plans will be alled *strategies*. In an extensive form game, each player i controls the collection H_i of nformation sets, which means that, whenever some information set $h \in H_i$ is reached y the game, player i should choose some available action $a \in A(h)$. It may therefore eem that a strategy for player i should assign an action $a \in A(h)$ to *each* information et $h \in H_i$. A problem with this definition is that, in many cases, it would contain redundant" information. Consider again the game in Figure 2.1, and consider the rofile of actions (b, c) for player 1. This profile represents the decision plan where layer 1 chooses action b at the beginning of the game, and chooses c if the game ould reach his second information set. However, if player 1 is confident to carry out his decision plan correctly, he knows that after choosing b his second information et can no longer be reached. Hence, the information that player 1 would choose

c at his second information set may be viewed redundant in this case. However, if player 1 plans to choose action a at the beginning, then he can no longer exclude the event that his second information will be reached, and hence a complete decision plan should in this case specify the action which he would choose in case his second information set is reached. In this example, we may thus distinguish three different *strategies* for player 1: $(a,c),(a,d)$ and b.

In order to formally introduce strategies, we need the following definition. Consider some collection $H_i^* \subseteq H_i$ of player i information sets, and a function s_i assigning to every $h \in H_i^*$ some available action $s_i(h) \in A(h)$. We say that an information set $h \in H_i$ is *avoided* by s_i if h cannot be reached if player i acts according to s_i. Formally, h is avoided by s_i if for every node $x \in h$ the path from the root to x crosses some information set $h' \in H_i^*$ at which the prescribed action $s_i(h')$ deviates from this path. Consider again the game of Figure 2.1. Let h_1 and h_2 be the first and second information set for player 1, respectively. Let $H_1^* = \{h_1\}$ and $s_1(h_1) = b$. Then, the path leading to the upper node in h_2 crosses $h_1 \in H_1^*$, at which the prescribed action $b = s_1(h_1)$ deviates from this path. The same holds for the path leading to the lower node in h_2. Therefore, h_2 is avoided by s_1.

Definition 2.1.3 *A strategy for player i is a function $s_i : H_i^* \to A$ where (1) $H_i^* \subseteq H_i$ is a collection of player i information sets, (2) $s_i(h) \in A(h)$ for all $h \in H_i^*$, (3) every $h \in H_i^*$ is not avoided by s_i and (4) every $h \in H_i \backslash H_i^*$ is avoided by s_i.*

Hence, a strategy for player i specifies player i's behavior at all information sets that are not avoided by the same strategy. Let S_i be the set of strategies for player i. Obviously, all sets S_i are finite since there are only finitely many information sets and actions in the game. The definition of a strategy applied here coincides with the notion of a *plan of action*, as discussed in Rubinstein (1991). The usual definition of a (pure) strategy that can be found in the literature specifies an action at *each* of player i information sets; also at those which are avoided by the same strategy. The particular definition used, however, will have no impact on the results presented in this book.

2.2 Decision-theoretic Framework

The main purpose of this book is to present, and analyze, several criteria of "rational decision making" for players in an extensive form game. In a one-person decision problem, rational decision making reflects the fact that the person has some preference relation on the set of possible decisions, and chooses a decision which is optimal given his preferences. If we wish to translate this principle to extensive form games, then a first necessary step would be to establish a formal decision-theoretic framework for such games. To this purpose, we first present an abstract model of one-person decision problems. Within this model we focus on those decision maker's preference relations that are of the so-called *subjective expected utility* type. In the following section, we apply the framework to the particular context of extensive form games. The model presented in this section is based upon work by Ramsey (1931), von Neumann and Morgenstern (1947), Savage (1954, 1972) and Anscombe and Aumann (1963).

Consider a decision maker who faces the problem of reaching a decision under uncertainty. Assume that the consequence of a decision depends upon uncertain events that are beyond the decision maker's control. We distinguish two types of uncertain events here. The first type contains random events for which the probabilities of the possible outcomes are objectively known. One may think of throwing a dice, or spinning a roulette wheel. Such events are called *objective unknowns*. The second type contains uncertain events to which no obvious objective probabilities can be assigned. Think, for instance, of events that are the result of a decision making process by other people. These events are called *subjective unknowns*.

Formally, let C be some finite set, representing the consequences that can possibly be induced by the decisions available. Let $\Delta(C)$ be the set of objective probability distributions on C, that is,

$$\Delta(C) = \{p : C \to [0,1] | \textstyle\sum_{c \in C} p(c) = 1\}.$$

Let Ω be some finite set of *states*, representing the subjective unknowns relevant for the decision problem. The fact that the consequence of a decision depends on the realization of objective and subjective unknowns is modeled by identifying each decision with some function $f : \Omega \to \Delta(C)$. Such functions, which assign to every state ω some objective probability distribution $f(\omega) \in \Delta(C)$ on the possible consequences, are called *acts*, and may be interpreted as follows. Suppose that we identify some decision d with an act f. Then, for a given state ω and consequence c, the number $f(\omega)(c)$ represents the objective probability that c will happen, if decision d is taken and the realized state is ω. Let $\mathcal{F} = \{f : \Omega \to \Delta(C)\}$ be the set of all acts.

In Savage (1954, 1972), an act is defined as a function which assigns to every state ω some *deterministic* outcome $f(\omega)$, and can thus be seen as a special case within our approach in which no objective unknowns are involved. In Anscombe and Aumann (1963), acts in the sense of Savage are called *horse lotteries*. In the special case where there is only one state, an act can be identified with an objective probability distribution on the set of consequences. Such acts are called *roulette lotteries* in Anscombe and Aumann (1963). The general definition of an act used here, which involves both objective and subjective unknowns, corresponds to the definition of *compound horse lotteries* in Anscombe and Aumann (1963).

The decision maker is assumed to have some preference relation $\succeq$ on the set of acts $\mathcal{F}$. Since every decision available may be identified with some act in $\mathcal{F}$, this implies that, in particular, the decision maker holds a preference relation on the set of possible decisions. Formally, a preference relation $\succeq$ on some set A is a subset $R \subseteq A \times A$, where $(a,b) \in R$ is to be interpreted as "the decision maker strictly prefers a to b, or is indifferent between both". Instead of $(a,b) \in R$ we write $a \succeq b$. If $a \succeq b$ and $b \succeq a$, we say that the decision maker is indifferent between a and b, written as $a \sim b$. If $a \succeq b$ but not $b \succeq a$, then we say that he strictly prefers a over b, denoted by $a \succ b$. Basic assumptions we impose on $\succeq$ are (A.1) $a \succeq a$ for all $a \in A$, (A.2) $a \succeq b$ and $b \succeq c$ imply $a \succeq c$ for all $a, b, c \in A$ and (A.3) for every $a, b \in A$ either $a \succeq b$ or $b \succeq a$. Conditions (A.1), (A.2) and (A.3) are called reflexiveness, transitivity and completeness, respectively.

Definition 2.2.1 *We say that the preference relation* $\succeq$ *on* $\mathcal{F}$ *is of the subjective* *expected utility type if there exists some probability distribution $\mu \in \Delta(\Omega)$ and some function $u : C \to \mathbb{R}$ such that for every $f, g \in \mathcal{F}$ it holds: $f \succeq g$ if and only if*

$$\sum_{\omega \in \Omega} \mu(\omega) \sum_{c \in C} f(\omega)(c)\ u(c) \geq \sum_{\omega \in \Omega} \mu(\omega) \sum_{c \in C} g(\omega)(c)\ u(c).$$

Here, μ is called a *subjective probability distribution* on the set of states, whereas u is called a *utility function* on the set of consequences. We say that the preference relation is *represented* by the functions μ and u. For a given μ, u, let $E_{\mu,u} : \mathcal{F} \to \mathbb{R}$ be the function defined by

$$E_{\mu,u}(f) = \sum_{\omega \in \Omega} \mu(\omega) \sum_{c \in C} f(\omega)(c)\ u(c).$$

We say that $E_{\mu,u}(f)$ is the *subjective expected utility* of act f generated by μ and u. A preference relation of the subjective expected utility type thus ranks the acts according to the subjective expected utility they induce. In the literature, conditions on the preference relation $\succeq$ can be found which guarantee that $\succeq$ is of the subjective expected utility type. For the sake of brevity, we do not discuss such conditions here. The interested reader may consult Myerson (1991), among many others.

A natural question which arises is whether the functions μ and u, representing the preference relation $\succeq$, are uniquely determined. The following theorem states that, under some regularity condition on the preference relation, the subjective probability distribution μ is unique, whereas the utility function u is unique up to positive affine transformations.

Theorem 2.2.2 *Let the preference relation $\succeq$ be of the subjective expected utility type, and let there be some $f, g \in \mathcal{F}$ with $f \succ g$. Let $\succeq$ be represented by some subjective probability distribution $\mu \in \Delta(\Omega)$ and some utility function $u : C \to \mathbb{R}$. Let $\tilde{\mu}$ and $\tilde{u}$ be such that they also represent $\succeq$. Then, $\tilde{\mu} = \mu$ and there exist numbers $A > 0$, $B \in \mathbb{R}$ such that $\tilde{u}(c) = A\ u(c) + B$ for all $c \in C$.*

The proof for this result is omitted, and can be found, for instance, in Myerson (1991). Given that the function μ is uniquely determined by $\succeq$, and u is "almost" uniquely determined, it would be interesting to see whether some intuitive interpretation can be assigned to the subjective probability distribution μ and the utility function u. The following procedure, which can be used to "extract" the functions μ and u from a decision maker with preferences of the subjective expected utility type, may shed some light on this. Suppose that a decision maker has a preference relation $\succeq$ of the subjective expected utility type. Assume that there is a most preferred act f^1 and a least preferred act f^0, that is, $f^1 \succeq f \succeq f^0$ for all $f \in \mathcal{F}$. In order to exclude uninteresting cases, assume that $f^1 \succ f^0$. For every $\lambda \in [0, 1]$, let $\lambda f^1 + (1 - \lambda) f^0$ be the act f given by

$$f(\omega)(c) = \lambda f^1(\omega)(c) + (1 - \lambda) f^0(\omega)(c)$$

for every $\omega \in \Omega, c \in C$. For a given $c \in C$, let f^c be the act given by

$$f^c(\omega)(c') = \begin{cases} 1, & \text{if } c' = c \\ 0, & \text{otherwise} \end{cases}$$

for all $\omega \in \Omega$. Hence, f^c always leads to consequence c for sure. For every $c \in C$, let $u(c) \in [0,1]$ be such that

$$f^c \sim u(c) f^1 + (1 - u(c)) f^0. \tag{2.1}$$

For a given state $\omega \in \Omega$, let the act f^ω be given by

$$f^\omega(\omega') = \begin{cases} f^1, & \text{if } \omega' = w \\ f^0, & \text{otherwise} \end{cases}$$

The act f^ω thus coincides with the most preferred act at state ω, and coincides with the least preferred act at all other states. For all $\omega \in \Omega$, let $\mu(\omega) \in [0,1]$ be such that

$$f^\omega \sim \mu(\omega) f^1 + (1 - \mu(\omega)) f^0. \tag{2.2}$$

Then, it can be shown that $\mu \in \Delta(\Omega)$ and, moreover, the functions u and μ obtained by this procedure represent the preference relation $\succeq$. Equation (2.1) suggests that the utility $u(c)$ of some consequence c may be interpreted as the "ranking" of act f^c between the most preferred and least preferred act. More precisely, it is the minimal weight λ one has to attach to f^1 in $\lambda f^1 + (1-\lambda) f^0$ in order to make $\lambda f^1 + (1-\lambda) f^0$ as least as good as f^c. In view of (2.2), the subjective probability $\mu(\omega)$ may be interpreted as a "measure of likelihood" for the event that the true state is ω. Namely, if $\mu(\omega)$ increases then, according to (2.2), f^ω becomes "more similar" to the most preferred act f^1. Since f^ω and f^1 coincide only at state ω, this suggest that the state ω becomes "more likely".

2.3 Conjectures about Opponents' Behavior

In this section we apply the decision-theoretic model developed in the previous section to the particular context of extensive form games. Formally speaking, we wish to represent the decision problem for each player i by some tuple $(D_i, C_i, \Omega_i, \mathcal{F}_i, g_i, \succeq_i)$ where D_i is the set of possible decisions, C_i is the set of possible consequences, Ω_i is the set of subjective unknowns which may influence the consequence of each decision, $\mathcal{F}_i$ is the set of acts $f_i : \Omega_i \to \Delta(C_i)$, $g_i : D_i \to \mathcal{F}_i$ is the function assigning to every decision $d_i \in D_i$ the act $g_i(d_i) \in \mathcal{F}_i$ with which d_i is to be identified, and $\succeq_i$ is player i's preference relation on $\mathcal{F}_i$. The latter preference relation induces, in particular, a preference relation on the possible decisions D_i.

We assume that the objects of choice for player i are his strategies in S_i, and hence we set $D_i = S_i$. In an extensive form game, the possible consequences of a strategic interaction among the players are represented by the set Z of terminal nodes in the game tree. Therefore, we choose $C_i = Z$ for every player i. The consequence of each strategy $s_i \in S_i$, that is, the terminal node reached by choosing s_i, depends on two uncertain events: the strategies chosen by the other players, and the realization of the chance moves during the game. Since no obvious objective probabilities can be assigned to the former event, we choose $\Omega_i = \times_{j \neq i} S_j$ as the set of subjective unknowns for player i. Here, $\times_{j \neq i} S_j$ is the set of strategy profiles $(s_j)_{j \neq i}$ that may possible be chosen by player i's opponents $j \in I \backslash \{i\}$. In the sequel we write $S_{-i} = \times_{j \neq i} S_j$ and the opponents' strategy profiles in S_{-i} are denoted by s_{-i}. On the other hand, each player is assumed to know the objective probabilities of each chance move, which means that chance moves are interpreted as objective unknowns for each player i. Every strategy s_i is now to be identified with some act $f_i : S_{-i} \to \Delta(Z)$ through the function $g_i : S_i \to \mathcal{F}_i$. In order to specify this function g_i, we need to know, for each strategy s_i, and each opponents' strategy profile $s_{-i} \in S_{-i}$, the objective probability distribution on Z induced by s_i and s_{-i}. To this purpose, we introduce the following notation.

Let the strategy s_i and the opponents' strategy profile s_{-i} be given. Write $s = (s_i, s_{-i}) = (s_j)_{j \in I}$. For a given player $j \in I$, information set $h \in H_j$ and action $a \in A(h)$, define

$$p_{(s_i,s_{-i})}(a) = \begin{cases} 1, & \text{if } h \text{ is not avoided by } s_j \text{ and } s_j(h) = a \\ 0, & \text{otherwise.} \end{cases}$$

For every terminal node $z \in Z$ in the game tree, let A_z be the set of actions on the path from the root x_0 to z, and let C_z be the set of outcomes of chance moves on this path. As an illustration, consider again the game in Figure 2.1. Let z be the third terminal node from above. Then, A_z contains the actions a for player 1, e for player 2 and c for player 1, whereas C_z contains the lower outcome of the chance move. For every terminal node z, define

$$\mathbb{P}_{(s_i,s_{-i})}(z) = \prod\nolimits_{a \in A_z} p_{(s_i,s_{-i})}(a) \prod\nolimits_{c \in C_z} \tau(c),$$

where $\tau(c)$ is the objective probability that outcome c is chosen at the corresponding chance move. It may easily be verified that $\sum_{z \in Z} \mathbb{P}_{(s_i,s_{-i})}(z) = 1$, and hence $\mathbb{P}_{(s_i,s_{-i})}$ constitutes a probability distribution on Z. Intuitively, $\mathbb{P}_{(s_i,s_{-i})}(z)$ is the objective probability that terminal node z is reached if players act according to (s_i, s_{-i}). In the example of Figure 2.1, consider the strategy profile (s_1, s_{-1}) in which player 1 chooses the strategy (a, d), player 2 chooses f and player 3 chooses g. Then, the induced probability distribution $\mathbb{P}_{(s_1,s_{-1})} \in \Delta(Z)$ assigns probability 0.5 to the second and fifth terminal node from above, and assigns probability zero to all other terminal nodes.

Now, the function $g_i : S_i \to \mathcal{F}_i$ is defined as follows. For a given strategy s_i, $g_i(s_i)$ is the act $g_i(s_i) : S_{-i} \to \Delta(Z)$ which assigns to every opponents' strategy profile s_{-i} the objective probability distribution $\mathbb{P}_{(s_i,s_{-i})} \in \Delta(Z)$ induced by s_i and s_{-i}.

We assume that each player i holds a preference relation $\succeq_i$ on $\mathcal{F}_i$ which is of the subjective expected utility type. Then, by definition, there is some subjective probability distribution $\mu^i \in \Delta(S_{-i})$ and some utility function $u_i : Z \to \mathbb{R}$ such that for every two acts $f_i, f_i' \in \mathcal{F}_i$, $f_i \succeq_i f_i'$ if and only if

$$\sum_{s_{-i} \in S_{-i}} \mu^i(s_{-i}) \sum_{z \in Z} f_i(s_{-i})(z)\, u_i(z) \geq \sum_{s_{-i} \in S_{-i}} \mu^i(s_{-i}) \sum_{z \in Z} f_i'(s_{-i})(z)\, u_i(z).$$

We know, by Theorem 2.2.2, that the subjective probability distribution μ^i is uniquely defined, and that u_i is unique up to positive affine transformations. Let $g_i(S_i) \subseteq \mathcal{F}_i$ be the set of those acts that correspond to strategies in S_i. Then, the restriction of the preference relation $\succeq_i$ to $g_i(S_i)$ may be identified with a preference relation $\succeq_i$ on S_i defined as follows: $s_i \succeq_i s_i'$ if and only if

$$\sum_{s_{-i} \in S_{-i}} \mu^i(s_{-i}) \sum_{z \in Z} \mathbb{P}_{(s_i, s_{-i})}(z)\ u_i(z) \geq \sum_{s_{-i} \in S_{-i}} \mu^i(s_{-i}) \sum_{z \in Z} \mathbb{P}_{(s_i', s_{-i})}(z)\ u_i(z).$$

This simply follows from the observation that a strategy s_i is being identified with the act f_i assigning to every s_{-i} the objective probability distribution $f_i(s_{-i}) = \mathbb{P}_{(s_i, s_{-i})}$. For given μ^i, u_i let

$$E_{\mu^i, u_i}(s_i) = \sum_{s_{-i} \in S_{-i}} \mu^i(s_{-i}) \sum_{z \in Z} \mathbb{P}_{(s_i, s_{-i})}(z)\ u_i(z)$$

be the *subjective expected utility* for strategy s_i, induced by μ^i and u_i. As an abuse of notation, we write $u_i(s_i, \mu^i) := E_{\mu^i, u_i}(s_i)$. Player i's preferences over his strategies can thus be characterized by: $s_i \succeq_i s_i'$ if and only if $u_i(s_i, \mu^i) \geq u_i(s_i', \mu^i)$.

Throughout this book, it is assumed that the utility functions $u_i : Z \to \mathbb{R}$ are *common knowledge* among the players. By the latter, we mean that all players know that each player i has a preference relation of the subjective expected utility type with the particular utility function $u_i : Z \to \mathbb{R}$ above, all players know that all players know this, all players know that all players know that all players know this, and so on. To be more precise, it is assumed common knowledge that each player i has the particular utility function $u_i : Z \to \mathbb{R}$ above, *or some positive affine transformation of it.* The reason is that for any given subjective probability distribution μ^i, any positive affine transformation $\tilde{u}_i$ of u_i would, together with μ^i, induce the same preference relation as μ^i and u_i would.

In the previous section we have seen a procedure which, for a decision maker with a preference relation $\succeq_i$ of the subjective expected utility type, "extracts" a utility function u_i and the unique subjective probability distribution μ^i that together represent $\succeq_i$. In the present context, the procedure for extracting u_i could be restated in the following way. Let z^1 be player i's "most preferred terminal node" and z^0 his "least preferred terminal node". Formally, a most preferred and least preferred terminal node are not defined since player i only has preferences over acts in $\mathcal{F}_i$, and not over terminal nodes. However, each terminal node z can be identified with the act f^z that for each s_{-i} leads to z with probability one, and therefore player i's preference relation on $\mathcal{F}_i$ induces a preference relation on the set of terminal nodes Z. For each terminal node z, we ask player i the question "for which $u_i(z) \in [0,1]$ would you be indifferent between the act f^z (always ending up in z) and the act $u_i(z) f^{z^1} + (1 - u_i(z)) f^{z^0}$?" Let us call this question $Q_i(z)$. If player i answers all these questions $Q_i(z)$ correctly, then the numbers $u_i(z)$, together with some appropriate subjective probability distribution μ^i, would represent $\succeq_i$. Given the common knowledge assumption above, we could therefore say that there is common knowledge (1) about the fact that every player holds preferences of the subjective expected utility type, and (2) about the answers that players give to the questions $Q_i(z)$. In view of this assumption, we may treat the utility functions u_i as exogenously given parameters, and may therefore as well include them in the description of the game. This leads to the following definition of an extensive form game.

Definition 2.3.1 *An extensive form game is a pair $\Gamma = (S, u)$ where S is an extensive form structure, and $u = (u_i)_{i \in I}$ is a profile of utility functions $u_i : Z \to \mathbb{R}$.*

Two extensive form games $\Gamma = (S, u)$ and $\Gamma' = (S', u')$ are called *equivalent* if they have the same extensive form structure, and if for every player i the utility function u'_i is a positive affine transformation of u_i. This definition is obvious, since for any profile $(\mu^i)_{i \in I}$ of subjective probability distributions $\mu^i \in \Delta(S_{-i})$, u and u' would induce the same profile of preferences for the players.

In contrast to the utility functions u_i, the subjective probability distributions μ^i are *not necessarily known* to the opponents, and can therefore not be included in the description of the game. As such, for a given player i the subjective probability distributions held by the other players could be modeled as *subjective unknowns* for this player. In fact, Chapter 7 provides a series of rationality concepts which is based upon this approach.

As we have seen in the previous section, the numbers $\mu^i(s_{-i})$ can be interpreted as some "measure of likelihood" that player i attaches to the strategy profile s_{-i}. In view of this, we speak, somewhat informally, about $\mu^i(s_{-i})$ as the "probability" that player i attaches to the event that his opponents will choose the strategy profile s_{-i}. The subjective probability distribution $\mu^i \in \Delta(S_{-i})$ is called player i's *conjecture* about the opponents' strategies.

We shall now impose some additional conditions on the conjectures μ^i, which will be used in Chapters 3 to 6, but not in the last chapter. First of all, it is assumed that the probability distribution $\mu^i \in \Delta(S_{-i})$ can be written as the product of its marginal distributions on S_j. For every player $j \neq i$, let $\mu^i_j \in \Delta(S_j)$ be the marginal distribution on S_j, given by

$$\mu^i_j(s_j) = \sum_{s'_{-i} \in S_{-i} : s'_j = s_j} \mu^i(s'_{-i})$$

for all $s_j \in S_j$.

Assumption 1. For every player i, the conjecture μ^i is such that for every $s_{-i} = (s_j)_{j \neq i} \in S_{-i}$ it holds: $\mu^i(s_{-i}) = \prod_{j \neq i} \mu^i_j(s_j)$.

We refer to μ^i_j as player i's conjecture about player j's strategy choice. The second assumption is that the conjectures held by different players about player k's strategy should coincide.

Assumption 2. For all triples $i, j, k \in I$ of pairwise different players, we have $\mu^i_k = \mu^j_k$.

Of course, assumptions 1 and 2 are trivially satisfied if the game involves only two players. By the assumptions above, we may define, for each player i, the probability distribution $\mu_i \in \Delta(S_i)$, representing the players' common conjecture about player i's strategy choice. Hence, $\mu_i = \mu^j_i$ for all players $j \neq i$. Such common conjectures are called *mixed* conjectures about player i.

Definition 2.3.2 *A mixed conjecture about player i is a probability distribution $\mu_i \in \Delta(S_i)$.*

The definition of a mixed conjecture about player i is mathematically equivalent to the definition of a *mixed strategy* for player i, which is normally found in the literature (conditional on the fact that one adopts our definition of a strategy). The interpretations of both concepts are different, however. A mixed strategy for player i is usually viewed as an explicit randomization over strategies in S_i carried out by player i. In our interpretation, the probability distribution μ_i does not represent a randomization which is actually carried out, but merely represents some common measure of likelihood that player i's opponents attach to player i's strategies. Or, if you wish, the probability distribution μ_i is "in the mind of player i's opponents", instead of being "in player i's mind".

2.4 Local Conjectures

Under the assumptions 1 and 2 of the previous section, the players' conjectures about the opponents' behavior may be summarized by a profile of mixed conjectures $(\mu_i)_{i\in I}$ where $\mu_i \in \Delta(S_i)$ is to be understood as the common conjecture of player i's opponents about player i's strategy choice. For practical purposes, it turns out to be convenient to represent these conjectures μ_i by profiles of *local* subjective probability distributions, to which we shall refer as *local mixed conjectures*. Formally, a *local mixed conjecture* λ_i about player i specifies for every information set $h \in H_i$ not avoided by λ_i some probability distribution λ_{ih} over the available actions at h. To be precise about what we mean by "information sets not avoided by λ_i", consider a collection $H_i^* \subseteq H_i$ of player i information sets, and some profile $\lambda_i = (\lambda_{ih})_{h\in H_i^*}$ of local subjective probability distributions, where $\lambda_{ih} \in \Delta(A(h))$ for every $h \in H_i^*$. We say that an information set $h \in H_i$ is *avoided* by λ_i if for every node $x \in h$, the path π from the root to x crosses some information set $h' \in H_i^*$ at which $\lambda_{ih'}$ assigns probability zero to the unique action at h' which lies on the path π.

Definition 2.4.1 *A local mixed conjecture about player i is a profile $\lambda_i = (\lambda_{ih})_{h\in H_i^*}$ of local subjective probability distributions λ_{ih} where (1) $H_i^* \subseteq H_i$ is a collection of player i information sets, (2) $\lambda_{ih} \in \Delta(A(h))$ for every $h \in H_i^*$, (3) every $h \in H_i^*$ is not avoided by λ_i and (4) every $h \in H_i \backslash H_i^*$ is avoided by λ_i.*

For a given information set $h \in H_i^*$ and action $a \in A(h)$, the number $\lambda_{ih}(a)$ may be interpreted as the subjective probability assigned by player i's opponents to the event that player i chooses action a at information set h. Every mixed conjecture μ_i induces, in a natural way, a local mixed conjecture λ_i as follows. For every information set $h \in H_i$, let $S_i(h)$ be the set of player i strategies that do not avoid h (in the sense of Section 2.1). For every $h \in H_i$ and every action $a \in A(h)$, let $S_i(h,a)$ be the set of strategies in $S_i(h)$ that choose action a at h. Say that an information set $h \in H_i$ is *avoided* by the mixed conjecture μ_i if $\mu_i(s_i) = 0$ for all $s_i \in S_i(h)$. Hence, μ_i only attaches positive probability to player i strategies that avoid h. Let $H_i(\mu_i)$ be the set of player i information sets that are not avoided by μ_i.

Definition 2.4.2 *Let μ_i be a mixed conjecture about player i. Then, the local mixed conjecture $\lambda_i = (\lambda_{ih})_{h\in H_i^*}$ induced by μ_i is given by (1) $H_i^* = H_i(\mu_i)$, (2) for every*

$h \in H_i(\mu_i)$ and every action $a \in A(h)$,

$$\lambda_{ih}(a) = \frac{\mu_i(S_i(h,a))}{\mu_i(S_i(h))}.$$

Here, $\mu_i(S_i(h,a))$ denotes the sum $\sum_{s_i \in S_i(h,a)} \mu_i(s_i)$. Similarly for $\mu_i(S_i(h))$. Note that for every $h \in H_i(\mu_i)$ we have, by definition, that $\mu_i(S_i(h)) > 0$. The local mixed conjecture λ_i induced by μ_i is thus obtained by first selecting those player i information sets that are not avoided by μ_i, and at each of those information sets h the subjective probability assigned to an action a is chosen equal to the conditional subjective probability that player i chooses some strategy in $S_i(h,a)$, *conditional* on the event that player i chooses a strategy in $S_i(h)$.

As an illustration, consider again the example in Figure 2.1. Let $\mu_1 = 0.2(a,c) + 0.3(a,d) + 0.5b$ be some mixed conjecture about player 1. Let h_1 and h_2 be the first and the second information set for player 1, respectively. Then, it is clear that $H_1(\mu_1) = \{h_1, h_2\}$, $S_1(h_1) = S_1$, $S_1(h_2) = \{(a,c),(a,d)\}$ and $\mu_1(S_1(h_2)) = 0.5$. The induced local mixed conjecture is then given by $\lambda_1 = (0.5a + 0.5b, 0.4c + 0.6d)$. If one would choose $\mu_1 = b$, then $H_1(\mu_1) = \{h_1\}$ since h_2 is avoided by μ_1, and the induced local mixed conjecture would be $\lambda_1 = b$.

The question remains whether one can always substitute the mixed conjecture μ_i be the induced local mixed conjecture λ_i while "preserving the players' preference relations". In order to answer this question we should first formalize what we mean by the phrase that "substituting μ_i by λ_i preserves the players' preference relations". Recall that every player i holds a preference relation $\succeq_i$on the set of acts $\mathcal{F}_i$ which is of the subjective expected utility type. Every preference relation $\succeq_i$ is induced by a subjective probability distribution $\mu^i \in \Delta(S_{-i})$ and a utility function $u_i : Z \to \mathbb{R}$, where $\mu^i(s_{-i}) = \prod_{j \neq i} \mu_j(s_j)$ for every $s_{-i} = (s_j)_{j \neq i} \in S_{-i}$. Now, suppose that we would substitute every mixed conjecture μ_j by some local mixed conjecture λ_j. Let $\lambda_{-i} = (\lambda_j)_{j \neq i}$. Then, we may define for every strategy s_i player i's subjective expected utility given (λ_{-i}, u_i) as follows.

Let $H_i(s_i)$ be those player i information sets that are not avoided by s_i, and for $j \neq i$, let $H_j(\lambda_j)$ be those player j information sets not avoided by λ_j. For every player $j \in I$, every information set $h \in H_j$ and every action $a \in A(h)$, define $p_{(s_i,\lambda_{-i})}(a)$ by

$$p_{(s_i,\lambda_{-i})}(a) = \begin{cases} 1, & \text{if } j = i,\ h \in H_i(s_i) \text{ and } s_i(h) = a, \\ \lambda_{jh}(a), & \text{if } j \neq i \text{ and } h \in H_j(\lambda_j), \\ 0, & \text{otherwise.} \end{cases}$$

For every terminal node $z \in Z$, let the probability $\mathbb{P}_{(s_i,\lambda_{-i})}(z)$ be given by

$$\mathbb{P}_{(s_i,\lambda_{-i})}(z) = \prod\nolimits_{a \in A_z} p_{(s_i,\lambda_{-i})}(a) \prod\nolimits_{c \in C_z} \tau(c),$$

where, as before, A_z is the set of actions on the path to z, and C_z is the set of outcomes of chance moves on this path. Player i's subjective expected utility of choosing s_i, given λ_{-i} and u_i, is then given by

$$E_{\lambda_{-i},u_i}(s_i) = \sum_{z \in Z} \mathbb{P}_{(s_i,\lambda_{-i})}(z)\ u_i(z).$$

In the obvious way, these subjective expected utilities induce for each player i a preference relation on S_i. Let $\mu_{-i} = (\mu_j)_{j\neq i}$ and $\mu_{-i}(s_{-i}) = \prod_{j\neq i} \mu_j(s_j)$ for all $s_{-i} \in S_{-i}$. Recall, from the previous section, that the subjective expected utility of s_i, given μ_{-i} and u_i, is

$$E_{\mu_{-i},u_i}(s_i) = \sum_{s_{-i}\in S_{-i}} \mu_{-i}(s_{-i}) \sum_{z\in Z} \mathbb{P}_{(s_i,s_{-i})}(z)\, u_i(z).$$

Definition 2.4.3 *A profile $(\mu_i)_{i\in I}$ of mixed conjectures and a profile $(\lambda_i)_{i\in I}$ of local mixed conjectures are called* equivalent *if for every player i, every utility function $u_i : Z \to \mathbb{R}$ and every strategy $s_i \in S_i$ we have that $E_{\lambda_{-i},u_i}(s_i) = E_{\mu_{-i},u_i}(s_i)$.*

Hence, $(\mu_i)_{i\in I}$ and $(\lambda_i)_{i\in I}$ are equivalent if, for each player i, they always induce the same preference relation on the set of strategies S_i. Since it is only this preference relation that matters at the end, there is no need to distinguish between $(\mu_i)_{i\in I}$ and $(\lambda_i)_{i\in I}$. We will now show that for games with perfect recall, one can always substitute the mixed conjectures μ_i by the induced local mixed conjectures λ_i without altering the preference relations for the players. This is an adapted version of Kuhn's Theorem (Kuhn, 1953).

Theorem 2.4.4 *Let S be an extensive form structure satisfying perfect recall, and $(\mu_i)_{i\in I}$ a profile of mixed conjectures. For every player i, let λ_i be the local mixed conjecture induced by μ_i. Then, $(\mu_i)_{i\in I}$ and $(\lambda_i)_{i\in I}$ are equivalent.*

Proof. Consider an extensive form structure S satisfying perfect recall, and a profile $(\mu_i)_{i\in I}$ of mixed conjectures. Let $(\lambda_i)_{i\in I}$ be the induced profile of local mixed conjectures. Let $(u_i)_{i\in I}$ be a profile of utility functions. Fix some player i and a strategy $s_i \in S_i$. In order to show that $E_{\lambda_{-i},u_i}(s_i)$ and $E_{\mu_{-i},u_i}(s_i)$ coincide, it suffices to show that for every terminal node z,

$$\mathbb{P}_{(s_i,\lambda_{-i})}(z) = \sum_{s_{-i}\in S_{-i}} \mu_{-i}(s_{-i})\, \mathbb{P}_{(s_i,s_{-i})}(z). \tag{2.3}$$

By definition,

$$\begin{aligned} \mathbb{P}_{(s_i,\lambda_{-i})}(z) &= \prod\nolimits_{a\in A_z} p_{(s_i,\lambda_{-i})}(a) \prod\nolimits_{c\in C_z} \tau(c), \\ \mathbb{P}_{(s_i,s_{-i})}(z) &= \prod\nolimits_{a\in A_z} p_{(s_i,s_{-i})}(a) \prod\nolimits_{c\in C_z} \tau(c) \end{aligned}$$

and hence, for showing (2.3) it is sufficient to prove that

$$\prod\nolimits_{a\in A_z} p_{(s_i,\lambda_{-i})}(a) = \sum_{s_{-i}\in S_{-i}} \mu_{-i}(s_{-i}) \prod\nolimits_{a\in A_z} p_{(s_i,s_{-i})}(a). \tag{2.4}$$

For every player j, let $A_{z,j}$ be the set of player j actions on the path to z, and let $S_j(z)$ be the set of player j strategies that prescribe all the actions in $A_{z,j}$. Then, by definition,

$$\prod\nolimits_{a\in A_z} p_{(s_i,s_{-i})}(a) = \begin{cases} 1, & \text{if } (s_i, s_{-i}) \in \times_{j\in I} S_j(z) \\ 0, & \text{otherwise,} \end{cases}$$

which implies that

$$\sum_{s_{-i}\in S_{-i}} \mu_{-i}(s_{-i}) \prod\nolimits_{a\in A_z} p_{(s_i,s_{-i})}(a) = \begin{cases} \prod_{j\neq i}\mu_j(S_j(z)), & \text{if } s_i\in S_i(z) \\ 0, & \text{otherwise.}\end{cases} \tag{2.5}$$

Suppose that $s_i \notin S_i(z)$. Then, $\sum_{s_{-i}\in S_{-i}} \mu_{-i}(s_{-i}) \prod_{a\in A_z} p_{(s_i,s_{-i})}(a) = 0$. Since $s_i \notin S_i(z)$, there is some player i information set h on the path π to z such that the prescribed action $s_i(h)$ deviates from this path. But then, $\prod_{a\in A_z} p_{(s_i,\lambda_{-i})}(a) = 0$ and hence (2.4) holds. Assume that $s_i \in S_i(z)$. Then, by (2.5),

$$\sum_{s_{-i}\in S_{-i}} \mu_{-i}(s_{-i}) \prod\nolimits_{a\in A_z} p_{(s_i,s_{-i})}(a) = \prod_{j\neq i}\mu_j(S_j(z)).$$

On the other hand,

$$\prod\nolimits_{a\in A_z} p_{(s_i,\lambda_{-i})}(a) = \prod\nolimits_{j\neq i}\prod\nolimits_{a\in A_{z,j}} p_{(s_i,\lambda_{-i})}(a),$$

since $p_{(s_i,\lambda_{-i})}(a) = 1$ for all $a \in A_{z,i}$. The latter follows from the fact that $s_i \in S_i(z)$. In order to prove (2.4) it thus remains to show that

$$\prod\nolimits_{a\in A_{z,j}} p_{(s_i,\lambda_{-i})}(a) = \mu_j(S_j(z)) \tag{2.6}$$

for all $j \neq i$. Choose some fixed player $j \neq i$. Let π be the path from the root to z, and let $h_1, h_2, ..., h_K$ be the player j information sets on π such that h_k comes before h_{k+1} for all k. For every $k = 1, .., K$ let $a_k \in A(h_k)$ be the unique action at h_k that lies on π. Since S satisfies perfect recall, there is for every h_k a unique sequence of player j actions that leads to h_k, namely the actions $a_1, ..., a_{k-1}$.

Suppose first that $\mu_j(S_j(z)) = 0$. Then, there is some $k \in \{1, ..., K\}$ such that $\mu_j(S_j(h_k)) > 0$ but $\mu_j(S_j(h_k, a_k)) = 0$. Then, by definition, $\lambda_{jh_k}(a_k) = 0$ and hence $\prod_{a\in A_{z,j}} p_{(s_i,\lambda_{-i})}(a) = 0$. In this case, (2.6) holds.

Assume now that $\mu_j(S_j(z)) > 0$. By perfect recall, $S_j(z) = S_j(h_K, a_K)$ and hence $\mu_j(S_j(h_K, a_K)) > 0$. Moreover, by perfect recall, $S_j(h_k, a_k) = S_j(h_{k+1})$ for all $k \in \{1, ..., K-1\}$. By construction of λ_j, we have that

$$\lambda_{jh_k}(a_k) = \frac{\mu_j(S_j(h_k, a_k))}{\mu_j(S_j(h_k))}$$

for all $k \in \{1, ..., K-1\}$. Note that $\mu_j(S_j(h_k)) > 0$ for all these k since $S_j(h_k) \supseteq S_j(h_K, a_K)$ and $\mu_j(S_j(h_K, a_K)) > 0$. We also have that $\lambda_{jh_k}(a_k) > 0$ for all k since $\mu_j(S_j(h_k, a_k)) > 0$ for all k. This implies that $h_k \in H_j(\lambda_j)$ for all k. Since $A_{z,j} = \{a_1, ..., a_K\}$ and $h_k \in H_j(\lambda_j)$ for all k, it follows that

$$\begin{aligned} \prod\nolimits_{a\in A_{z,j}} p_{(s_i,\lambda_{-i})}(a) &= \prod\nolimits_{k=1}^K \lambda_{jh_k}(a_k) = \prod\nolimits_{k=1}^K \frac{\mu_j(S_j(h_k, a_k))}{\mu_j(S_j(h_k))} \\ &= \left[\prod\nolimits_{k=1}^{K-1} \frac{\mu_j(S_j(h_{k+1}))}{\mu_j(S_j(h_k))}\right] \frac{\mu_j(S_j(h_K, a_K))}{\mu_j(S_j(h_K))} \\ &= \frac{\mu_j(S_j(h_K, a_K))}{\mu_j(S_j(h_1))} = \mu_j(S_j(z)) \end{aligned}$$

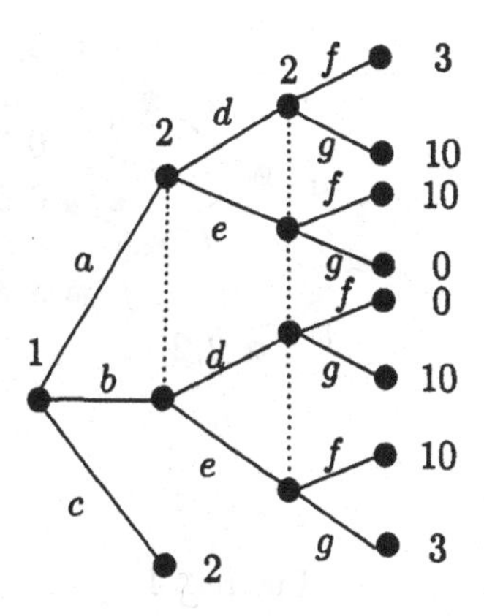

Figure 2.3

since $S_j(h_K, a_K) = S_j(z)$ and $S_j(h_1) = S_j$. As such, (2.6) holds also if $\mu_j(S_j(z)) > 0$. This completes the proof of the theorem. ■

The theorem above no longer holds if perfect recall is not satisfied. Consider, for instance, the extensive form structure in Figure 2.3. It is clear that perfect recall is violated in this example since player 2, at his second information set, does not remember which action he took at his first information set. Let player 1's utility function $u_1 : Z \to \mathbb{R}$ be as the numbers at the terminal nodes indicate. Consider the mixed conjecture about player 2 given by $\mu_2 = 0.5(d, f) + 0.5(e, g)$. The induced subjective expected utilities for player 1's strategies a, b and c are $E_{\mu_2,u_1}(a) = E_{\mu_2,u_1}(b) = 1.5$, and $E_{\mu_2,u_1}(c) = 2$. Hence, player 1 strictly prefers c over a and b, and is indifferent between a and b. We show that this preference relation cannot be induced by a local mixed conjecture λ_2 about player 2, given the utility function u_1. Choose an arbitrary local mixed conjecture $\lambda_2 = (\alpha d + (1-\alpha)e, \beta f + (1-\beta)g)$. Suppose that λ_2 and u_1 would induce the preference relation above. Since player 1 is indifferent between a and b we have that $E_{\lambda_2,u_1}(a) = E_{\lambda_2,u_1}(b)$ which implies that $\alpha + \beta = 1$. But then, it may be verified that $E_{\lambda_2,u_1}(a) + E_{\lambda_2,u_1}(b) = 34(\alpha - 0.5)^2 + 46/4 > 6$, from which we may conclude that either $E_{\lambda_2,u_1}(a) > 3$ or $E_{\lambda_2,u_1}(b) > 3$. However, this means that either player 1 strictly prefers a over c, or strictly prefers b over c, which is a contradiction. Hence, μ_2 cannot be substituted by a local mixed conjecture λ_2 without altering player 1's preference relation over his strategies.

Now, suppose that the local mixed conjecture λ_i is given. Then, one can easily "recover" a mixed conjecture μ_i such that λ_i is induced by μ_i. We refer to this μ_i as the mixed conjecture *induced by* λ_i. Recall that $H_i(\lambda_i)$ is the collection of player i information sets not avoided by λ_i and $H_i(s_i)$ is the collection of player i information sets not avoided by the strategy s_i.

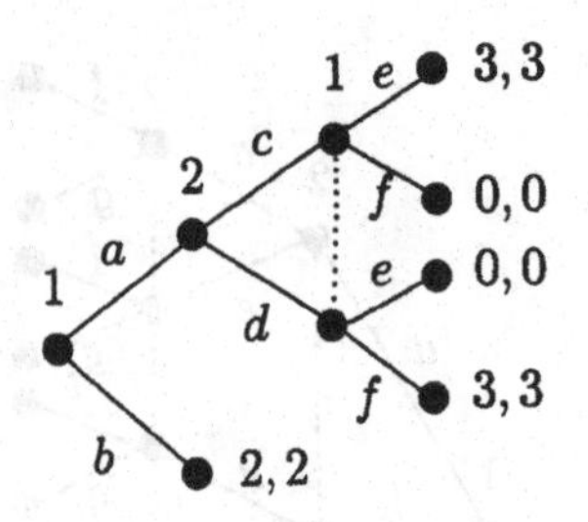

Figure 2.4

Definition 2.4.5 *Let λ_i be a local mixed conjecture. Then, the mixed conjecture induced by λ_i is the mixed conjecture $\mu_i \in \Delta(S_i)$ given by*

$$\mu_i(s_i) = \begin{cases} \prod_{h \in H_i(s_i)} \lambda_{ih}(s_i(h)), & \text{if } H_i(s_i) \subseteq H_i(\lambda_i) \\ 0, & \text{otherwise.} \end{cases}$$

It may be checked that if μ_i is the mixed conjecture induced by λ_i, then λ_i is the mixed conjecture induced by μ_i, and vice versa. In view of Theorem 2.4.4, whenever $(\lambda_i)_{i \in I}$ is given, we may substitute it by the induced profile of mixed conjectures $(\mu_i)_{i \in I}$ without altering the players' preferences.

2.5 Revision of Conjectures

From now on, we always assume that the extensive form structure satisfies perfect recall. In view of Theorem 2.4.4, the players' conjectures about the opponents' behavior may then either be represented by a profile of mixed conjectures $(\mu_i)_{i \in I}$ or, equivalently, by the profile of induced local mixed conjectures $(\lambda_i)_{i \in I}$. However, it may happen that one or more of these conjectures are *contradicted* during the play of the game. Consider, for instance, the game in Figure 2.4. The numbers at the terminal nodes denote the utilities $u_1(z)$ and $u_2(z)$ of the players. Suppose that $\mu_1 = b$, which means that player 2 believes that player 1 chooses strategy b with probability one. If the game would reach player 2's information set, then player 2 must conclude that his conjecture μ_1 is false, since it is evident that player 1 has chosen action a. In order to evaluate the consequences of his strategies c and d at this point of the game, player 2 should thus revise his conjecture about player 1's strategy. This is important since at this stage of the game, the final outcome depends not only on player 2's choice between c and d but also on the action chosen by player 1 at his second information set. Speaking in terms of preferences, player 2 is asked to revise his preference relation on his set of strategies after observing that player 1's behavior has contradicted his initial conjecture μ_1, but it is assumed that player 2's utility function u_2 remains unchanged during this revision; only the conjecture μ_1 is revised. Suppose that player 2's "new" conjecture after observing action a is $\mu_1' = 0.7(a, e) + 0.3(a, f)$. In this case, player 2's expected utility by choosing c is 2.1 whereas his expected utility by choosing d is 0.9. Hence, player 2 should choose c. Note that with respect to his *initial* conjecture μ_1, player 2 would be indifferent between c and d, since μ_1 attached probability zero to the event that player 2's information set would be reached. However, once player 2's information set is reached this conjecture can no longer be maintained, and with respect to the *revised* conjecture c is strictly preferred to d.

In this example, it is convenient to represent the initial conjecture μ_1 and the new conjecture μ_1' about player 1 by their induced local mixed conjectures $\lambda_1 = b$ and $\lambda_1' = (a, 0.7e + 0.3f)$. Here, the second component of λ_1', that is, $0.7e + 0.3f$, is the local subjective probability distribution that player 2 assigns to player 1's actions at player 1's second information set. The first component of λ_1', which is a, is in some sense redundant since a is the unique action for player 1 which leads to his second information set, and could thus as well be skipped without loosing information. By putting the "relevant" local probability distributions of λ_1 and λ_1' in one and the same object, we obtain the vector of local probability distributions $\sigma_1 = (b, 0.7e + 0.3f)$, which is to be read as follows: at the beginning of the game, player 2 assigns probability one to action b. However, if this initial conjecture is contradicted, that is, if player 1 has chosen a instead, player 2 assigns probability 0.7 to the event that player 1 will choose action e at his last information set, and assigns probability 0.3 to the event that player 1 will choose action f there. The vector σ_1 is called a *behavioral conjecture* about player 1.

Formally, a behavioral conjecture σ_i about player i should specify, for *each* of player i's information sets $h \in H_i$, some local probability distribution σ_{ih} on the available actions at h.

Definition 2.5.1 *A behavioral conjecture about player i is a vector $\sigma_i = (\sigma_{ih})_{h \in H_i}$ where $\sigma_{ih} \in \Delta(A(h))$ for every $h \in H_i$.*

For every $h \in H_i$ and every action $a \in A(h)$, the probability $\sigma_{ih}(a)$ is to be interpreted as the players' common subjective probability assigned to the event that player i would choose action a at h. The crucial difference with a local mixed conjecture λ_i is that the latter only specifies a local probability distribution for every information set $h \in H_i$ that is *not avoided* by λ_i, whereas σ_i specifies a local probability distribution also at those information sets that are actually avoided by σ_i. In the example above, for instance, player 1's second information set is avoided by the behavioral conjecture $\sigma_1 = (b, 0.7e + 0.3f)$, however σ_1 specifies a local probability distribution also at this information set. The reason is that a behavioral conjecture σ_i should reflect the players' conjecture about player i's behavior *at each of player i's information sets*, including those which actually contradict the players' initial conjecture about player i's behavior.

Every behavioral conjecture profile $\sigma = (\sigma_i)_{i \in I}$ thus specifies, in particular, the players' *initial* conjectures about the opponents' behavior. Formally, for a given behavioral conjecture σ_i, let λ_i be the local mixed conjecture induced by σ_i, obtained by taking the restriction of σ_i on those information sets $h \in H_i$ that are not avoided by σ_i. Let μ_i be the mixed conjecture induced by λ_i, as defined in the previous section.

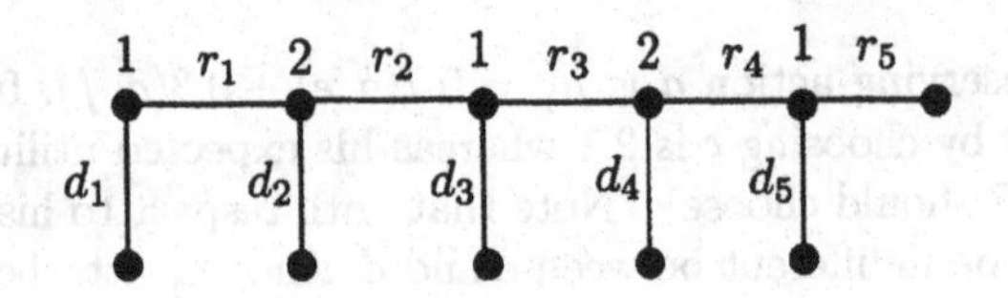

Figure 2.5

Then, we say that μ_i is the *initial mixed conjecture induced by* σ_i. For a given strategy s_i we write $\sigma_i(s_i)$ as to denote the subjective probability assigned to s_i by the mixed conjecture μ_i induced by σ_i. Formally, we have that

$$\sigma_i(s_i) = \prod\nolimits_{h \in H_i(s_i)} \sigma_{ih}(s_i(h))$$

for all $s_i \in S_i$.

In general, a behavioral conjecture σ_i induces, for every information set $h \in H_i$, a mixed conjecture $\mu_{i|h} \in \Delta(S_i(h))$ which assigns positive probability only to strategies in $S_i(h)$. Recall that $S_i(h)$ is the set of player i strategies that do not avoid h. The mixed conjecture $\mu_{i|h}$ reflects the common conjecture that player i's opponents have about player i's behavior if information set h is reached. This conjecture may be different from the initial mixed conjecture induced by σ_i, since the event of reaching information set h may not be compatible with this initial conjecture. We refer to $\mu_{i|h}$ as the *revised mixed conjecture* induced by σ_i at h.

Formally, $\mu_{i|h}$ is defined as follows. Consider an information set $h \in H_i$. By perfect recall, there is a unique sequence $h_1, h_2, ..., h_K$ of player i information sets that leads to h. Here, h_k follows h_{k-1} for every k, h follows h_K, and there are no player i information sets between h_{k-1} and h_k or between h_K and h. At every h_k, let $a_k \in A(h_k)$ be the unique action at h_k that leads to h. Let σ_i be a behavioral conjecture which possibly avoids h. By $\sigma_{i|h}$ we denote the behavioral conjecture which at every h_k assigns probability one to the action a_k leading to h, and which coincides with σ_i at all $h \in H_i \backslash \{h_1, ..., h_K\}$. The behavioral conjecture $\sigma_{i|h}$ may be seen as the revised behavioral conjecture induced at h, since at h it is evident that player i has chosen the actions $a_1, ..., a_K$. The mixed conjecture induced by $\sigma_{i|h}$ is called $\mu_{i|h}$.

Since $\sigma_{i|h}$ assigns probability one to all player i actions leading to h, it follows that $\mu_{i|h}$ attaches positive probability only to strategies in $S_i(h)$, and hence $\mu_{i|h} \in \Delta(S_i(h))$. For every information set $h \in H_i$ and every $s_i \in S_i(h)$, we write $\sigma_{i|h}(s_i)$ as to denote the subjective probability assigned to s_i by the revised mixed conjecture $\mu_{i|h}$ at h.

It may occur that in a game not only the players' *initial* conjecture about player i is contradicted at a certain stage of the game, but also that their *revised* conjecture about player i is contradicted later on. Consider the example in Figure 2.5. Suppose that player 2 holds the behavioral conjecture $\sigma_1 = (d_1, d_3, d_5)$ about player 1. Then, player 2's initial conjecture is that player 1 chooses strategy d_1 with probability one. However, if player 2's first information set is reached, he knows that this initial con-

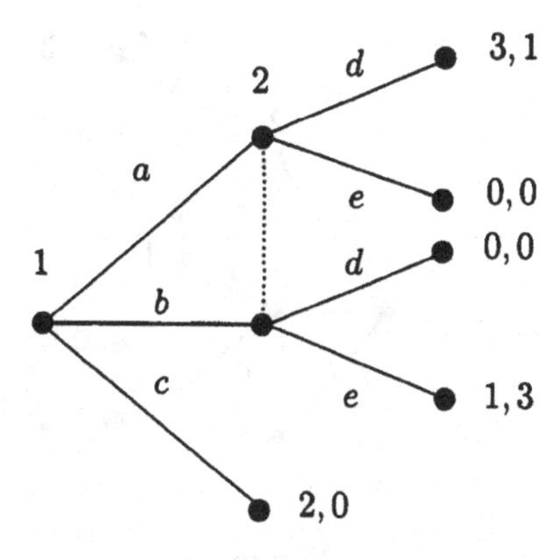

Figure 2.6

jecture has been contradicted. According to σ_1, player 2's revised conjecture is that player 1 chooses (r_1, d_3) with probability one. Note that r_1 is the unique action that leads to player 1's second information set. However, if player 2's second information is reached, also this revised conjecture has been contradicted. The third component in σ_1 then states that player 2's newly revised conjecture is that player 1 chooses strategy (r_1, r_3, d_5) with probability one. Here, (r_1, r_3) is the unique sequence of player 1 actions that leads to player 1's final information set.

A behavioral conjecture σ_i about player i thus specifies, *at each of player i's information sets*, the opponents' common conjecture about player i's play. However, it may also be necessary for a player j to revise his conjecture about the opponents' past play *at some of his own information sets*. Consider the game in Figure 2.6. Assume that player 2 holds the behavioral conjecture $\sigma_1 = c$. If player 2's information set is reached, then player 2 knows that his initial conjecture about player 1's behavior has been contradicted, and should thus revise this conjecture. To undertake such a revision at his own information set is important for player 2, since the fact whether action e or d is optimal for player 2, or both, depends crucially on the revised conjecture about player 1's past behavior. For instance, if player 2's revised conjecture is that player 1 has chosen a with probability 0.9 and b with probability 0.1, then d would be optimal. If the revised conjecture would assign probability 0.1 to a and probability 0.9 to b, then e would be optimal. Such revised conjectures about the opponents' past play that players undertake *at their own information sets* are called *beliefs*.

Formally, a *belief* for player i should specify at each of player i's information sets a conjecture about the opponents' past play that has led to this information set. At a given information set, this conjecture somehow reflects player i's "personal theory" about what happened in the past. Suppose that the game has reached some information set $h \in H_i$ and that player i is asked to form a conjecture about the opponents' past play. Since player i knows that one of the nodes in h has been

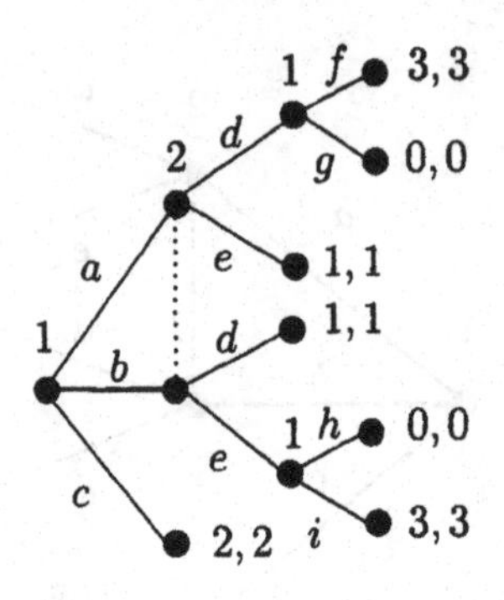

Figure 2.7

reached, and each node in h can be identified with the path, or past play, that leads to this node, such a conjecture about past play may be represented as a subjective probability distribution β_{ih} on the set of nodes in h. This leads to the following definition of a *belief*.

Definition 2.5.2 *A belief for player i is a vector $\beta_i = (\beta_{ih})_{h \in H_i}$ where $\beta_{ih} \in \Delta(h)$ for every $h \in H_i$. A profile $(\beta_i)_{i \in I}$ of beliefs is called a belief system.*

The combination of a behavioral conjecture profile $(\sigma_i)_{i \in I}$ and a belief system $(\beta_i)_{i \in I}$ specifies, for each player i and each information set $h \in H_i$, player i's conjecture about past play up to h (given by β_{ih}) and player i's conjecture about future play after h (given by σ_{-i}). We refer to such combinations as *assessments*.

Definition 2.5.3 *An assessment is a pair (σ, β) where $\sigma = (\sigma_i)_{i \in I}$ is a behavioral conjecture profile and $\beta = (\beta_i)_{i \in I}$ is a belief system.*

Consider, as an illustration, the game in Figure 2.7. Let (σ, β) be the assessment where $\sigma_1 = (c, 0.2f + 0.8g, 0.8h + 0.2i)$, $\sigma_2 = 0.5d + 0.5e$ and $\beta_2 = (0.3, 0.7)$. Here, β_2 is the belief for player 2 which assigns probability 0.3 to the upper node in player 2's information set, and probability 0.7 to the lower node. The beliefs for player 1 are not specified since player 1 solely controls information sets with only one node, and hence his beliefs are the trivial ones that assign probability one to this single node at each of his information sets. Consider player 2's information set h_2. At this information set, player 2's initial conjecture about player 1's behavior, namely that he chooses c with probability one, has been contradicted. The revised conjecture about player 1's past behavior is given by β_2, and assigns probability 0.3 to the event that player 1 has chosen a, and probability 0.7 to the event that player 1 has chosen b. Player 2's conjecture at h_2 about player 1's future behavior is given by σ_1. The information provided by β_2 and σ_1 allows us to determine, for both strategies d and e, the expected utility they induce for player 2 at h_2. For instance, the expected utility generated by strategy d at h_2 is $0.3 \cdot (0.2 \cdot 3 + 0.8 \cdot 0) + 0.7 \cdot 1 = 0.88$. The expected utility generated by strategy e at h_2 is $0.3 \cdot 1 + 0.7 \cdot (0.8 \cdot 0 + 0.2 \cdot 3) = 0.72$. A formal definition of the expected utility at an information set may be found in the following section.

In general, an assessment (σ, β) allows us to determine, for every player i and every information set $h \in H_i$, the expected utility at h generated by the different player i strategies. Consequently, (σ, β) induces at every information set $h \in H_i$ player i's preference relation over his strategies, conditional on the event that h has been reached. An assessment (σ, β) may therefore be seen as a sufficiently complete description of the players' conjectures about the opponents' behavior at every instance of the game.

It seems natural to require, however, that the conjectures about the opponents' past play contained in β be "consistent" with the conjectures about opponents' future play contained in σ. Suppose, for instance, that player 2 in the game of Figure 2.7 would hold the behavioral conjecture $\sigma_1 = (0.2a+0.3b+0.5c, 0.2f+0.8g, 0.8h+0.2i)$. Then, if player 2's information set is reached, player 2's conditional belief about player 1's past behavior assigns probability 0.4 to a and probability 0.6 to b. Hence, the only "consistent" beliefs in this case are $\beta_2 = (0.4, 0.6)$. This consistency condition is defined formally in the following section.

2.6 Best Responses to Conjectures

Suppose that the players' conjectures about the opponents' behavior at every stage of the game is given by an assessment (σ, β). A fundamental question which arises is: which behavior by the players may be viewed optimal, given this assessment? In this section we shall distinguish three types of optimal behavior, namely *best responses, sequential best responses* and *local best responses*.

Before discussing these notions, we introduce some additional notation. For a given assessment (σ, β), let $\mu = (\mu_i)_{i \in I}$ be the profile of initial mixed conjectures induced by σ, as defined in the previous section. For a strategy $s_i \in S_i$, let $u_i(s_i, \sigma_{-i}) := E_{\mu_{-i}, u_i}(s_i)$ denote the expected utility of s_i, given player i's conjecture σ_{-i} about the opponents' behavior. Note that $u_i(s_i, \sigma_{-i})$ only depends upon the *initial* mixed conjectures induced by σ, and not upon the way conjectures are revised in σ.

Definition 2.6.1 *A strategy s_i is called a* **best response** *against σ if $u_i(s_i, \sigma_{-i}) = \max_{s_i' \in S_i} u_i(s_i', \sigma_{-i})$.*

In the obvious way, we may define best responses against *mixed* conjecture profiles μ as well. By construction, if s_i is a best response against σ, then it is a best response against every behavioral conjecture profile σ' which induces the same initial mixed conjectures as σ. As such, the notion of best response selects strategies that are optimal "at the beginning of the game". However, a best response may no longer be optimal once the initial conjectures in σ have been contradicted by the play of the game. Consider, for instance, the game in Figure 2.8. Given the behavioral conjecture profile $\sigma = (b, d)$, the strategy d is a best response for player 2. However, if player 2's information set is reached, player 2's initial conjecture has been contradicted, and at that instance d is no longer optimal. Strategies that are not only best responses at the beginning of the game, but remain responses after the initial conjectures have

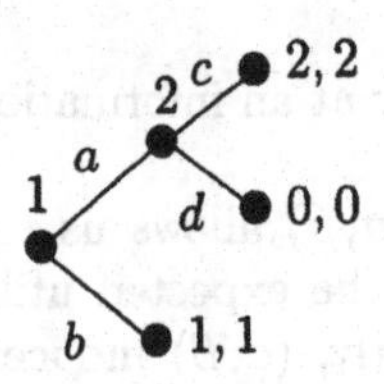

Figure 2.8

been contradicted, are called sequential best responses. To formally define the notion of sequential best response, we need to define the expected utility generated by a strategy at a certain information set. Let the assessment (σ, β) be given, and let $h \in H_i$ be an information set for player i. For every node $x \in h$, let $Z(x)$ be the set of terminal nodes following x. For every $z \in Z(x)$, let $A_{z,x}$ be the set of actions on the path from x to z, and let $C_{z,x}$ be the set of chance moves on this path. Recall that $S_i(h)$ is the set of player i strategies that do not avoid h. For a given strategy $s_i \in S_i(h)$ and an action $a \in A_{z,x}$ available at some information set $h' \in H_j$, let

$$p_{(s_i,\sigma_{-i})}(a) = \begin{cases} 1, & \text{if } j = i,\ h' \in H_i(s_i) \text{ and } s_i(h) = a, \\ \sigma_{jh'}(a), & \text{if } j \neq i \\ 0, & \text{otherwise.} \end{cases}$$

be the probability that action a is chosen, given (s_i, σ_{-i}). For a terminal node $z \in Z(x)$,

$$\mathbb{P}_{(s_i,\sigma_{-i})}(z|\ x) = \prod\nolimits_{a \in A_{z,x}} p_{(s_i,\sigma_{-i})}(a) \prod\nolimits_{c \in C_{z,x}} \tau(c)$$

denotes the probability that z will be reached if the game were to start at x. Let

$$u_i((s_i, \sigma_{-i})|\ x) = \sum\nolimits_{z \in Z(x)} \mathbb{P}_{(s_i,\sigma_{-i})}(z|\ x)\ u_i(z)$$

be the expected utility of choosing strategy s_i if the game would be at node $x \in h$. Given the assesment (σ, β), every node $x \in h$ is assigned probability $\beta_{ih}(x)$ by player i. By

$$u_i((s_i, \sigma_{-i})|\ h, \beta_{ih}) = \sum\nolimits_{x \in h} \beta_{ih}(x)\ u_i((s_i, \sigma_{-i})|\ x)$$

we denote the expected utility of choosing s_i at information set h, given the beliefs β_{ih}, and given the conjecture σ_{-i} about the opponents' future behavior.

Definition 2.6.2 *A strategy s_i is called a sequential best response against the assessment (σ, β) if at every information set $h \in H_i(s_i)$ we have that $u_i((s_i, \sigma_{-i})|\ h, \beta_{ih}) = \max_{s'_i \in S_i(h)} u_i((s'_i, \sigma_{-i})|\ h, \beta_{ih})$.*

Recall that $H_i(s_i)$ is the collection of player i information sets that are not avoided by s_i. A sequential best response thus constitutes a best response at each of player i's information sets that are not avoided by it. In Section 4.3 we will see that, without imposing additional conditions on the assessment, a sequential best response against (σ, β) need not exist. However, under suitable conditions on the assessment one can guarantee the existence of sequential best responses for every player.

Under some "consistency conditions" on the beliefs β, each sequential best response is a best response in the sense defined above. The consistency condition basically states that at every information set $h \in H_i$ the beliefs β_{ih} should be compatible with player i's conjecture σ_{-i} about the opponents' behavior, whenever h is reached with positive probability under σ. In this case, we say that the assessment is Bayesian consistent. For a given node x, let

$$\mathbb{P}_\sigma(x) = \prod_{a \in A_x} \sigma(a) \prod_{c \in C_x} \tau(c)$$

be the probability that x is reached under σ. Here, A_x denotes the set of actions on the path to x, and C_x is the set of outcomes of chance moves on this path. As an abuse of notation, we denote by $\sigma(a)$ the probability that σ assigns to the action a. For a given information set h, let $\mathbb{P}_\sigma(h) = \sum_{x \in h} \mathbb{P}_\sigma(x)$ be the probability that some node in h is reached.

Definition 2.6.3 *An assessment (σ, β) is called Bayesian consistent if for every player i and every information set $h \in H_i$ with $\mathbb{P}_\sigma(h) > 0$ it holds that $\beta_{ih}(x) = \mathbb{P}_\sigma(x)/\mathbb{P}_\sigma(h)$ for all nodes $x \in h$.*

In other words, if h is reached with positive probability under σ, then the beliefs at h should be derived from σ by Bayesian updating. By perfect recall, every path leading to h contains the same set of player i actions. But then, it may be verified easily that the ratio $\mathbb{P}_\sigma(x)/\mathbb{P}_\sigma(h)$ does not depend on σ_i. We may thus define $\mathbb{P}_{\sigma_{-i}}(x|\ h) := \mathbb{P}_{(s_i,\sigma_{-i})}(x)/\mathbb{P}_{(s_i,\sigma_{-i})}(h)$ where s_i is an arbitrary strategy for which $\mathbb{P}_{(s_i,\sigma_{-i})}(h) > 0$. We now show that under Bayesian consistency, every sequential best response constitutes a best response.

Lemma 2.6.4 *Let (σ, β) be a Bayesian consistent assessment and s_i a sequential best response against (σ, β). Then, s_i is a best response against σ.*

Proof. Let (σ, β) be a Bayesian consistent assessment and s_i^* a sequential best response against (σ, β). Assume that s_i^* is not a best response against σ. Then, there is some $s_i \in S_i$ with $u_i(s_i^*, \sigma_{-i}) < u_i(s_i, \sigma_{-i})$. Let H_i^0 be the set of player i information sets that are not preceded by any other player i information set. Let Z_0 be the set of terminal nodes that are not preceded by any player i information set. Then, for every strategy s_i,

$$\begin{aligned} u_i(s_i, \sigma_{-i}) &= \sum_{h \in H_i^0} \sum_{x \in h} \mathbb{P}_{(s_i,\sigma_{-i})}(x)\ u_i((s_i, \sigma_{-i})|\ x) + \\ &+ \sum_{z \in Z_0} \mathbb{P}_{(s_i,\sigma_{-i})}(z)\ u_i(z). \end{aligned}$$

Since every information set $h \in H_i^0$ is not preceded by any player i information set, the probability $\mathbb{P}_{(s_i,\sigma_{-i})}(h)$ does depend upon s_i. Hence, we may write $\mathbb{P}_{\sigma_{-i}}(h) := \mathbb{P}_{(s_i,\sigma_{-i})}(h)$ for every $h \in H_i^0$. Let H_i^{0*} be the set of information sets $h \in H_i^0$ with $\mathbb{P}_{\sigma_{-i}}(h) > 0$. Moreover, every terminal node $z \in Z_0$ is not preceded by any player i information set. We thus have that $\mathbb{P}_{(s_i,\sigma_{-i})}(z)$ does not depend upon s_i for every $z \in Z_0$. We write $\mathbb{P}_{\sigma_{-i}}(z) := \mathbb{P}_{(s_i,\sigma_{-i})}(z)$ for every $z \in Z_0$. It follows that

$$\begin{aligned} u_i(s_i,\sigma_{-i}) &= \sum\nolimits_{h\in H_i^{0*}} \mathbb{P}_{\sigma_{-i}}(h) \sum\nolimits_{x\in h} \frac{\mathbb{P}_{(s_i,\sigma_{-i})}(x)}{\mathbb{P}_{\sigma_{-i}}(h)}\, u_i((s_i,\sigma_{-i})|\, x) + \\ &\quad + \sum\nolimits_{z\in Z_0} \mathbb{P}_{\sigma_{-i}}(z)\, u_i(z). \end{aligned}$$

By definition, $\mathbb{P}_{(s_i,\sigma_{-i})}(x)/\mathbb{P}_{\sigma_{-i}}(h) = \mathbb{P}_{\sigma_{-i}}(x|\, h)$ for every $h \in H_i^{0*}$ and every $x \in h$. Moreover, since (σ,β) is Bayesian consistent, $\mathbb{P}_{\sigma_{-i}}(x|\, h) = \beta_{ih}(x)$ for all such h, x. We thus obtain

$$\begin{aligned} u_i(s_i,\sigma_{-i}) &= \sum\nolimits_{h\in H_i^{0*}} \mathbb{P}_{\sigma_{-i}}(h) \sum\nolimits_{x\in h} \beta_{ih}(x)\, u_i((s_i,\sigma_{-i})|\, x) + \\ &\quad + \sum\nolimits_{z\in Z_0} \mathbb{P}_{\sigma_{-i}}(z)\, u_i(z) \\ &= \sum\nolimits_{h\in H_i^{0*}} \mathbb{P}_{\sigma_{-i}}(h)\, u_i((s_i,\sigma_{-i})|\, h,\beta_{ih}) + \sum\nolimits_{z\in Z_0} \mathbb{P}_{\sigma_{-i}}(z)\, u_i(z). \end{aligned}$$

Since this holds for every $s_i \in S_i$ and $u_i(s_i^*,\sigma_{-i}) < u_i(s_i,\sigma_{-i})$ for some s_i, it follows that

$$u_i((s_i^*,\sigma_{-i})|\, h,\beta_{ih}) < u_i((s_i,\sigma_{-i})|\, h,\beta_{ih})$$

for some $h \in H_i^{0*}$. However, this contradicts the assumption that s_i^* is a sequential best response against (σ,β). ■

In order to define the notion of local best response, consider an assessment (σ,β) and an information set $h \in H_i$. Let σ_{-h} be the restriction of σ on information sets other than h. For every action a at h, let $u_i((a,\sigma_{-h})|\, h,\beta_{ih})$ be the expected utility of choosing action a at h, given the conjecture σ_{-h} about the behavior at information sets other than h (including the conjecture about player i's behavior at such information sets), and the beliefs β_{ih}.

Definition 2.6.5 *Let (σ,β) be an assessment, $h\in H_i$, and a an action at h. Then, we say that a is a local best response against (σ,β) if*

$$u_i((a,\sigma_{-h})|\, h,\beta_{ih}) = \max_{a'\in A(h)} u_i((a',\sigma_{-h})|\, h,\beta_{ih}).$$

In contrast to the notions of best response and sequential best response, local best response is a criterion which judges the optimality of *actions* rather than strategies.

Chapter 3

Backward Induction and Nash Equilibrium

In the previous chapter, we have seen that the conjectures held by the players about the opponents' behavior, and the revision of these conjectures during the game, may be represented by means of an assessment (σ, β). We have also seen alternative ways to specify optimal behavior against such assessments, formalized by the notions of *best response* and *sequential best response*. Roughly speaking, a criterion of rational decision making in extensive form games should give an answer to the following two questions: (1) for a given assessment (σ, β), which strategies may be viewed acceptable?, and (2) which assessments (σ, β) may be viewed reasonable? Throughout this monograph, we shall assume that for a given assessment, players are expected to choose either a best response or a sequential best response against this assessment, depending on the concept, and hence the first question is answered by this assumption. The second question, however, is the most interesting one, and the remainder of this book is concerned with presenting and exploring rationality criteria that give different answers to the question which assessments are "reasonable". In this chapter, we discuss a first series of rationality concepts that put restrictions on the behavioral conjecture profiles σ that may be held in a given assessment. All of these concepts are based on two influential ideas in the theory of extensive form games: *backward induction* and *Nash equilibrium.* The idea in backward induction is that at every stage of the game, the players' initial or revised conjectures about the opponents' behavior should only assign positive probability to actions that are optimal given the assessment. In the concept of Nash equilibrium, the players' initial conjectures should only assign positive probability to opponents' strategies that are best responses against σ. These two concepts are presented in Sections 2.1 and 2.2, respectively. In the remaining sections, we present the concepts of *subgame perfect equilibrium, perfect equilibrium, quasi-perfect equilibrium* and *proper equilibrium,* which put restrictions on the players' conjectures that are more stringent than those imposed by backward induction and Nash equilibrium.

3.1 Backward Induction

In this section we focus on a special class of games, namely that in which at each point in time the player knows perfectly which actions and chance moves have occurred until that moment. Formally, this means that every information set contains exactly one node. Such games are called *games with perfect information.* Typical examples are recreational games like tic-tac-toe, chess, checkers or backgammon. Since the players' beliefs β_i are trivial in a game with perfect information, the players' conjectures about the opponents' behavior at any point in the game are completely described by a behavioral conjecture profile $\sigma = (\sigma_i)_{i\in I}$.

We shall now present a rationality criterion, called *backward induction,* which puts restrictions on the conjectures that players may hold in a behavioral conjecture profile σ. More precisely, backward induction imposes that at every node x in the game, σ should only assign positive probability to those actions at x which are local best responses against σ, in the sense of Section 2.6.

Definition 3.1.1 *Let Γ be a game with perfect information and $\sigma = (\sigma_i)_{i\in I}$ a behavioral conjecture profile. Then, σ is said to satisfy backward induction if for every player i and every decision node x controlled by player i, $\sigma_{ix}(a) > 0$ only if a is a local best response against σ. A strategy s_i is called a backward induction strategy if there is a behavioral conjecture profile σ satisfying backward induction such that s_i is a sequential best response against σ.*

Here, "local best response against σ" and "sequential best response against σ" should be read as local (sequential) best response against (σ, β), where β is the trivial belief system in a game with perfect information. Since (σ, β) is trivially Bayesian consistent in a game with perfect information, Lemma 2.6.4 implies that a backward induction strategy is always a *best response* against some σ satisfying backward induction.

Behavioral conjecture profiles satisfying backward induction can easily be found by means of the following iterative procedure, which is called the *backward induction procedure.* Roughly speaking, the procedure consists of "solving the game backwards". Consider first the set X^0 of "final" decision nodes, that is, decision nodes which are not followed by any other decision node. Choose a node $x \in X^0$ controlled by player i. For every action a available at x the expected utility $u_i(a|\ x)$ of choosing a at x is well-defined, since no player moves after action a. Let $a(x)$ be an action at x with $u_i(a(x)|\ x) = \max_{a\in A(x)} u_i(a|\ x)$. Here, $A(x)$ denotes the set of actions available at x. Define $\sigma_{ix} = a(x)$, by which we mean that σ_{ix} assigns probability one to action $a(x)$. We do so for every node in X^0. Afterwards, consider the set X^1 of decision nodes $x \notin X^0$ for which every decision node following x is in X^0. Choose some $x \in X^1$, and assume that x is controlled by player i. For every action a at x, let $u_i(a|\ x)$ be the expected utility of choosing a at x, given that the conjecture about future behavior is specified by the local probability distributions σ_{iy} for $y \in X^0$ defined above. Again, this is well-defined. Choose some action $a(x)$ with $u_i(a(x)|\ x) = \max_{a\in A(x)} u_i(a|\ x)$ and set $\sigma_{ix} = a(x)$. After doing so for every $x \in X^1$, turn to the set X^2 of decision nodes $x \notin X^0 \cup X^1$ for which every decision node following x is in $X^0 \cup X^1$. Choose

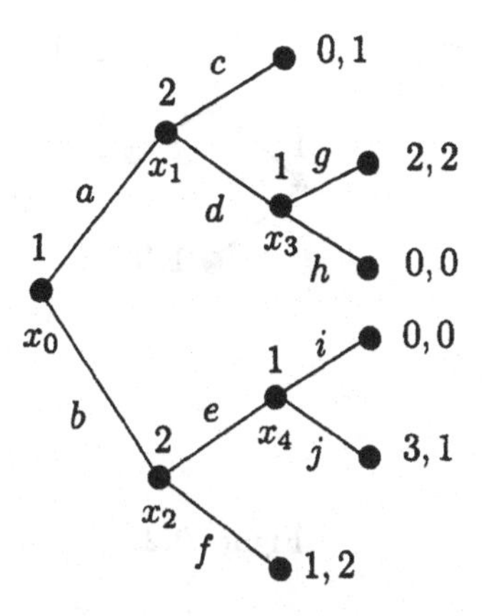

Figure 3.1

some node $x \in X^2$ and assume that x is controlled by player i. For every action a at x let $u_i(a|\ x)$ be the expected utility of choosing a at x, given that the conjecture about future behavior is specified by the local probability distributions σ_{iy} defined above. Define an optimal action $a(x)$ as above and set $\sigma_{ix} = a(x)$, and so on. By proceding like this, one eventually covers all decision nodes in the game, and the result is some behavioral conjecture profile $\sigma = (\sigma_i)_{i \in I}$. By construction, at every decision node x controlled by player i the local probability distribution σ_{ix} assigns probability one to some action $a(x)$ which is a local best response against σ. Hence, the behavioral conjecture profile σ obtained by this backward induction procedure satisfies backward induction. Moreover, the so obtained σ assigns at every decision node probability one to some particular action. Behavioral conjecture profiles with this property are called *pure*. The two observations above thus lead to the following insight.

Lemma 3.1.2 *For every game with perfect information there is some pure behavioral conjecture profile σ which satisfies backward induction.*

As an illustration of the backward induction procedure, consider the game in Figure 3.1. In this game, $X^0 = \{x_3, x_4\}, X^1 = \{x_1, x_2\}$ and $X^2 = \{x_0\}$. It may be easily be checked that $a(x_3) = g$ and $a(x_4) = j$ which implies that $\sigma_{1x_3} = g$ and $\sigma_{1x_4} = j$. Consequently, $u_2(d|\ x_1) = 2$, $u_2(e|\ x_2) = 1$ and hence $a(x_1) = d$, $a(x_2) = f$. We therefore set $\sigma_{2x_1} = d$ and $\sigma_{2x_2} = f$. Then, $u_1(a|\ x_0) = 2$, $u_1(b|\ x_0) = 1$ which implies that $a(x_0) = a$ and $\sigma_{1x_0} = a$. The unique behavioral conjecture profile which results from the backward induction procedure is thus $\sigma = ((a, g, j), (d, f))$.

Suppose now that the game with perfect information Γ is such that there are no chance moves, and that for every player i and every pair of different terminal nodes $z, z' \in Z$ we have that $u_i(z) \neq u_i(z')$. In this case, we say that the game Γ is *in generic*

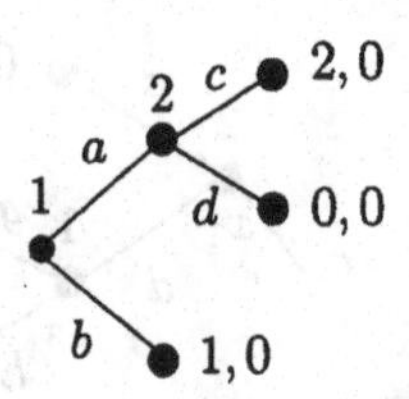

Figure 3.2

position. For such a game, it is easily seen that there is a *unique* behavioral conjecture profile that satisfies backward induction. Suppose namely that σ satisfies backward induction for such a game. Consider a final decision node $x \in X^0$, controlled by some player i. Since every action at x leads to a different terminal node, and player i's utilities $u_i(z)$ are different for different terminal nodes, it follows that there is a unique optimal action $a(x)$ at node x. Hence, it should hold that $\sigma_{ix} = a(x)$ for every $x \in X^0$. Consider now some decision node $x \in X^1$ controlled by some player i. Since at every node $y \in X^0$ the local probability distribution σ_{iy} necessarily puts probability one to the unique optimal action $a(y)$, it follows that for every action a at x there is some terminal node $z(x,a)$ such that $u_i((a,\sigma_{-x})|\ x) = u_i(z(x,a))$. Here, $z(x,a)$ is simply the terminal node that would be reached if player i chooses a at x and players would choose the actions $a(y)$ at nodes $y \in X^0$. Since $z(x,a) \neq z(x,a')$ for two different actions a, a' at x, it follows that there is a unique action $a(x) \in A(x)$ with $u_i((a(x),\sigma_{-x})|\ x) = \max_{a \in A(x)} u_i((a,\sigma_{-x})|\ x)$. Hence, for every $x \in X^1$ the local probability distribution σ_{ix} should attach probability one to $a(x)$. By repeating this argument for nodes in $X^2, X^3, ...$ we may conclude that at every node x there is a unique optimal action $a(x)$ and σ_{ix} should thus attach probability one to this action $a(x)$. We therefore obtain the following result.

Lemma 3.1.3 *Let Γ be a game with perfect information in generic position. Then, there is a unique behavioral conjecture profile σ which satisfies backward induction. Moreover, this σ is pure.*

If the game is not in generic position, there may exist several behavioral conjecture profiles satisfying backward induction. Consider, for instance, the game in Figure 3.2. Clearly, the game is not in generic position, since player 2's utilities at all terminal nodes coincide. In this game, $(a,c),(b,d)$ and $(0.5a+0.5b, 0.5c+0.5d)$ all satisfy backward induction, among others.

The following theorem states an important property of backward induction: if σ satisfies backward induction, then every behavioral conjecture σ_i assigns only positive probability to those player i strategies that are best responses against σ. Recall that, for a given strategy s_i, we denote by $\sigma_i(s_i)$ the probability assigned to s_i by the initial mixed conjecture μ_i induced by σ_i.

Theorem 3.1.4 *Let Γ be a game with perfect information and $\sigma = (\sigma_i)_{i\in I}$ a behavioral conjecture profile which satisfies backward induction. Then, for every player i, we have that $\sigma_i(s_i) > 0$ only if s_i is a best response against σ.*

Proof. Let $\sigma = (\sigma_i)_{i\in I}$ be a behavioral conjecture profile satisfying backward induction. Let s_i^* be a strategy for player i with $\sigma_i(s_i^*) > 0$. We will show that s_i^* is a sequential best response against (σ, β), where β is the trivial belief system in the game. Since (σ, β) is trivially Bayesian consistent, Lemma 2.6.4 would then guarantee that s_i^* is a best response against σ. Let X_i be the collection of decision nodes controlled by player i, and let $X_i(s_i^*)$ be those decision nodes in X_i that are not avoided by s_i^*. In order to prove that s_i^* is a sequential best response against (σ, β), we must show that

$$u_i((s_i^*, \sigma_{-i})|\, x) = \max_{s_i\in S_i(x)} u_i((s_i, \sigma_{-i})|\, x) \tag{3.1}$$

for all $x \in X_i$, where $S_i(x)$ is the set of player i strategies that do not avoid x. In order to prove (3.1) we show that

$$u_i((s_i^*, \sigma_{-i})|\, x) = u_i(\sigma|\, x) = \max_{s_i\in S_i(x)} u_i((s_i, \sigma_{-i})|\, x) \tag{3.2}$$

for every $x \in X_i(s_i^*)$. Here, $u_i(\sigma|\, x) = \sum_{a\in A(x)} \sigma_{ix}(a)\, u_i((a, \sigma_{-x})|\, x)$ is the expected utility induced by σ at x. We prove (3.2) by induction on the number of player i nodes that follow x. We say that x is followed by at most k player i nodes if every path starting at x crosses at most k other player i nodes. Suppose first that x is not followed by any player i node. Then, the expected utility at x does not depend on player i's behavior at nodes other than x, hence, for every $s_i \in S_i(x)$, $u_i((s_i, \sigma_{-i})|\, x) = u_i((s_i(x), \sigma_{-x})|\, x)$. Since σ satisfies backward induction, we have that $\sigma_{ix}(a) > 0$ only if $u_i((a, \sigma_{-x}|\, x) = \max_{a'\in A(x)} u_i((a', \sigma_{-x}|\, x)$. Hence, $u_i(\sigma|\, x) = \max_{a\in A(x)} u_i((a, \sigma_{-x}|\, x)$, which implies that $u_i(\sigma|\, x) \geq u_i((s_i(x), \sigma_{-x})|\, x) = u_i((s_i, \sigma_{-i})|\, x)$ for every $s_i \in S_i(x)$. Now, let a^* be the action chosen by s_i^* at x. Since $\sigma_i(s_i^*) > 0$ we have that $\sigma_{ix}(a^*) > 0$. Since σ satisfies backward induction, $u_i((a^*, \sigma_{-x})|\, x) = \max_{a\in A(x)} u_i((a, \sigma_{-x}|\, x) = u_i(\sigma|\, x)$. On the other hand, since x is not followed by any other player i information set, $u_i((s_i^*, \sigma_{-i})|\, x) = u_i((a^*, \sigma_{-x})|\, x)$ and hence $u_i((s_i^*, \sigma_{-i})|\, x) = u_i(\sigma|\, x)$. We have thus shown (3.2).

Now, let $k \geq 1$ and suppose that (3.2) holds for all nodes $x \in X_i(s_i^*)$ that are followed by at most $k-1$ player i nodes. Assume that $x \in X_i(s_i^*)$ is followed by at most k player i nodes. Let X_i^* be the set of player i nodes y that follow x and for which there is no player i node between x and y. Let $Z_0(x)$ be the set of terminal nodes that follow x and that are not preceded by any node in X_i^*. For a given strategy $s_i \in S_i(x)$, let $X_i^*(s_i)$ be those player i nodes in X_i^* that are not avoided by s_i. If player i chooses s_i and the game would start at x, then every terminal node that can possibly be reached is either preceded by a node $y \in X_i^*(s_i)$, or belongs to $Z_0(x)$.

Hence,

$$u_i((s_i, \sigma_{-i})| \ x) = \sum_{y \in X_i^*(s_i)} \mathbb{P}_{(s_i,\sigma_{-i})}(y| \ x) \ u_i((s_i, \sigma_{-i})| \ y) + + \sum_{z \in Z_0(x)} \mathbb{P}_{(s_i,\sigma_{-i})}(z| \ x) \ u_i(z), \tag{3.3}$$

where $\mathbb{P}_{(s_i,\sigma_{-i})}(y| \ x)$ is the probability that node y would be reached if the game would start at x. Similarly, for every action a at x we have

$$u_i((a, \sigma_{-x})| \ x) = \sum_{y \in X_i^*(a)} \mathbb{P}_{(a,\sigma_{-x})}(y| \ x) \ u_i((a, \sigma_{-x})| \ y) + + \sum_{z \in Z_0(x)} \mathbb{P}_{(a,\sigma_{-x})}(z| \ x) \ u_i(z), \tag{3.4}$$

where $X_i^*(a)$ is the set of player i nodes in X_i^* that are not avoided if the game would start at x, and a is chosen at x. Let a be the action chosen by $s_i \in S_i(x)$ at x. For every $y \in X_i^*(s_i)$, the probability $\mathbb{P}_{(s_i,\sigma_{-i})}(y| \ x)$ does not depend on player i's behavior at nodes other than x, since there is no player i node between x and y. The same property holds for the probabilities $\mathbb{P}_{(s_i,\sigma_{-i})}(z| \ x)$ where $z \in Z_0(x)$, since there is no player i node between x and z. Hence, $\mathbb{P}_{(s_i,\sigma_{-i})}(y| \ x) = \mathbb{P}_{(a,\sigma_{-x})}(y| \ x)$ for all $y \in X_i^*(s_i)$ and $\mathbb{P}_{(s_i,\sigma_{-i})}(z| \ x) = \mathbb{P}_{(a,\sigma_{-x})}(z| \ x)$ for all $z \in Z_0(x)$. By construction, every node $y \in X_i^*(s_i)$ is followed by at most $k-1$ other player i nodes, and hence by induction assumption we know that $u_i((s_i, \sigma_{-i})| \ y) \leq u_i(\sigma| \ y)$ at all such nodes y. Since the expected utility at y does not depend upon player i's behavior at x, it holds that $u_i(\sigma| \ y) = u_i((a, \sigma_{-x})| \ y)$ and hence $u_i((s_i, \sigma_{-i})| \ y) \leq u_i((a, \sigma_{-x})| \ y)$ at all $y \in X_i^*(s_i)$. It is also clear that $X_i^*(s_i) = X_i^*(a)$ since there are no player i nodes between x and X_i^*. Combining these insights with (3.3) leads to

$$\begin{aligned} u_i((s_i, \sigma_{-i})| \ x) &\leq \sum_{y \in X_i^*(a)} \mathbb{P}_{(a,\sigma_{-x})}(y| \ x) \ u_i((a, \sigma_{-x})| \ y) + \\ &\quad + \sum_{z \in Z_0(x)} \mathbb{P}_{(a,\sigma_{-x})}(z| \ x) \ u_i(z) \\ &= u_i((a, \sigma_{-x})| \ x). \end{aligned} \tag{3.5}$$

Now, let a' be some action at x with $\sigma_{ix}(a') > 0$. Since σ satisfies backward induction, we have that a' is local best response against σ, hence $u_i((a', \sigma_{-x})| \ x) = \max_{a'' \in A(x)} u_i((a'', \sigma_{-x})| \ x)$. Since this holds for every a' with $\sigma_{ix}(a') > 0$, we have that $u_i(\sigma| \ x) = \max_{a'' \in A(x)} u_i((a'', \sigma_{-x})| \ x)$. By (3.5) it follows that $u_i((s_i, \sigma_{-i})| \ x) \leq u_i(\sigma| \ x)$ for all $s_i \in S_i(x)$.

We finally show that $u_i((s_i^*, \sigma_{-i})| \ x) = u_i(\sigma| \ x)$, which would imply (3.2). By (3.3) we know that

$$u_i((s_i^*, \sigma_{-i})| \ x) = \sum_{y \in X_i^*(s_i^*)} \mathbb{P}_{(s_i^*,\sigma_{-i})}(y| \ x) \ u_i((s_i^*, \sigma_{-i})| \ y) + + \sum_{z \in Z_0(x)} \mathbb{P}_{(s_i^*,\sigma_{-i})}(z| \ x) \ u_i(z). \tag{3.6}$$

By induction assumption, we know for every $y \in X_i^*(s_i^*)$ that $u_i((s_i^*, \sigma_{-i})| \, y) = u_i(\sigma| \, y)$. Let a^* be the action chosen by s_i^* at x. Since $u_i(\sigma| \, y)$ does not depend on the behavior at x, we have $u_i(\sigma| \, y) = u_i((a^*, \sigma_{-x})| \, y)$ and hence $u_i((s_i^*, \sigma_{-i})| \, y) = u_i((a^*, \sigma_{-x})| \, y)$ for every $y \in X_i^*(s_i^*)$. By the same argument as used above for (s_i, σ_{-i}), we may conclude that $\mathbb{P}_{(s_i^*, \sigma_{-i})}(y| \, x) = \mathbb{P}_{(a^*, \sigma_{-x})}(y| \, x)$ for all $y \in X_i^*(s_i^*)$, $\mathbb{P}_{(s_i^*, \sigma_{-i})}(z| \, x) = \mathbb{P}_{(a^*, \sigma_{-x})}(z| \, x)$ for all $z \in Z_0(x)$ and $X_i^*(s_i^*) = X_i^*(a^*)$. Combining these insights with (3.6) we obtain

$$
\begin{aligned}
u_i((s_i^*, \sigma_{-i})| \, x) &= \sum_{y \in X_i^*(a^*)} \mathbb{P}_{(a^*, \sigma_{-x})}(y| \, x) \, u_i((a^*, \sigma_{-x})| \, y) + \\
&\quad + \sum_{z \in Z_0(x)} \mathbb{P}_{(a^*, \sigma_{-x})}(z| \, x) \, u_i(z) \\
&= u_i((a^*, \sigma_{-x})| \, x).
\end{aligned}
$$

By assumption, $\sigma_i(s_i^*) > 0$ and hence $\sigma_{ix}(a^*) > 0$. We know, from above, that then $u_i((a^*, \sigma_{-x})| \, x) = \max_{a \in A(x)} u_i((a, \sigma_{-x})| \, x) = u_i(\sigma| \, x)$. It thus follows that $u_i((s_i^*, \sigma_{-i})| \, x) = u_i(\sigma| \, x)$, which was to show. By induction, it follows that (3.2) holds for every $x \in X_i(s_i^*)$.We have thus shown that s_i^* is a sequential best response against (σ, β), which implies that s_i^* is a best response against σ. This completes the proof. ■

Suppose that σ satisfies backward induction in a game with perfect information. We then know, by the theorem above, that the induced initial conjecture about player i assigns positive probability only to those strategies that are best responses for player i, given his conjecture about the opponents' behavior. However, this property does not only hold for the *initial* conjecture about player i induced by σ, but also for the *revised* conjectures about player i. To formalize this property, we need the following definitions. For a given node x, let Γ_x be the part of the game that follows x. We refer to Γ_x as the *subgame* starting at x, since Γ_x may be interpreted as a game itself, with root x. For every player i, let σ_i^x be the restriction of σ_i on the player i nodes in Γ_x. Then, σ_i^x may be interpreted as the common conjecture that i's opponents hold about player i's behavior in Γ_x. Denote by $\sigma^x = (\sigma_i^x)_{i \in I}$ be the induced profile of behavioral conjectures in Γ_x. Let S_i^x be the set of player i strategies in the subgame Γ_x . Since Γ_x is a game, the phrase "a strategy $s_i \in S_i^x$ is a best response against σ^x in Γ_x" is well-understood. For every strategy $s_i \in S_i^x$, we denote by $\sigma_i^x(s_i)$ the probability assigned to s_i by σ_i^x in the subgame Γ_x. Note that $\sigma_i^x(s_i)$ may be generated by the initial mixed conjecture induced by σ_i, but also by some *revised* mixed conjecture about player i.

Theorem 3.1.5 *Let Γ be a game with perfect information and $\sigma = (\sigma_i)_{i \in I}$ a behavioral conjecture profile which satisfies backward induction. Then, for every node x and every strategy $s_i \in S_i^x$ we have that $\sigma_i^x(s_i) > 0$ only if s_i is a best response against σ^x in the subgame Γ_x.*

This result follows immediately from Theorem 3.1.4. Namely, consider a behavioral conjecture profile σ satisfying backward induction and an arbitrary subgame Γ_x.

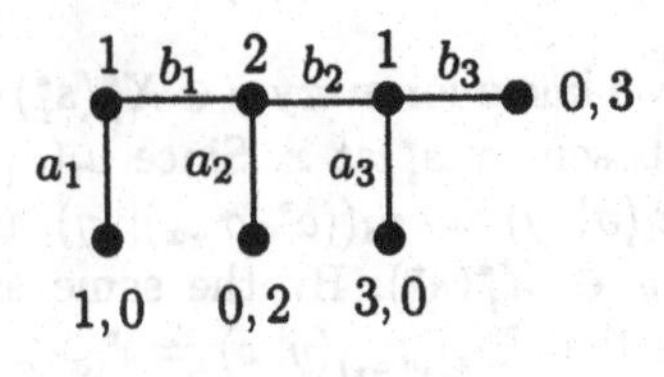

Figure 3.3

Then, σ^x satisfies backward induction in the subgame Γ_x. By applying Theorem 3.1.4 to σ^x and Γ_x, we obtain the desired result.

In view of Theorem 3.1.5, the concept of backward induction assumes that at every subgame, all players believe that their opponents will choose best responses against σ in the sequel. In particular, if a subgame can only be reached through a non-best response by player i, then i's opponents should still believe that player i will choose a best response against σ in this subgame. It is this property of backward induction that has sometimes been critized in the literature. Consider, for instance, the game in Figure 3.3. The unique behavioral conjecture profile satisfying backward induction is $\sigma = ((a_1, a_3), a_2)$. Suppose now that player 2's information set is reached. Then, player 2 knows that player 1 has not chosen a best response against σ. Backward induction, at this stage, requires player 2 to "ignore" this fact, and to believe that player 1 will still choose optimally in the remainder of the game, that is, will choose a_3 at the last node. The question is whether this requirement is reasonable. If player 2 observes that player 1 has chosen non-optimally against σ, then an alternative approach would be to allow player 2 to believe that player 1 is not rational. In this case, player 2 could believe that player 1 chooses b_3 instead of a_3. Given this belief revision process by player 2, it would be optimal for player 2 to choose b_2 instead of a_2. By this argument, also strategy b_2 can be viewed as a "reasonable" choice for player 2. For a more detailed discussion of this issue, the reader is referred to Sections 7.5 and 7.6 and the references therein. See also Selten (1978) and Rosenthal (1981) for a discussion of examples in which backward induction leads to "unintuitive" outcomes.

3.2 Nash Equilibrium

As we have mentioned in Section 2.3, the players' utility functions $u_i : Z \to \mathbb{R}$ are assumed to be common knowledge. The only uncertainty that remains about the opponents' preference relations are thus the subjective probability distributions $(\mu_i)_{i \in I}$ held by the players. Assume now that these subjective probability distributions would also be common knowledge, that is, there would be common knowledge about the exact preference relations held by the players. Then, player i would be informed about the conjecture μ_i held by the opponents about player i, and each of player i's opponents would be informed about the conjecture μ_{-i} held by player i about the other players' behavior. Consequently, if the opponents believe that player i chooses a best response, then the opponents' common conjecture μ_i about player i should assign positive probability only to those player i strategies s_i that are best responses against

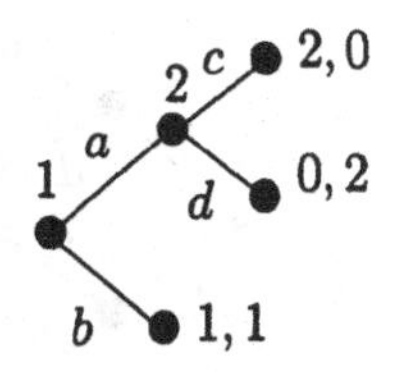

Figure 3.4

μ_{-i}. Mixed conjecture profiles with this property are called Nash equilibria (Nash 1950, 1951).

Definition 3.2.1 *A mixed conjecture profile $\mu = (\mu_i)_{i \in I}$ is called a Nash equilibrium if for every player i, $\mu_i(s_i) > 0$ only if s_i is a best response against μ. A strategy s_i is called a Nash equilibrium strategy if there is a Nash equilibrium μ such that s_i is a best response against μ.*

Similarly, we say that a behavioral conjecture profile $\sigma = (\sigma_i)_{i \in I}$ is a Nash equilibrium if for every player i, $\sigma_i(s_i) > 0$ only if s_i is a best response against σ. In contrast to backward induction, the players' *revision* of conjectures does not play a role in the concept of Nash equilibrium. Namely, if σ is a Nash equilibrium, then every behavioral conjecture profile σ' which induces the same initial mixed conjectures as σ is a Nash equilibrium as well.

By definition, if a strategy s_i is assigned positive probability in some Nash equilibrium μ, then s_i is a Nash equilibrium strategy. The converse, however, is not true. Consider the game in Figure 3.4.In this game, $\mu^* = (b, 0.5c + 0.5d)$ is a Nash equilibrium. Since a is a best response against σ, we have that a is a Nash equilibrium strategy. However, there is no Nash equilibrium μ which assigns positive probability to a. Suppose, namely, that $\mu = (\mu_1, \mu_2)$ would be a Nash equilibrium with $\mu_1(a) > 0$. Then, d is player 2's unique best response against μ, and hence $\mu_2 = d$. Consequently, b is player 1's unique best response against μ, which implies that $\mu_1 = b$. However, this contradicts the assumption that $\mu_1(a) > 0$.

We now turn to the question whether a Nash equilibrium always exists. The game in Figure 3.5 shows that a game may fail to have a Nash equilibrium in *pure* behavioral conjectures.In this game there is a unique Nash equilibrium in behavioral conjectures, namely $\sigma = (0.5a + 0.5b, 0.5c + 0.5d)$, which is not pure. We will show, however, that a Nash equilibrium always exists in every extensive form game. Before doing so, we present Kakutani's fixed point theorem (Kakutani, 1941), which plays a key role in the existence proof. We need the following definitions. Let A and B be subsets of some Euclidean space $\mathbb{R}^d$. A *correspondence* $F : A \twoheadrightarrow B$ is a mapping assigning to every $a \in A$ some subset $F(a) \subseteq B$. The correspondence F is called *upper-hemicontinuous* if for every sequence $(a^k)_{k \in \mathbb{N}}$ in A converging to some $a \in A$ and every sequence $(b^k)_{k \in \mathbb{N}}$ in B converging to some $b \in B$ with $b^k \in F(a^k)$ for all k, it holds that $b \in F(a)$. Let $F : A \twoheadrightarrow A$ be a correspondence from a set A to the same set A. A point $a^* \in A$ is called a *fixed point* of F if $a^* \in F(a^*)$.

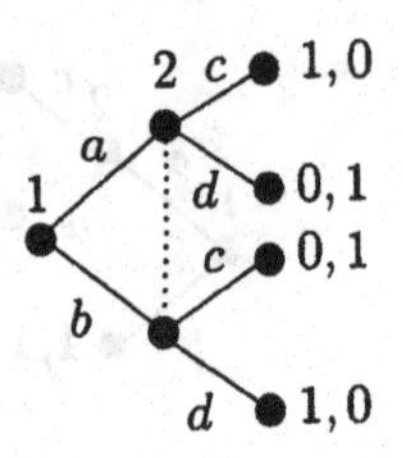

Figure 3.5

Theorem 3.2.2 *(Kakutani's Fixed Point Theorem). Let $A \subseteq \mathbb{R}^d$ be a nonempty, convex and compact set. Let $F : A \twoheadrightarrow A$ be an upper-hemicontinuous correspondence such that $F(a)$ is a nonempty, convex and compact subset of A for all $a \in A$. Then, F has a fixed point.*

With this result at hand, we are able to prove the following theorem, which is due to Nash (1950, 1951).

Theorem 3.2.3 *For every extensive form game, there exists a Nash equilibrium $\mu = (\mu_i)_{i\in I}$.*

Proof. Let Γ be an extensive form game. Let $M = \times_{i\in I}\Delta(S_i)$ be the set of mixed conjecture profiles. Obviously, M is a nonempty, convex and compact subset of some Euclidean space $\mathbb{R}^d$. For every mixed conjecture profile $\mu = (\mu_i)_{i\in I}$ and every player i let $\tilde{B}_i(\mu) = \{s_i \in S_i | \ s_i$ best response against $\mu\}$ and let $B_i(\mu) = \Delta(\tilde{B}_i(\mu)) \subseteq \Delta(S_i)$. Obviously, $B_i(\mu)$ is nonempty, convex and compact for every μ. Let B be the correspondence from M to M assigning to every $\mu \in M$ the set $B(\mu) = \times_{i\in I}B_i(\mu) \subseteq M$. Since $B_i(\mu)$ is nonempty, convex and compact for all i, the set $B(\mu)$ is a nonempty, convex and compact subset of M for all μ. It remains to show that the correspondence B is upper-hemicontinuous. Suppose that $(\mu^n)_{n\in\mathbb{N}}$ is a sequence of mixed conjecture profiles converging to μ. Let the sequence $(\bar{\mu}^n)_{n\in\mathbb{N}}$ in M be such that $\bar{\mu}^n \in B(\mu^n)$ for all n and let $(\bar{\mu}^n)_{n\in\mathbb{N}}$ converge to $\bar{\mu}$. We must show that $\bar{\mu} \in B(\mu) = \times_{i\in I}\Delta(\tilde{B}_i(\mu))$. Hence, we must prove for every player i that $\bar{\mu}_i \in \Delta(\tilde{B}_i(\mu))$. Suppose not. Then, there is some s_i which is not a best response against μ with $\bar{\mu}_i(s_i) > 0$. Hence, there is some strategy $r_i \in S_i$ with $u_i(s_i, \mu_{-i}) < u_i(r_i, \mu_{-i})$. Consequently, for n large enough, $u_i(s_i, \mu^n_{-i}) < u_i(r_i, \mu^n_{-i})$, which implies that s_i is not a best response against μ^n if n is large enough. Since $\bar{\mu}^n_i \in \Delta(\tilde{B}_i(\mu^n))$ for all n, it follows that $\bar{\mu}^n_i(s_i) = 0$ for n large enough. Since $(\bar{\mu}^n_i)_{n\in\mathbb{N}}$ converges to $\bar{\mu}_i$, it follows that $\bar{\mu}_i(s_i) = 0$, which is a contradiction. We may thus conclude that the correspondence B is upper-hemicontinuous. By Kakutani's fixed point theorem, we know that B has some fixed point $\mu^* = (\mu^*_i)_{i\in I}$. Hence, by definition, $\mu^*_i \in B_i(\mu^*)$ for all players i. In other words, $\mu^*_i(s_i) > 0$ only if $s_i \in \tilde{B}_i(\mu^*)$. But then, by definition, μ^* is a Nash equilibrium. ■

Of course, the theorem above implies that every extensive form game has a Nash equilibrium $\sigma = (\sigma_i)_{i\in I}$ in behavioral conjectures as well. We shall now present a technical characterization of Nash equilibria which will be useful for proofs. For a given behavioral conjecture profile $\sigma = (\sigma_i)_{i \in I}$, we refer to $u_i(\sigma) = \sum_{s_i \in S_i} \sigma_i(s_i)\, u_i(s_i, \sigma_{-i})$ as the expected utility for player i induced by σ. Nash equilibria may now be characterized as follows.

Lemma 3.2.4 *A behavioral conjecture profile $\sigma = (\sigma_i)_{i \in I}$ is a Nash equilibrium if and only if for every player i, $u_i(\sigma) = \max_{\sigma_i'} u_i(\sigma_i', \sigma_{-i})$.*

Proof. Let $\sigma = (\sigma_i)_{i \in I}$ be a Nash equilibrium. Then, $\sigma_i(s_i) > 0$ only if $u_i(s_i, \sigma_{-i}) = \max_{s_i'} u_i(s_i', \sigma_{-i})$, which implies that $u_i(\sigma) = \max_{\sigma_i'} u_i(\sigma_i', \sigma_{-i})$. On the other hand, let σ be such that $u_i(\sigma) = \max_{\sigma_i'} u_i(\sigma_i', \sigma_{-i})$ for every player i. Suppose that σ is not a Nash equilibrium. Hence, there is some $s_i, s_i' \in S_i$ with $u_i(s_i, \sigma_{-i}) < u_i(s_i', \sigma_{-i})$ but $\sigma_i(s_i) > 0$. But then, every $\tilde{\sigma}_i$ with $u_i(\tilde{\sigma}_i, \sigma_{-i}) = \max_{\sigma_i'} u_i(\sigma_i', \sigma_{-i})$ should assign probability zero to s_i. Hence, $u_i(\sigma) < \max_{\sigma_i'} u_i(\sigma_i', \sigma_{-i})$, which is a contradiction. ■

In the remainder of this section we apply the Nash equilibrium concept to games with perfect information and zero-sum games, and discuss its relationship with backward induction and max-min strategies, respectively. At the end of the section, we combine the obtained results in order to prove Zermelo's theorem.

3.2.1 Games with Perfect Information

Let Γ be a game with perfect information and σ a behavioral conjecture profile satisfying backward induction. By Theorem 3.1.4 we know that σ is then a Nash equilibrium. Or, even stronger, Theorem 3.1.5 states that for every subgame Γ_x, the induced behavioral conjecture profile σ^x constitutes a Nash equilibrium in Γ_x. Since Lemma 3.1.2 guarantees that we can always find a *pure* behavioral conjecture profile satisfying backward induction, it follows that a game with perfect information always contains a Nash equilibrium in *pure* behavioral conjectures. These facts are gathered in the following lemma.

Lemma 3.2.5 *Every game with perfect information contains a Nash equilibrium $\sigma = (\sigma_i)_{i \in I}$ in pure behavioral conjectures. Moreover, this Nash equilibrium σ can be chosen such that in every subgame Γ_x the induced behavioral conjecture profile σ^x constitutes a Nash equilibrium in Γ_x.*

In a game with perfect information, not every Nash equilibrium satisfies backward induction, however. Consider, for instance, the game in Figure 3.6. The behavioral conjecture profile $\sigma = (b, d)$ is a Nash equilibrium, but it does not satisfy backward induction. The fundamental difference between both concepts is that Nash equilibrium does not require players to believe that their opponents choose optimally at every subgame, whereas backward induction does. Note that in this example, d is a best response for player 2 against σ, but it is no longer a best response in the subgame that starts at player 2's decision node.

3.2.2 Zero-Sum Games

In this subsection we focus on two-player games in which both players have opposite interests. Think, for instance, of recreational games for two players in which either

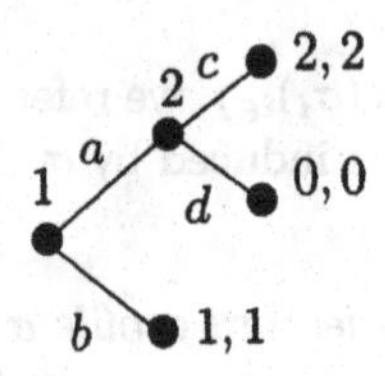

Figure 3.6

one of both players wins at the end, or the game ends in a draw. Or, more generally, think of situations where a fixed amount has to be distributed among the players at the end. In contrast to the other sections in this book, suppose that the preference relations for both players are not represented by subjective expected utility functions, as introduced in Sections 2.2 and 2.3, but are instead of the *max-min type.* Moreover, we shall assume now that the objects of choice for both players are *randomized strategies* instead of strategies. Here, a *randomized strategy* for player i is a probability distribution $\rho_i \in \Delta(S_i)$ on the set of strategies. Despite the fact that randomized strategies are mathematically equivalent to mixed conjectures, their interpretation is crucially different. A randomized strategy ρ_i reflects a situation where player i chooses an objective randomization device that selects each of his strategies s_i with the objective probability $\rho_i(s_i)$. One may think of drawing a coloured ball from an urn, where each colour corresponds to a strategy in S_i. In a mixed conjecture μ_i, the probabilities $\mu_i(s_i)$ are subjective probabilities held by player i's opponent instead of objective probabilities chosen by player i. Let R_i be the set of randomized strategies for player i. For every pair of randomized strategies $\rho = (\rho_i, \rho_j)$, let $\mathbb{P}_\rho \in \Delta(Z)$ be the objective probability distribution on the set of terminal nodes induced by ρ. We assume that both players i hold a preference relation $\succeq_i$ on their set of randomized strategies R_i which is of the *max-min type.* By the latter we mean that there is a utility function $u_i : Z \to \mathbb{R}$ such that for every $\rho_i, \rho_i' \in R_i$ we have: $\rho_i \succeq_i \rho_i'$ if and only if

$$\min_{\rho_j \in R_j} \sum\nolimits_{z \in Z} \mathbb{P}_{(\rho_i, \rho_j)}(z)\ u_i(z) \geq \min_{\rho_j \in R_j} \sum\nolimits_{z \in Z} \mathbb{P}_{(\rho_i', \rho_j)}(z)\ u_i(z).$$

For every randomized strategy pair $\rho = (\rho_i, \rho_j)$ we refer to $\sum_{z \in Z} \mathbb{P}_{(\rho_i, \rho_j)}(z)\ u_i(z)$ as the *expected utility* induced by (ρ_i, ρ_j), and denote it by $u_i(\rho_i, \rho_j)$. Let $v_i(\rho_i) = \min_{\rho_j \in R_j} u_i(\rho_i, \rho_j)$ be the minimum expected utility that player i may obtain when choosing ρ_i. Hence, both players are assumed to choose randomized strategies that maximize the minimum expected utility $v_i(\rho_i)$. We further assume that the utility functions $u_1, u_2 : Z \to \mathbb{R}$ for players 1 and 2 can be chosen such that $u_1(z) + u_2(z) = 0$ for every terminal node z. This property reflects the fact that both players have opposite interests in the game. If this property is satisfied, we say that the game is a *zero-sum game*. Let $v_i = \max_{\rho_i \in R_i} v_i(\rho_i)$ be the maximum expected utility that player i can guarantee for himself. We say that v_i is the *reservation value* for player i. A randomized strategy ρ_i with $v_i(\rho_i) = v_i$ is called a *max-min strategy* for player i.

The following theorem, which is due to von Neumann (1928), establishes an important property of max-min strategies in zero-sum games: if both players choose max-min strategies ρ_i and ρ_j, then ρ_i is a best response against ρ_j, and ρ_j is a best response against ρ_i. Moreover, the other direction is also true: if ρ_i and ρ_j constitute mutual best responses, then both ρ_i and ρ_j are max-min strategies. For the proof of this result we use the fact that a Nash equilibrium in mixed conjectures always exists in zero-sum games. By interpreting the mixed conjectures of the Nash equilibrium as randomized strategies, the result follows easily. *En route* to this result we prove two additional properties of zero-sum games, namely that the sum of the reservation values for the players is zero, and that every pair of max-min strategies yields the same pair of expected utilites.

Theorem 3.2.6 *Let Γ be a two-person zero sum game. Then,*
(1) (ρ_i, ρ_j) is a pair of max-min strategies if and only if $u_i(\rho_i, \rho_j) = \max_{\rho_i' \in R_i} u_i(\rho_i', \rho_j)$ and $u_j(\rho_i, \rho_j) = \max_{\rho_j' \in R_j} u_j(\rho_i, \rho_j')$,
(2) for every pair of max-min strategies (ρ_i, ρ_j) we have $u_i(\rho_i, \rho_j) = v_i$ and $u_j(\rho_i, \rho_j) = v_j$ and
(3) $v_i + v_j = 0$.

Proof. The extensive form structure, together with the utility functions $u_i : Z \to \mathbb{R}$ constitute an extensive form game Γ' in the sense defined in Section 2.3. In Γ', the players are assumed to have preferences of the subjective expected utility type, and $u_i : Z \to \mathbb{R}$ represents the utility function for player i used to represent these preferences. By Theorem 3.2.3 we know that Γ' has a Nash equilibrium in mixed conjectures (μ_i^*, μ_j^*). Now, let ρ_i^*, ρ_j^* be those randomized strategies whose objective probabilities coincide with the subjective probabilities in μ_i^*, μ_j^*, that is, $\rho_i^*(s_i) = \mu_i^*(s_i)$ for all s_i, and the same for ρ_j^*. Since (μ_i^*, μ_j^*) is a Nash equilibrium, we have that $\mu_i^*(s_i) > 0$ only if s_i is a best response against μ_j^*. However, this implies that $u_i(\rho_i^*, \rho_j^*) = \max_{\rho_i} u_i(\rho_i, \rho_j^*)$. Similarly, $u_j(\rho_i^*, \rho_j^*) = \max_{\rho_j} u_j(\rho_i^*, \rho_j)$. Since $u_i(\rho_i^*, \rho_j) = -u_j(\rho_i^*, \rho_j)$, it follows that $u_i(\rho_i^*, \rho_j^*) = \min_{\rho_j} u_i(\rho_i^*, \rho_j)$, and hence $u_i(\rho_i^*, \rho_j^*) \leq v_i$. Using the same argument for player j, it can be shown that $u_j(\rho_i^*, \rho_j^*) \leq v_j$. Since $u_i(\rho_i^*, \rho_j^*) + u_j(\rho_i^*, \rho_j^*) = 0$, we have $v_i + v_j \geq 0$.

On the other hand, let (ρ_i, ρ_j) be a pair of max-min strategies. Then, $u_i(\rho_i, \rho_j) \geq v_i$ and $u_j(\rho_i, \rho_j) \geq v_j$, implying that $v_i + v_j \leq u_i(\rho_i, \rho_j) + u_j(\rho_i, \rho_j) = 0$. Since we have seen that $v_i + v_j \geq 0$, it follows that $v_i + v_j = 0$. Hence we have shown (3). Since for the pair of max-min strategies (ρ_i, ρ_j) we know that $u_i(\rho_i, \rho_j) \leq v_i$, $u_j(\rho_i, \rho_j) \leq v_j$, $u_i(\rho_i, \rho_j) + u_j(\rho_i, \rho_j) = 0$ and $v_i + v_j = 0$, it follows that $u_i(\rho_i, \rho_j) = v_i$ and $u_j(\rho_i, \rho_j) = v_j$, which proves (2).

We have seen above that there is a pair (ρ_i^*, ρ_j^*) with $u_i(\rho_i^*, \rho_j^*) = \max_{\rho_i} u_i(\rho_i, \rho_j^*)$ and $u_j(\rho_i^*, \rho_j^*) = \max_{\rho_j} u_j(\rho_i^*, \rho_j)$. Since we have seen above that $u_i(\rho_i^*, \rho_j^*) \leq v_i$, $u_j(\rho_i^*, \rho_j^*) \leq v_j$, $u_i(\rho_i^*, \rho_j^*) + u_j(\rho_i^*, \rho_j^*) = 0$ and $v_i + v_j = 0$, it follows that $u_i(\rho_i^*, \rho_j^*) = v_i$ and $u_j(\rho_i^*, \rho_j^*) = v_j$. Moreover, we have seen that $u_i(\rho_i^*, \rho_j^*) = \min_{\rho_j} u_i(\rho_i^*, \rho_j)$ and hence $\min_{\rho_j} u_i(\rho_i^*, \rho_j) = v_i$, which implies that ρ_i^* is a max-min strategy. Similarly, it follows that ρ_j^* is a max-min strategy. This completes one direction of (1). Suppose, on the other hand, that (ρ_i, ρ_j) is a pair of max-min strategies. Then, we know by (2) that $u_j(\rho_i, \rho_j) = v_j$. Since $v_j = \min_{\rho_i'} u_j(\rho_i', \rho_j)$, it follows that $\max_{\rho_i'} u_i(\rho_i', \rho_j) = -v_j = v_i$.

On the other hand, we know by (2) that $u_i(\rho_i, \rho_j) = v_i$. We may thus conclude that $\max_{\rho'_i} u_i(\rho'_i, \rho_j) = u_i(\rho_i, \rho_j)$. Similarly, $\max_{\rho'_j} u_j(\rho_i, \rho'_j) = u_j(\rho_i, \rho_j)$. We have therefore shown (1). This completes the proof. ■

3.2.3 Zermelo's Theorem

We finally combine the results obtained in the previous two subsections to prove Zermelo's Theorem (Zermelo, 1913) for the game of chess. Note that chess is a typical example of a two-person zero-sum game with perfect information, and hence both Subsections 3.2.1 and 3.2.2 apply to this game. Suppose that we adopt the following stopping rule for the game of chess: if the same configuration on the chess board occurs more than twice, the game ends in a draw. Since there are only finitely many possible configurations on the chess board, the game ends after finitely many moves. As such, the game can be represented by a finite game-tree. We assume that the utility functions $u_i : Z \to \mathbb{R}$ for the players are as follows. If white wins, white has utility 1 and black has utility -1. If black wins, the utilities are reversed. If the game ends in a draw, both players get utility zero. By construction, Γ is then a zero-sum game. Zermelo's Theorem states that in the game of chess, either some of the players has a winning strategy, or both players have a strategy that guarantees at least a draw to them. Here, we say that s_w is a *winning strategy* for white if for each of black's strategies s_b the strategy profile (s_w, s_b) leads to a win for white. We say that the strategy s_w guarantees at least a draw to white if for each of black's strategies s_b the strategy profile (s_w, s_b) either leads to a draw, or leads to a win for white. The proof we offer here is quite different from Zermelo's original proof. For a detailed discussion of Zermelo's approach to the problem, the reader is referred to Schwalbe and Walker (2001). The appendix of that paper contains a translation to english of Zermelo's paper.

Theorem 3.2.7 (*Zermelo's Theorem*). *Consider the game of chess described above. Then, one of the following three statements is true:*
(1) White has a winning strategy s_w^*.
(2) Black has a winning strategy s_b^*.
(3) White and black have strategies s_w^* *and* s_b^* *which guarantee at least a draw to them.*

Proof. Since this game of chess is a finite extensive form game with perfect information, we know by Lemma 3.2.5 that there is a Nash equilibrium $\sigma = (\sigma_w, \sigma_b)$ in pure behavioral conjectures. Let (μ_w, μ_b) be the initial mixed conjectures induced by (σ_w, σ_b). Then, both μ_w and μ_b assign probability one to one particular strategy, say s_w^* and s_b^*. Identify s_w^* and s_b^* with the randomized strategies that put probability one on s_w^* and s_b^*, respectively. Since (μ_w, μ_b) is a Nash equilibrium, we have that $u_w(s_w^*, s_b^*) = \max_{s_w} u_w(s_w, s_b^*)$ and $u_b(s_w^*, s_b^*) = \max_{s_b} u_b(s_w^*, s_b)$, and hence $u_w(s_w^*, s_b^*) = \max_{\rho_w} u_w(\rho_w, s_b^*)$ and $u_b(s_w^*, s_b^*) = \max_{\rho_b} u_b(s_w^*, \rho_b)$. Here, ρ_w and ρ_b denote the randomized strategies for the players. The latter equalities follow from the fact that $\max_{\rho_w} u_w(\rho_w, s_b^*)$ can always be achieved by using a randomized strategy ρ_w that puts probability one on one particular strategy. By Theorem 3.2.6 it then follows that (s_w^*, s_b^*) is a pair of max-min strategies, $u_w(s_w^*, s_b^*) = v_w$, $u_b(s_w^*, s_b^*) = v_b$, and $v_w + v_b = 0$. Here, v_w and v_b are the reservation values for the players, as defined in the previous subsection. Obviously, (s_w^*, s_b^*) either leads to a win for white, or a win for black, or a draw. (1) If (s_w^*, s_b^*) leads to a win for white, then $u_w(s_w^*, s_b^*) = 1$, which implies that $v_w = 1$. But this means that white can guarantee a win. Since s_w^* is a max-min strategy, it follows that s_w^* guarantees a win for white. (2) If (s_w^*, s_b^*) leads to a win for black, than we can conclude by the same argument that s_b^* guarantees a win for black. (3) If (s_w^*, s_b^*) leads to a draw, then $u_w(s_w^*, s_b^*) = u_b(s_w^*, s_b^*) = 0$, which implies that $v_w = v_b = 0$. Hence, both players can guarantee a draw. Since s_w^*, s_b^* are max-min strategies, both s_w^* and s_b^* guarantee a draw. ■

The reason why chess is still a game of intellectual interest is the fact that we do not know which of the three statements in Zermelo's Theorem is true. The problem is that the game-tree of chess is simply too big to compute a Nash equilibrium, which would then reveal which of the three statements holds. Of course, Zermelo's Theorem does not only hold for the game of chess, but also for other recreational games with perfect information, like tic-tac-toe, checkers or backgammon. For the game of tic-tac-toe, for instance, we know that statement (3) holds. This game is small enough such that we can explicitly compute a Nash equilibrium by backward induction.

3.3 Subgame Perfect Equilibrium

In this section we introduce the concept of *subgame perfect equilibrium* (Selten, 1965), which may be viewed as a combination of backward induction and Nash equilibrium. By Theorem 3.1.5 we know that in games with perfect information, a behavioral conjecture profile satisfying backward induction induces, in every subgame, a Nash equilibrium of that subgame. The idea of subgame perfect equilibrium is to extend this property to games without perfect information. To this purpose, we should first define what we mean by *subgames* in games without perfect information. Consider an extensive form game Γ (possibly with imperfect information) and its game-tree T. Let x be a non-terminal node and T_x the subtree which starts at x. Let Γ_x be the subtree T_x together with the utilities at its terminal nodes. We call Γ_x a *subgame* if every information set in Γ is either completely contained in Γ_x or completely contained in the complement of Γ_x. Intuitively, this means that every player in Γ_x knows that the game has started at x, since every node in Γ_x can be distinguished from nodes not belonging to Γ_x. In particular, the node x itself should be a singleton information set. A subgame Γ_x can thus be interpreted as a game itself, with root x. For a given behavioral conjecture profile σ, let σ^x be the restriction of σ to the subgame Γ_x, i.e. $\sigma^x = (\sigma_{ih})_{i\in I, h\in H_i(x)}$, where $H_i(x)$ are the player i information sets in Γ_x.

Definition 3.3.1 *A behavioral conjecture profile σ is called a subgame perfect equilibrium if for every subgame Γ_x, the induced behavioral conjecture profile σ^x is a Nash equilibrium in Γ_x.*

By construction, every subgame perfect equilibrium is a Nash equilibrium since the whole game Γ is a subgame. In games with perfect information, we know by

Theorem 3.1.5 that every behavioral conjecture profile satisfying backward induction is a subgame perfect equilibrium. In fact, the other direction is also true. Consider namely a game Γ with perfect information and a subgame perfect equilibrium σ in this game. Choose a final decision node x controlled by some player i, that is, x is only followed by terminal nodes. Since Γ_x is a subgame and σ^x is a Nash equilibrium in Γ_x, σ assigns positive probability only to those actions at x that are optimal for i. By working backwards from the end of the game to the beginning, one can show that for every node x, σ assigns positive probability only to those actions at x that are local best responses against σ, and hence σ satisfies backward induction. The following equivalence result is therefore obtained.

Lemma 3.3.2 *In a game with perfect information, a behavioral conjecture profile σ is a subgame perfect equilibrium if and only if it satisfies backward induction.*

In the following theorem, we prove that a subgame perfect equilibrium always exists in every extensive form game.

Theorem 3.3.3 *Every game in extensive form has at least one subgame perfect equilibrium.*

Proof. We construct a behavioral conjecture profile σ as follows. Let $M(\Gamma)$ be the collection of minimal subgames in Γ, that is, subgames that contain no smaller subgames. By construction, all subgames in $M(\Gamma)$ are disjoint. For every subgame $\Gamma_x \in M(\Gamma)$, choose a corresponding Nash equilibrium σ^x. The existence of such a Nash equilibrium σ^x is guaranteed by the previous section. Let Γ^1 be the "reduced" game obtained as follows. Take the game Γ, replace each subgame $\Gamma_x \in M(\Gamma)$ by a terminal node z_x and choose the utilities at z_x equal to the expected utilities of the players in σ^x. Formally, let $u_i(\sigma|\ x) = \sum_{s_i \in S_i^x} \sigma_i^x(s_i)\ u_i((s_i, \sigma_{-i}^x)|\ x)$ be player i's expected utility generated by σ^x in the subgame Γ_x. For every "new" terminal node z_x, define the utility $u_i(z_x)$ by $u_i(z_x) = u_i(\sigma|\ x)$. In the reduced game Γ^1, take the collection $M(\Gamma^1)$ of minimal subgames in Γ^1, and for every subgame $\Gamma_x \in M(\Gamma^1)$ choose a Nash equilibrium σ^x. In the obvious way, we define the reduced game Γ^2 induced by Γ^1, and so on, until the game Γ^r can not be reduced any more. By putting all these Nash equilibria σ^x of such mimimal subgames together, we obtain a behavioral conjecture profile σ in the "big game" Γ. We show that σ is a subgame perfect equilibrium in Γ.

For $r = 0, 1, 2, ...$ we define *subgames of degree r* by the following recursive definition. A subgame of degree 0 is a minimal subgame of Γ. For $r \geq 1$, a subgame of degree r is a subgame Γ_x such that (1) every proper subgame of Γ_x is of degree $\leq r-1$ and (2) Γ_x contains a proper subgame of degree $r-1$. Here, by proper subgame we mean a subgame of Γ_x which is not Γ_x itself. We show by induction on r that σ constitutes a Nash equilibrium in every subgame of degree r.

If $r = 0$, the statement is true by construction. Now, let $r \geq 1$ and let the statement be true for every $r' < r$. Let Γ_x be a subgame of degree r and let σ^x be the restriction of σ to Γ_x. In view of Lemma 3.2.4, showing that σ^x is a Nash equilibrium in Γ_x is equivalent to showing that $u_i(\sigma|\ x) = \max_{\sigma_i'} u_i((\sigma_i', \sigma_{-i})|\ x)$ for all players i and all behavioral conjectures σ_i'. Let $Y(x)$ be the set of nodes y in Γ_x such that (1) Γ_y is a proper subgame of Γ_x, and (2) there is no node y' between x and y such that $\Gamma_{y'}$ is a proper subgame of Γ_x. Then,

$$u_i(\sigma | x) = \sum_{y \in Y(x)} \mathbb{P}_\sigma(y | x) \, u_i(\sigma | y), \tag{3.7}$$

where $\mathbb{P}_\sigma(y| x)$ is the probability under σ that y is reached if the game starts at x. Since every subgame $\Gamma_y, y \in Y(x)$, is a subgame of degree $\leq r-1$, we know by induction assumption that

$$u_i(\sigma | y) = \max_{\sigma_i'} u_i((\sigma_i', \sigma_{-i}) | y) \tag{3.8}$$

for all $y \in Y(x)$. Let Γ^k be the reduced game in which Γ_x is a minimal subgame. Then, every node $y \in Y(x)$ is a terminal node z_y in Γ^k with utilities $u_i(z_y) = u_i(\sigma | y)$. So, with (3.7),

$$u_i(\sigma | x) = \sum_{y \in Y(x)} \mathbb{P}_\sigma(y | x) \, u_i(z_y).$$

Since, by construction, σ^x constitutes a Nash equilibrium in this minimal subgame of Γ^k, we have by Lemma 3.2.4,

$$u_i(\sigma | x) = \max_{\sigma_i''} u_i((\sigma_i'', \sigma_{-i}) | x) = \max_{\sigma_i''} \sum_{y \in Y(x)} \mathbb{P}_{(\sigma_i'', \sigma_{-i})}(y | x) \, u_i(z_y).$$

Together with (3.8), it follows that

$$\begin{aligned} u_i(\sigma | x) &= \max_{\sigma_i''} \sum_{y \in Y(x)} \mathbb{P}_{(\sigma_i'', \sigma_{-i})}(y | x) \, \max_{\sigma_i'} u_i((\sigma_i', \sigma_{-i}) | y) \\ &= \max_{\sigma_i'} \sum_{y \in Y(x)} \mathbb{P}_{(\sigma_i', \sigma_{-i})}(y | x) \, u_i((\sigma_i', \sigma_{-i}) | y) \\ &= \max_{\sigma_i'} u_i((\sigma_i', \sigma_{-i}) | x). \end{aligned}$$

Since this holds for every player i, we have that σ^x is a Nash equilibrium in Γ_x. By induction, it follows that σ^x is a Nash equilibrium in Γ_x for every subgame Γ_x. ■

As with backward induction, in a subgame perfect equilibrium σ players are required to believe, in every subgame, that their opponents will choose best responses against σ in this subgame. In particular, they should believe so even if this subgame could only have been reached via non-best responses by the same opponents against σ. Consider, for instance, the game in Figure 3.7. In the subgame which starts at player 2's decision node, the unique Nash equilibrium is $(0.5e + 0.5f, 0.5c + 0.5d)$. Consequently, the game has a unique subgame perfect equilibrium, namely $\sigma = ((b, 0.5e + 0.5f), 0.5c + 0.5d)$. Suppose now that player 2's decision node is reached. Then, player 2 knows that player 1 has not chosen a best response against σ. In spite of this, the concept of subgame perfect equilibrium requires σ to constitute a Nash

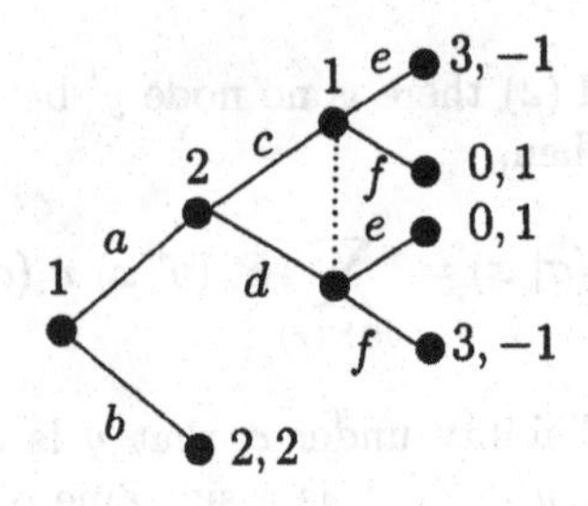

Figure 3.7

equilibrium also in the subgame which remains after player 2's decision node. As with backward induction, the question may be raised why σ should still constitute a Nash equilibrium in this subgame if the event of reaching this subgame has contradicted player 2's conjecture that player 1 will choose optimally against σ. One could say that the concept of subgame perfect equilibrium requires the players to "ignore" the history that has led to a particular subgame, and that each subgame should be treated as if the game would actually start here.

3.4 Perfect Equilibrium

In the concepts of Nash equilibrium and subgame perfect equilibrium, players are required to attach positive probability only to those opponents' strategies that constitute best responses. However, players may still assign positive probability to opponents' strategies that are *not local best responses*. Consider for instance the game in Figure 3.8.In this game, $\sigma = (c, e)$ is a subgame perfect equilibrium (and hence a Nash equilibrium), but e is not a local best response for any beliefs β for player 2. Note that the conjecture $\sigma_2 = e$ cannot be ruled out by subgame perfect equilibrium since there is no subgame which starts at player 2's information set. The reason why e is a best response against σ but not a local best response lies in the fact that player 2's conjecture $\sigma_1 = c$ excludes the event that player 2's information set will be reached. Consequently, player 2 is indifferent between his two actions. However, if player 2 would attach a small positive probability to the actions a and b, then e would no longer be a best response, and the "unreasonable" conjecture $\sigma_2 = e$ could thus be eliminated. In other words, if player 1 believes that player 2 attaches positive probability to each of player 1's actions, then player 1 should believe (with probability one) that player 2 chooses d. This is the basic idea underlying the concept of *perfect equilibrium* (Selten, 1975) which will be presented in this section.

The idea above may be generalized as follows. Say that a behavioral conjecture σ_i about player i is *strictly positive* if $\sigma_{ih}(a) > 0$ for every $h \in H_i$ and every action

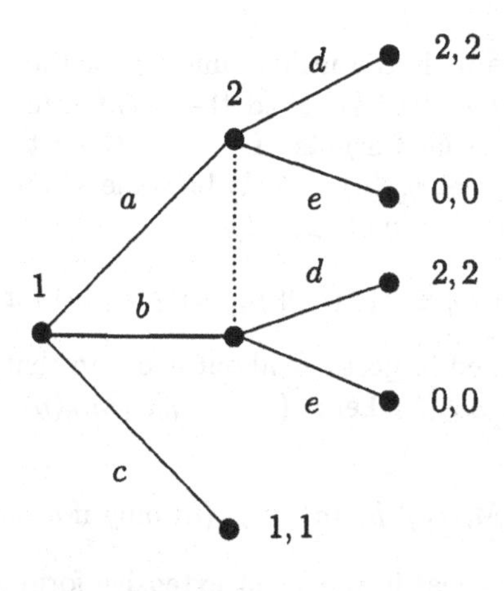

Figure 3.8

$a \in A(h)$. A strictly positive behavioral conjecture thus takes into account all possible actions that may be chosen by player i. A behavioral conjecture profile $\sigma = (\sigma_i)_{i \in I}$ is called strictly positive if each σ_i is strictly positive. For a given strictly positive behavioral conjecture profile σ every information set is reached with positive probability, that is, $\mathbb{P}_\sigma(h) > 0$ for every h. But then, conform Section 2.6, there is a *unique* belief system β such that (σ, β) is a Bayesian consistent assessment. We refer to β as the belief system induced by σ. We say that the Bayesian consistent assessment (σ, β) is strictly positive if σ is strictly positive.

Definition 3.4.1 *A behavioral conjecture profile $\sigma = (\sigma_i)_{i \in I}$ is called a perfect equilibrium if there is a sequence $(\sigma^n, \beta^n)_{n \in \mathbb{N}}$ of strictly positive Bayesian consistent assesments such that (1) $(\sigma^n)_{n \in \mathbb{N}}$ converges to σ and (2) $\sigma_{ih}(a) > 0$ only if a is a local best response against (σ^n, β^n) for every n.*

Hence, in the concept of perfect equilibrium, for the players to attach positive probability to an action a it is not sufficient that a is optimal against σ; the action a should, in addition, remain optimal if σ is replaced by a slightly perturbed conjecture $\tilde{\sigma}$ that takes all possible actions into account. In the example of Figure 3.8, (c, e) is not a perfect equilibrium since e is never a local best response against any strictly positive assessment. We shall now prove that a perfect equilibrium always exists in every extensive form game.

Theorem 3.4.2 *In every extensive form game there is at least one perfect equilibrium.*

Proof. Let Γ be an extensive form game. We transform Γ into a new game $A(\Gamma)$ in which every information set is controlled by a different player, and in which the "new player" controlling information set h has the same preferences as the "old player" controlling h. For every "old player" $i \in I$ and information set $h \in H_i$, we denote by ih the "new player" controlling h. The utility function for ih is given by $u_{ih}(z) = u_i(z)$

for every $z \in Z$, where u_i is the utility function of the "old player" controlling h. The so constructed game $A(\Gamma)$ is called the *agent extensive form* of Γ. The "new players" ih in $A(\Gamma)$ are called agents. Let J be the set of agents. For every agent ih and action $a \in A(h)$, let $\eta_{ih}(a) \in (0,1)$ be some small positive number such that $\sum_{a \in A(h)} \eta_{ih}(a) < 1$. Let $\eta = (\eta_{ih})_{ih \in J}$. By

$$M_{ih}(\eta) = \{\mu_{ih} \in \Delta(A(h)) | \; \mu_{ih}(a) \geq \eta_{ih}(a) \text{ for all } a \in A(h)\}$$

we denote the set of mixed conjectures about agent ih that assign at least probability $\eta_{ih}(a)$ to every action $a \in A(h)$. Let $M(\eta) = \times_{ih \in J} M_{ih}(\eta)$. For every mixed conjecture profile $\mu \in M(\eta)$, let

$$B_{ih}(\mu, \eta) = \{\tilde{\mu}_{ih} \in M_{ih}(\eta) | \; \tilde{\mu}_{ih}(a) > \eta_{ih}(a) \text{ only if } a \text{ best response against } \mu\}.$$

Here, we mean best response in the agent extensive form $A(\Gamma)$. Hence, a mixed conjecture in $B_{ih}(\mu, \eta)$ assigns minimum probability $\eta_{ih}(a)$ to every action a that is not a best response against μ. Let $B(\mu, \eta) = \times_{ih \in J} B_{ih}(\mu, \eta) \subseteq M(\eta)$. Consider the correspondence $B^\eta : M(\eta) \twoheadrightarrow M(\eta)$ assigning to every $\mu \in M(\eta)$ the set $B(\mu, \eta)$. We show, by Kakutani's fixed point theorem, that B^η has a fixed point. It is clear that $M(\eta)$ is a nonempty, convex and compact subset of some space $\mathbb{R}^d$. We show that $B^\eta(\mu)$ is nonempty, convex and compact. It suffices to show that, for every ih, the set $B_{ih}(\mu, \eta)$ is nonempty, convex and compact. It may be verified easily by the reader that $B_{ih}(\mu, \eta)$ is nonempty and compact. Let $\mu^1_{ih}, \mu^2_{ih} \in B_{ih}(\mu, \eta)$ and let $\lambda \in [0, 1]$. We show that $\tilde{\mu}_{ih} = \lambda \mu^1_{ih} + (1 - \lambda)\mu^2_{ih} \in B_{ih}(\mu, \eta)$. Let a not be a best response against μ. Then, $\mu^1_{ih}(a) = \mu^2_{ih}(a) = \eta_{ih}(a)$, and hence $\tilde{\mu}_{ih}(a) = \eta_{ih}(a)$, which implies that $\tilde{\mu}_{ih} \in B_{ih}(\mu, \eta)$. Hence, $B_{ih}(\mu, \eta)$ is convex. The reader may also verify that the correspondence B^η is upper-hemicontinuous. As such, Kakutani's fixed point theorem guarantees that B^η has some fixed point $\mu \in M(\eta)$. Hence, by construction, $\mu_{ih}(a) > \eta_{ih}(a)$ only if a is a best response against μ.

Now, let $(\eta^n)_{n \in \mathbb{N}}$ be some sequence of such strictly positive vectors η^n with $\lim_{n \to \infty} \eta^n_{ih}(a) = 0$ for every a. We know that for every η^n there is some fixed point $\mu^n \in M(\eta^n)$ of B^{η^n}. For every n, let $\tilde{B}_{ih}(\mu^n)$ be the set of actions at h that are best responses against μ. Since there are only finitely many actions in the game, there is some subsequence of $(\mu^n)_{n \in \mathbb{N}}$ such that for every ih, the set $\tilde{B}_{ih}(\mu^n)$ remains constant in this subsequence. Without loss of generality, we may assume that $\tilde{B}_{ih}(\mu^n)$ remains constant in the sequence $(\mu^n)_{n \in \mathbb{N}}$. Let M be the set of (unconstrained) mixed conjecture profiles in the agent extensive form $A(\Gamma)$. Since $\mu^n \in M$ for every n and M is a compact set, there is some subsequence of $(\mu^n)_{n \in \mathbb{N}}$ which converges to some $\mu \in M$. Without loss of generality, assume that $(\mu^n)_{n \in \mathbb{N}}$ converges to μ. Now, let a be some action which is not a best response against μ^n for some n. Then, by the assumption above, a is not a best response against μ^n for all n. Since μ^n is a fixed point of B^{η^n}, we then have that $\mu^n_{ih}(a) = \eta^n_{ih}(a)$ for every n, which implies that $\mu_{ih}(a) = 0$. We thus have the following property: if a is not a best response against μ^n for some n, then $\mu_{ih}(a) = 0$. Clearly, every μ^n may be identified with a behavioral conjecture profile σ^n of the original game Γ, and μ may be identified with some σ. For every n, let β^n be the belief system induced by σ^n. Then, we have that $(\sigma^n, \beta^n)_{n \in \mathbb{N}}$ is a sequence of strictly positive Bayesian consistent assessments, with $(\sigma^n)_{n \in \mathbb{N}}$ converging to σ. Moreover, if an action a is not a local best response against some (σ^n, β^n), then a is not a best response against μ^n, and hence $\mu_{ih}(a) = \sigma_{ih}(a) = 0$. But then, by definition, σ is a perfect equilibrium. ■

From the definition of perfect equilibrium it is not immediately clear that every perfect equilibrium is a Nash equilibrium. Note that the concept of perfect equilibrium requires players to assign zero probability to ***suboptimal actions by the opponents against σ^n***, but does not explicitly require them to assign zero probability to ***suboptimal strategies by the opponents against σ***. The following theorem, however, will show that the latter property is implied by the former. Moreover, this holds for every *subgame*, which implies that every perfect equilibrium is a subgame perfect equilibrium.

Theorem 3.4.3 *Every perfect equilibrium σ is a subgame perfect equilibrium.*

Proof. Let σ be a perfect equilibrium in Γ. We first prove that σ is a Nash equilibrium in Γ. Since σ is a perfect equilibrium, there is some sequence $(\sigma^n, \beta^n)_{n\in\mathbb{N}}$ of strictly positive Bayesian consistent assessments such that (1) $(\sigma^n)_{n\in\mathbb{N}}$ converges to σ and (2) $\sigma_{ih}(a) > 0$ only if a is a local best response against (σ^n, β^n) for every n. Since the set of all belief systems is a compact set, there is some subsequence of $(\beta^n)_{n\in\mathbb{N}}$ which converges to some belief system β. Without loss of generality, we assume that $(\beta^n)_{n\in\mathbb{N}}$ converges to β. Then, it may easily be verified that the assessment (σ, β) is Bayesian consistent. Now, let s_i^* be a strategy with $\sigma_i(s_i^*) > 0$. We must show that s_i^* is a best response against σ. In view of Lemma 2.6.4, it suffices to show that s_i^* is a sequential best response against (σ, β). Hence, we must prove that

$$u_i((s_i^*, \sigma_{-i})|\, h, \beta_{ih}) = \max_{s_i\in S_i(h)} u_i(s_i, \sigma_{-i}|\, h, \beta_{ih}) \tag{3.9}$$

for every $h \in H_i(s_i^*)$. At every information set $h \in H_i(s_i^*)$, let

$$u_i(\sigma|\, h, \beta_{ih}) = \sum_{a\in A(h)} \sigma_{ih}(a)\, u_i((a, \sigma_{-h})|\, h, \beta_{ih})$$

be the expected utility at h, generated by σ and β_{ih}. We shall prove that

$$u_i((s_i^*, \sigma_{-i})|\, h, \beta_{ih}) = u_i(\sigma|\, h, \beta_{ih}) = \max_{s_i\in S_i(h)} u_i((s_i, \sigma_{-i})|\, h, \beta_{ih}) \tag{3.10}$$

for every $h \in H_i(s_i^*)$. We prove (3.10) by induction on the number of player i information sets that follow h. We say that a player i information set is followed by at most k other player i information sets if every path which starts at some node in h crosses at most k other player i information sets. Suppose first that h is not followed by any other information set of player i. Let a be some action at h with $\sigma_{ih}(a) > 0$. By construction, a is then a local best response against σ^n for every n, which means that

$$u_i((a, \sigma_{-h}^n)|\, h, \beta_{ih}^n) = \max_{a'\in A(h)} u_i((a', \sigma_{-h}^n)|\, h, \beta_{ih}^n)$$

for every n. However, this implies that

$$u_i((a,\sigma_{-h})|\ h,\beta_{ih}) = \max_{a'\in A(h)} u_i((a',\sigma_{-h})|\ h,\beta_{ih}). \tag{3.11}$$

Since h is not followed by any other player i information set, the expression $u_i(s_i,\sigma_{-i}|\ h,\beta_{ih})$ does not depend on the actions prescribed by s_i at information sets other than h. As such, (3.11) implies that $u_i((a,\sigma_{-h})|\ h,\beta_{ih}) = \max_{s_i\in S_i(h)} u_i((s_i,\sigma_{-i})|\ h,\beta_{ih})$. Since this holds for all actions a with $\sigma_{ih}(a) > 0$, it follows that $u_i(\sigma|\ h,\beta_h) = \max_{s_i\in S_i(h)} u_i((s_i,\sigma_{-i})|\ h,\beta_{ih})$. Now, let a^* be the action prescribed by s_i^* at h. Then, we know that $\sigma_{ih}(a^*) > 0$ and hence, by the argument above, $u_i((a^*,\sigma_{-h})|\ h,\beta_{ih}) = \max_{s_i\in S_i(h)} u_i((s_i,\sigma_{-i})|\ h,\beta_{ih})$. Since h is not followed by any other player i information set, we have that $u_i((s_i^*,\sigma_{-i})|\ h,\beta_{ih}) = u_i((a^*,\sigma_{-h})|\ h,\beta_{ih})$. We have thus shown (3.10).

Now, let $k \geq 1$ and assume that for every player i information set $h' \in H_i$ followed by at most $k-1$ other player i information sets, (3.10) holds. Let $h \in H_i$ be such that h is followed by at most k player i information sets. By definition, for every strategy $s_i \in S_i(h)$,

$$u_i((s_i,\sigma_{-i})|\ h,\beta_{ih}) = \sum_{x\in h} \beta_{ih}(x)\ u_i((s_i,\sigma_{-i})|\ x). \tag{3.12}$$

Choose a node $x \in h$. Let $Z(x)$ be the set of terminal nodes that follow x. Let $Z_0(x)$ be the set of terminal nodes in $Z(x)$ for which the path from x to z does not cross any other player i information set. By H_i^* we denote the collection of player i information sets h' that follow h and for which there is no player i information set between h and h'. By $H_i^*(s_i)$ we denote those player i information sets in H_i^* that are not avoided by s_i. For a given $h' \in H_i^*(s_i)$, let $Y(x,h')$ be the set of nodes in h' that follow x. Note that this set may be empty. If the game reaches x and player i chooses s_i, then either the game ends at some terminal node $z \in Z_0(x)$ or it crosses exactly one information set $h' \in H_i^*(s_i)$. It thus follows that

$$\begin{aligned} u_i((s_i,\sigma_{-i})|\ x) &= \sum_{h'\in H_i^*(s_i)} \sum_{y\in Y(x,h')} \mathbb{P}_{(s_i,\sigma_{-i})}(y|\ x)\ u_i((s_i,\sigma_{-i})|\ y) + \\ &\quad + \sum_{z\in Z_0(x)} \mathbb{P}_{(s_i,\sigma_{-i})}(z|\ x)\ u_i(z), \end{aligned} \tag{3.13}$$

where $\mathbb{P}_{(s_i,\sigma_{-i})}(y|\ x)$ is the probability of reaching y, conditional on the event that x is reached. For every n, let $\mathbb{P}_{(s_i,\sigma_{-i}^n)}(h)$ be the probability that h is reached under (s_i,σ_{-i}^n). Since $s_i \in S_i(h)$, we have that s_i chooses all player i actions which lead to h. Since σ_{-i}^n assigns positive probability to all actions, we may conclude that $\mathbb{P}_{(s_i,\sigma_{-i}^n)}(h) > 0$ for all n. Choose some information set $h' \in H_i^*$. We define $\mathbb{P}_{(s_i,\sigma_{-i}^n)}(h')$ in the same way. Since h' follows h we have $\mathbb{P}_{(s_i,\sigma_{-i}^n)}(h')/\mathbb{P}_{(s_i,\sigma_{-i}^n)}(h) \in [0,1]$ for all n. Hence, there is a subsequence of $(\mathbb{P}_{(s_i,\sigma_{-i}^n)}(h')/\mathbb{P}_{(s_i,\sigma_{-i}^n)}(h))_{n\in\mathbb{N}}$ which converges to some number in $[0,1]$. Without loss of generality, we assume that the sequence itself already converges to some number. Define

$$\mathbb{P}_{(s_i,\sigma_{-i})}(h'|h) := \lim_{n\to\infty} (\mathbb{P}_{(s_i,\sigma_{-i}^n)}(h')/\mathbb{P}_{(s_i,\sigma_{-i}^n)}(h))_{n\in\mathbb{N}}.$$

Claim 1. For every $x \in h$ and $y \in Y(x,h')$ it holds: $\beta_{ih}(x)\ \mathbb{P}_{(s_i,\sigma_{-i})}(y|\ x) = \beta_{ih}(y)\ \mathbb{P}_{(s_i,\sigma_{-i})}(h'|h)$.

Proof of claim 1. We distinguish two cases. Assume first that $s_i \in S_i(h')$, hence s_i chooses all player i actions that lead to h'. Then, for all n we have

$$\begin{aligned} \beta^n_{ih}(x)\ \mathbb{P}_{(s_i,\sigma^n_{-i})}(y|\ x) &= \frac{\mathbb{P}_{(s_i,\sigma^n_{-i})}(x)}{\mathbb{P}_{(s_i,\sigma^n_{-i})}(h)} \frac{\mathbb{P}_{(s_i,\sigma^n_{-i})}(y)}{\mathbb{P}_{(s_i,\sigma^n_{-i})}(x)} = \frac{\mathbb{P}_{(s_i,\sigma^n_{-i})}(y)}{\mathbb{P}_{(s_i,\sigma^n_{-i})}(h)} \\ &= \frac{\mathbb{P}_{(s_i,\sigma^n_{-i})}(y)}{\mathbb{P}_{(s_i,\sigma^n_{-i})}(h')} \frac{\mathbb{P}_{(s_i,\sigma^n_{-i})}(h')}{\mathbb{P}_{(s_i,\sigma^n_{-i})}(h)} \\ &= \beta^n_{ih'}(y)\ \frac{\mathbb{P}_{(s_i,\sigma^n_{-i})}(h')}{\mathbb{P}_{(s_i,\sigma^n_{-i})}(h)}. \end{aligned} \tag{3.14}$$

Note that $\mathbb{P}_{(s_i,\sigma^n_{-i})}(x) > 0$ since s_i chooses all player i actions leading to x, and σ^n_{-i} is strictly positive. Moreover, $\mathbb{P}_{(s_i,\sigma^n_{-i})}(h') > 0$ since $s_i \in S_i(h')$. Since (3.14) holds for all n, we may conclude that

$$\begin{aligned} \beta_{ih}(x)\ \mathbb{P}_{(s_i,\sigma_{-i})}(y|\ x) &= \lim_{n\to\infty} \beta^n_{ih}(x)\ \mathbb{P}_{(s_i,\sigma^n_{-i})}(y|\ x) = \lim_{n\to\infty} \beta^n_{ih'}(y)\ \frac{\mathbb{P}_{(s_i,\sigma^n_{-i})}(h')}{\mathbb{P}_{(s_i,\sigma^n_{-i})}(h)} \\ &= \beta_{ih'}(y)\ \mathbb{P}_{(s_i,\sigma_{-i})}(h'|h). \end{aligned}$$

Assume now that $s_i \notin S_i(h')$. Then, $\mathbb{P}_{(s_i,\sigma_{-i})}(y|\ x) = 0$. Moreover, $\mathbb{P}_{(s_i,\sigma^n_{-i})}(h') = 0$ for all n and hence $\mathbb{P}_{(s_i,\sigma_{-i})}(h'|h) = 0$. This completes the proof of the claim.

By combining the claim with (3.12) and (3.13) we obtain

$$\begin{aligned} u_i((s_i,\sigma_{-i})|h,\beta_{ih}) &= \sum_{x\in h}\sum_{h'\in H^*_i(s_i)} \mathbb{P}_{(s_i,\sigma_{-i})}(h'|h) \sum_{y\in Y(x,h')} \beta_{ih'}(y)\ u_i((s_i,\sigma_{-i})|y) + \\ &\quad + \sum_{x\in h}\beta_{ih}(x) \sum_{z\in Z_0(x)} \mathbb{P}_{(s_i,\sigma_{-i})}(z|\ x)\ u_i(z) \\ &= \sum_{h'\in H^*_i(s_i)} \mathbb{P}_{(s_i,\sigma_{-i})}(h'|h) \sum_{y\in h'} \beta_{ih'}(y)\ u_i((s_i,\sigma_{-i})|\ y) + \\ &\quad + \sum_{x\in h}\beta_{ih}(x) \sum_{z\in Z_0(x)} \mathbb{P}_{(s_i,\sigma_{-i})}(z|\ x)\ u_i(z) \\ &= \sum_{h'\in H^*_i(s_i)} \mathbb{P}_{(s_i,\sigma_{-i})}(h'|h)\ u_i((s_i,\sigma_{-i})|\ h',\beta_{ih'}) + \\ &\quad + \sum_{x\in h}\beta_{ih}(x) \sum_{z\in Z_0(x)} \mathbb{P}_{(s_i,\sigma_{-i})}(z|\ x)\ u_i(z). \end{aligned} \tag{3.15}$$

Similarly, we have for every action $a \in A(h)$,

$$\begin{aligned} u_i((a,\sigma_{-h})|\ h,\beta_{ih}) &= \sum_{h'\in H^*_i(a)} \mathbb{P}_{(a_i,\sigma_{-h})}(h'|h)\ u_i((a,\sigma_{-h})|\ h',\beta_{ih'}) + \\ &\quad + \sum_{x\in h}\beta_{ih}(x) \sum_{z\in Z_0(x)} \mathbb{P}_{(a,\sigma_{-h})}(z|\ x)\ u_i(z), \end{aligned} \tag{3.16}$$

where $\mathbb{P}_{(a,\sigma_{-h})}(h'|h) = \lim_{n\to\infty} \mathbb{P}_{(a,\sigma^n_{-h})}(h')/\mathbb{P}_{(a,\sigma^n_{-h})}(h)$. Note that $\mathbb{P}_{(a,\sigma^n_{-h})}(h) > 0$ for all n. By $H_i^*(a)$ we denote those player i information sets in H_i^* that are not avoided if player i chooses a at h.

Let a be the action chosen by s_i at h. Then, $H_i^*(s_i) = H_i^*(a)$. Since every $h' \in H_i^*(s_i)$ is followed by at most $k-1$ player i information sets, we know by induction assumption that $u_i((s_i,\sigma_{-i})|\ h',\beta_{ih'}) \leq u_i(\sigma|\ h',\beta_{ih'})$ for every $h' \in H_i^*(s_i)$. Since $u_i(\sigma|\ h',\beta_{ih'})$ does not depend on player i's behavior at h we have that $u_i(\sigma|\ h',\beta_{ih'}) = u_i((a,\sigma_{-h})|\ h',\beta_{ih'})$. For every terminal node $z \in Z_0(x)$, the probability $\mathbb{P}_{(s_i,\sigma_{-i})}(z|\ x)$ does not depend upon player i's choices at information sets other than h. Therefore, $\mathbb{P}_{(s_i,\sigma_{-i})}(z|\ x) = \mathbb{P}_{(a,\sigma_{-h})}(z|\ x)$ for every $x \in h$ and every $z \in Z_0(x)$. For every $h' \in H_i^*(s_i)$ there is no player i information set between h and h', and therefore $\mathbb{P}_{(s_i,\sigma_{-i})}(h'|h)$ does not depend upon player i's choices at information sets other than h. It thus follows that $\mathbb{P}_{(s_i,\sigma_{-i})}(h'|h) = \mathbb{P}_{(a,\sigma_{-h})}(h'|h)$. Combining these insights with (3.15) leads to

$$\begin{aligned} u_i((s_i,\sigma_{-i})|\ h,\beta_{ih}) \ &\leq \ \sum_{h'\in H_i^*(a)} \mathbb{P}_{(a,\sigma_{-h})}(h'|h)\ u_i((a,\sigma_{-h\_}|\ h',\beta_{ih'}) \\ &\quad + \sum_{x\in h} \beta_{ih}(x) \sum_{z\in Z_0(x)} \mathbb{P}_{(a,\sigma_{-h})}(z|\ x)\ u_i(z) \\ &= \ u_i((a,\sigma_{-h})|\ h,\beta_{ih}). \end{aligned} \tag{3.17}$$

Now, let a' be some action at h with $\sigma_{ih}(a') > 0$. Since σ is a perfect equilibrium, "justified" by the sequence $(\sigma^n)_{n\in\mathbb{N}}$, we know that a' is a local best response against σ^n for all n, hence $u_i((a',\sigma^n_{-h})|\ h,\beta^n_{ih}) = \max_{a''\in A(h)} u_i((a'',\sigma^n_{-h})|\ h,\beta^n_{ih})$ for all n. This implies that $u_i((a',\sigma_{-h})|\ h,\beta_{ih}) = \max_{a''\in A(h)} u_i((a'',\sigma_{-h})|\ h,\beta_{ih})$. Since this holds for every $a' \in A(h)$ with $\sigma_{ih}(a') > 0$, it follows that $u_i(\sigma|\ h,\beta_{ih}) = \max_{a''\in A(h)} u_i((a'',\sigma_{-h})|\ h,\beta_{ih})$. By (3.17) we may conclude that $u_i((s_i,\sigma_{-i})|\ h,\beta_{ih}) \leq u_i(\sigma|\ h,\beta_{ih})$ for all $s_i \in S_i(h)$.

We next show that $u_i((s_i^*,\sigma_{-i})|\ h,\beta_{ih}) = u_i(\sigma|\ h,\beta_{ih})$, which would imply (3.10). By (3.15) we know that

$$\begin{aligned} u_i((s_i^*,\sigma_{-i})|\ h,\beta_{ih}) \ &= \ \sum_{h'\in H_i^*(s_i^*)} \mathbb{P}_{(s_i^*,\sigma_{-i})}(h'|h)\ u_i((s_i^*,\sigma_{-i})|\ h',\beta_{ih'}) + \\ &\quad + \sum_{x\in h} \beta_{ih}(x) \sum_{z\in Z_0(x)} \mathbb{P}_{(s_i^*,\sigma_{-i})}(z|\ x)\ u_i(z). \end{aligned} \tag{3.18}$$

By induction assumption, we know that for every $h' \in H_i^*(s_i^*)$ it holds that $u_i((s_i^*,\sigma_{-i})|\ h',\beta_{ih'}) = u_i(\sigma|\ h',\beta_{ih'})$. Let a^* be the action chosen by s_i^* at h. Since $u_i(\sigma|\ h',\beta_{ih'})$ does not depend on player i's behavior at h, we have that $u_i(\sigma|\ h',\beta_{ih'}) = u_i((a^*,\sigma_{-h})|\ h',\beta_{ih'})$ and hence $u_i((s_i^*,\sigma_{-i})|\ h',\beta_{ih'}) = u_i((a^*,\sigma_{-h})|\ h',\beta_{ih'})$ for every $h' \in H_i^*(s_i^*)$. By the same argument as used above for (s_i,σ_{-i}), we may conclude that $\mathbb{P}_{(s_i^*,\sigma_{-i})}(h'|h) = \mathbb{P}_{(a^*,\sigma_{-h})}(h'|h)$ for all $h' \in H_i^*(s_i^*)$ and $\mathbb{P}_{(s_i^*,\sigma_{-i})}(z|\ x) = \mathbb{P}_{(a^*,\sigma_{-h})}(z|$

x) for all $x \in h$ and $z \in Z_0(x)$. Combining these insights with (3.18) we obtain

$$\begin{aligned} u_i((s_i^*, \sigma_{-i})|\, h, \beta_{ih}) &= \sum_{h' \in H_i^*(s_i^*)} \mathbb{P}_{(a^*, \sigma_{-h})}(h'|h)\, u_i((a^*, \sigma_{-h})|\, h', \beta_{ih'}) + \\ &\quad + \sum_{x \in h} \beta_{ih}(x) \sum_{z \in Z_0(x)} \mathbb{P}_{(a^*, \sigma_{-h})}(z|\, x)\, u_i(z) \\ &= u_i((a^*, \sigma_{-h})|\, h, \beta_{ih}). \end{aligned} \tag{3.19}$$

By assumption, $\sigma_i(s_i^*) > 0$ and hence $\sigma_{ih}(a^*) > 0$. We know, from above, that then $u_i((a^*, \sigma_{-h})|\, h, \beta_{ih}) = \max_{a \in A(h)} u_i((a, \sigma_{-h})|\, h, \beta_{ih}) = u_i(\sigma|\, h, \beta_{ih})$. By (3.19) it follows that $u_i((s_i^*, \sigma_{-i})|\, h, \beta_{ih}) = u_i(\sigma|\, h, \beta_{ih})$, which was to show. By induction, it follows that (3.10) holds for every $h \in H_i(s_i^*)$.We have thus shown that s_i^* is a sequential best response against (σ, β), which implies that s_i^* is a best response against σ. Since this holds for every s_i^* with $\sigma_i(s_i^*) > 0$, it follows that σ is a Nash equilibrium in Γ.

We finally prove that σ is not only a Nash equilibrium, but even a subgame perfect equilibrium. Recall that there is a sequence $(\sigma^n, \beta^n)_{n \in \mathbb{N}}$ of strictly positive Bayesian consistent assesments with $(\sigma^n)_{n \in \mathbb{N}}$ converging to σ such that $\sigma_{ih}(a) > 0$ only if a is a local best response against (σ^n, β^n) for every n. Let Γ_x be a subgame, let σ^x be the restriction of σ to Γ_x, and for every n let $(\sigma^{n,x}, \beta^{n,x})$ be the restriction of (σ^n, β^n) to Γ_x. Then, obviously, $(\sigma^{n,x})_{n \in \mathbb{N}}$ converges to σ^x and $\sigma^x_{ih}(a) > 0$ only if a is a local best response against $(\sigma^{n,x}, \beta^{n,x})$ for every n. Hence, σ^x is a perfect equilibrium in Γ_x. By the first part of the proof, we know that σ^x is then a Nash equilibrium in Γ_x. Hence, σ induces a Nash equilibrium in every subgame, and is thus a subgame perfect equilibrium. ■

3.5 Quasi-Perfect Equilibrium

By construction, the concept of perfect equilibrium requires players to assign probability zero to those actions that are not local best responses against any strictly positive behavioral conjecture profile. The intuition behind this property is that players, when choosing an action at a given information set, should not exclude any action that may be chosen at other information sets. More precisely, if player i has to choose at information set h, then he is believed to choose an action a which is not only a local best response against the original conjecture profile σ, but also against some $\tilde{\sigma}$ close to σ that assigns positive probability to all actions. Here, $\tilde{\sigma}$ may be interpreted as some slightly perturbed conjecture. A disturbing feature, however, is the fact that in the perturbed conjecture $\tilde{\sigma}$, player i's conjecture $\tilde{\sigma}_{-h}$ about the behavior at other information sets should also assign positive probability to all future actions *controlled by player i himself.* This restriction seems difficult to justify if we assume that players are confident to carry out correctly their own strategies. To illustrate this difficulty, consider the game in Figure 3.9. In this game, (a, c) is not a perfect equilibrium since a is no longer a local best response if player 1 believes that he will play d with some small probability at the second information set. The unique perfect equilibrium is (b, c). However, from player 1's viewpoint there is no need to distinguish between the

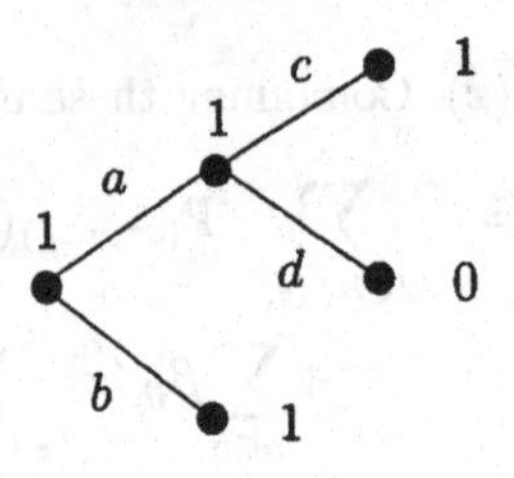

Figure 3.9

strategies (a, c) and b, since they yield the same utility given that he is certain to play c at his second information set.

A possible way to overcome this drawback is by stating that a player, at each of his information sets, is believed to choose a continuation strategy that is optimal against a slight perturbation of his conjecture about strategy choices *not controlled by himself.* This is the idea behind the concept of *quasi-perfect equilibrium,* introduced in van Damme (1984). In order to formally state the concept, we need the following definitions.

Let σ^*_{-i} be a strictly positive behavioral conjecture profile about player i's opponents. Consider an information set h controlled by player i. Let $s_i \in S_i(h)$ be a player i strategy that does not avoid h. Then, obviously, h is reached with positive probability under (s_i, σ^*_{-i}) and hence (s_i, σ^*_{-i}) induces a unique belief vector β_{ih} at h. Moreover, this belief vector does not depend upon the particular choice of $s_i \in S_i(h)$, and we may therefore say that β_{ih} is the *belief vector induced by* σ^*_{-i}. For an arbitrary strategy $s_i \in S_i(h)$, we define

$$u_i((s_i, \sigma^*_{-i})|\ h) := u_i((s_i, \sigma^*_{-i})|\ h, \beta_{ih})$$

where β_{ih} is the belief vector induced by the strictly positive σ^*_{-i}. Hence, $u_i((s_i, \sigma^*_{-i})|\ h)$ is the usual expected utility induced by s_i at h, given the induced belief vector β_{ih}. We say that $s_i \in S_i(h)$ is a *sequential best response against* σ^*_{-i} *at* h if $u_i((s_i, \sigma^*_{-i})|\ h) = \max_{s'_i \in S_i(h)} u_i((s'_i, \sigma^*_{-i})|\ h)$.

Recall, from Section 2.5, that $\sigma_{i|h}$ is the revised behavioral conjecture about player i, once h is reached. By definition, $\sigma_{i|h}$ only assigns positive probability to strategies $s_i \in S_i(h)$. We say that a behavioral conjecture profile σ is a *quasi-perfect equilibrium* if it can be approximated by a sequence of strictly positive behavioral conjecture profiles, such that at every information set $h \in H_i$ the revised conjecture $\sigma_{i|h}$ only assigns positive probability to strategies that constitute a sequential best response at h *against each conjecture profile in the sequence.*

Definition 3.5.1 *A behavioral conjecture profile σ is called a quasi-perfect equilibrium if there is a sequence $(\sigma^n)_{n\in\mathbb{N}}$ of strictly positive behavioral conjecture profiles such that (1) $(\sigma^n)_{n\in\mathbb{N}}$ converges to σ and (2) for every player i, every information set*

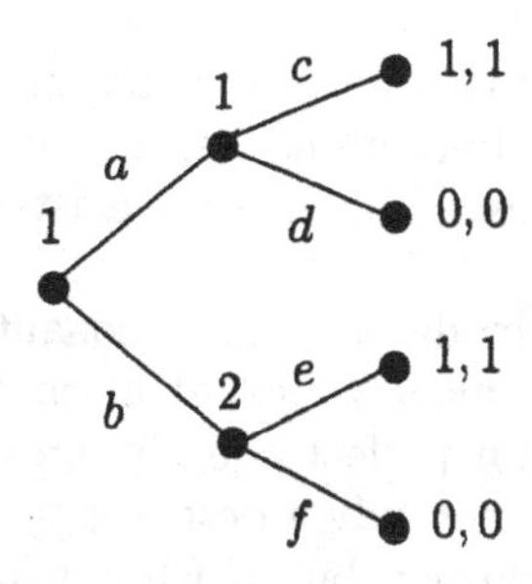

Figure 3.10

$h \in H_i$ and every $s_i \in S_i(h)$ it holds that $\sigma_{i|h}(s_i) > 0$ only if s_i is a sequential best response against σ^n_{-i} at h for every n.

Each player i is thus believed, at each of his information sets, to solely apply strategies that not only are optimal against the original conjecture profile σ_{-i}, but also against some slight perturbation σ^n_{-i} of σ_{-i} in which player i takes into account all possible actions by the opponents. In contrast to the concept of perfect equilibrium, the perturbation σ^n_{-i} no longer includes player i's own actions. In fact, there is no general logical relationship between perfect equilibrium and quasi-perfect equilibrium, that is, a perfect equilibrium need not be quasi-perfect, and a quasi-perfect equilibrium need not be perfect. In the example of Figure 3.9, for instance, (a, c) is a quasi-perfect equilibrium, but it is not perfect. Consider now the game if Figure 3.10. The behavioral conjecture profile $((b, c), e)$ is a perfect equilibrium. To see this, consider the sequence $(\sigma^n, \beta^n)_{n\in\mathbb{N}}$ of strictly positive Bayesian consistent assessments where $\sigma^n = ((\frac{1}{n}a + (1-\frac{1}{n})b, (1-\frac{1}{n})c + \frac{1}{n}d), (1-\frac{1}{2n})e + \frac{1}{2n}f)$. Then, $(\sigma^n)_{n\in\mathbb{N}}$ converges to $((b, c), e)$. For every n, playing b at the first information set is a local best reponse for player 1 against σ^n and hence it follows that $((b, c), e)$ is a perfect equilibrium. However, $((b, c), e)$ is not a quasi-perfect equilibrium. The reason is that for every strictly positive behavioral conjecture σ^n_2 about player 2, the expected utility for player 1 at his first information set from playing (a, c) is strictly higher than from playing b, and hence every quasi-perfect equilibrium should assign probability zero to b. As such, $((a, c), e)$ is the unique quasi-perfect equilibrium.

Despite the fact that there is no general logical relationship between perfect equilibrium and quasi-perfect equilibrium, they have an important property in common, namely that they both constitute a refinement of subgame perfect equilibrium.

Theorem 3.5.2 *Every quasi-perfect equilibrium is a subgame perfect equilibrium.*

The proof of this result is postponed until Section 4.2. In that section we shall show that every quasi-perfect equilibrium induces a sequential equilibrium, and that every sequential equilibrium constitutes a subgame perfect equilibrium. Another issue of interest is whether a quasi-perfect equilibrium always exists. We indeed show that every game contains a quasi-perfect equilibrium. The proof of existence is given in the following section, in which we introduce the concept of proper equilibrium. It is shown that every proper equilibrium induces a quasi-perfect equilibrium, and that a proper equilibrium always exists, implying the existence of quasi-perfect equilibria in every game.

We conclude this section by discussing a rationality concept that is closely related to quasi-perfect equilibrium, namely that of *normal form perfect equilibrium*. The idea is that in a normal form perfect equilibrium σ, every player i is believed to choose a strategy s_i which is not only a best response against his original conjecture σ_{-i} about the opponents' behavior, but which in addition *remains* a best response if σ_{-i} is replaced by a nearby conjecture $\tilde{\sigma}_{-i}$ that takes all possible opponents' actions into account.

Definition 3.5.3 *A behavioral conjecture profile σ is called a normal form perfect equilibrium if there is a sequence $(\sigma^n)_{n\in\mathbb{N}}$ of strictly positive behavioral conjecture profiles such that (1) $(\sigma^n)_{n\in\mathbb{N}}$ converges to σ and (2) $\sigma_i(s_i) > 0$ only if s_i is a best response against σ^n for all n.*

It is easily shown that every quasi-perfect equilibrium is a normal form perfect equilibrium.

Lemma 3.5.4 *Every quasi-perfect equilibrium is a normal form perfect equilibrium.*

Proof. Let σ be a quasi-perfect equilibrium. Then, there is a sequence $(\sigma^n)_{n\in\mathbb{N}}$ of strictly positive behavioral conjecture profiles such that (1) $(\sigma^n)_{n\in\mathbb{N}}$ converges to σ and (2) for every $h \in H_i$ and $s_i \in S_i(h)$ we have that $\sigma_{i|h}(s_i) > 0$ only if s_i is a sequential best response against σ^n_{-i} at h for all n. Now, let $s_i \in S_i$ be such that $\sigma_i(s_i) > 0$. We show that s_i is a best response against σ^n for all n, which would imply that σ is a normal form perfect equilibrium. Let $n \in \mathbb{N}$ be given, and let β^n be the belief system induced by σ^n. We shall prove that s_i is a sequential best response against (σ^n, β^n), which would then, by Lemma 2.6.4, imply that s_i is a best response against σ^n. Let $h \in H_i(s_i)$. We must prove that $u_i((s_i, \sigma^n_{-i})|\, h, \beta^n_{ih}) = \max_{s'_i \in S_i(h)} u_i((s'_i, \sigma^n_{-i})|\, h, \beta^n_{ih})$. Since $h \in H_i(s_i)$ and $\sigma_i(s_i) > 0$, we have that $\sigma_{i|h}(s_i) > 0$. But then, by assumption, s_i is a sequential best response against σ^n_{-i} at h, which is equivalent to $u_i((s_i, \sigma^n_{-i})|\, h, \beta^n_{ih}) = \max_{s'_i \in S_i(h)} u_i((s'_i, \sigma^n_{-i})|\, h, \beta^n_{ih})$. Since this holds for every $h \in H_i(s_i)$, we may conclude that s_i is a sequential best response against (σ^n, β^n), and hence s_i is a best response against σ^n. This completes the proof. ■

In the following section, it is shown that a quasi-perfect equilibrium always exists, which, by the lemma above, guarantees the existence of a normal form perfect equilibrium in every game. In contrast to perfect equilibrium and quasi-perfect equilibrium, not every normal form perfect equilibrium is a subgame perfect equilibrium. Consider the game in Figure 3.11. In this game, the unique behavioral conjecture profile satisfying backward induction, and hence the unique subgame perfect equilibrium, is $((b_1, b_3), b_2)$. However, $\sigma = ((a_1, a_3), a_2)$ is a normal form perfect equilibrium. Note that $\sigma_1(a_1) = 1$ and $\sigma_2(a_2) = 1$. Consider the sequence of strictly positive behavioral conjecture profiles $\sigma^n = (((1-\frac{1}{n})a_1 + \frac{1}{n}b_1, (1-\frac{1}{n})a_3 + \frac{1}{n}b_3), (1-\frac{1}{n})a_2 + \frac{1}{n}b_2)$. Then, a_1 is a best response against σ^n for all n, and a_2 is a best response against σ^n for all

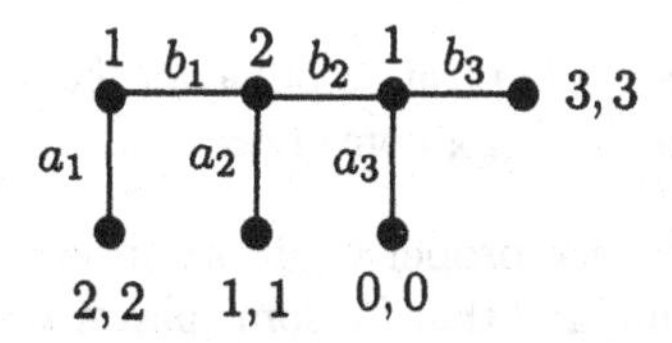

Figure 3.11

n, which implies that σ is a normal form perfect equilibrium. The crucial difference between subgame perfect equilibrium and normal form perfect equilibrium in this example is the following. In the normal form perfect equilibrium σ defined above, player 2, after observing that player 1 has chosen b_1, and thus has contradicted player 2's initial conjecture, does no longer believe that player 1 will choose optimally at his second information set. This is reflected by the fact that $\sigma_1 = (a_1, a_3)$, in which player 2 believes that player 1 chooses the suboptimal action a_3 after observing that his initial conjecture has been falsified. In the concept of subgame perfect equilibrium, however, player 2 is required to believe that player 1 chooses optimally at his second information set, that is, chooses b_3.

As was the case for quasi-perfect equilibrium, there is no general logical relationship between normal form perfect equilibrium and perfect equilibrium. In the game of Figure 3.9, (a, c) is a normal form perfect equilibrium, but is not perfect. In the game of Figure 3.10, $((b, c), e)$ is a perfect equilibrium, but not a normal form perfect equilibrium

3.6 Proper Equilibrium

In this section, we introduce the concept of *proper equilibrium,* which is due to Myerson (1978) and for which we prove existence. We then prove that every proper equilibrium induces a quasi-perfect equilibrium, implying the existence of quasi-perfect equilibria in every game. This relationship between proper equilibrium and quasi-perfect equilibrium is due to van Damme (1984). Kohlberg and Mertens (1986) have independently proved a weaker version of this theorem, stating that every proper equilibrium induces a sequential equilibrium; a concept which is weaker than quasi-perfect equilibrium.

We now turn to the definition of proper equilibrium. Let $\sigma = (\sigma_i)_{i \in I}$ be a behavioral conjecture profile and $\epsilon > 0$ some small strictly positive number. We say that σ is *ϵ-proper* if σ is strictly positive and if for every player i and every two strategies $s_i, t_i \in S_i$ with $u_i(s_i, \sigma_{-i}) < u_i(t_i, \sigma_{-i})$ it holds that $\sigma_i(s_i) \leq \epsilon \sigma_i(t_i)$. Hence, if strategy s_i performs worse than some other strategy t_i, then player i's opponents should deem s_i much less likely than t_i.

Definition 3.6.1 *A behavioral conjecture profile σ is called a proper equilibrium if there is some sequence $(\epsilon^n)_{n \in \mathbb{N}}$ of strictly positive numbers converging to zero, and*

some sequence $(\sigma^n)_{n\in\mathbb{N}}$ *of strictly positive behavioral conjecture profiles such that* σ^n *is* ϵ^n*-proper for every* n *and* $(\sigma^n)_{n\in\mathbb{N}}$ *converges to* σ.

It is easily shown that every proper equilibrium is a Nash equilibrium. Suppose that σ is a proper equilibrium, and that for some player i, some strategy s_i is not a best response against σ. In this case, there exists some other strategy t_i with $u_i(s_i, \sigma_{-i}) < u_i(t_i, \sigma_{-i})$. Since σ is proper, there is a sequence $(\epsilon^n)_{n\in\mathbb{N}}$ converging to zero, and a corresponding sequence $(\sigma^n)_{n\in\mathbb{N}}$ of ϵ^n-proper conjecture profiles converging to σ. If n is large enough, we have that $u_i(s_i, \sigma^n_{-i}) < u_i(t_i, \sigma^n_{-i})$ and hence $\sigma^n_i(s_i) \leq \epsilon^n \sigma^n_i(t_i)$. This implies that $\sigma_i(s_i) = \lim_{n\to\infty} \sigma^n_i(s_i) = 0$. Since this holds for all players i and all strategies s_i which are not best responses against σ, it follows that σ is a Nash equilibrium.

We shall now provide a characterization of proper equilibria, which will be useful in the sequel. Say that a sequence $(\sigma^n)_{n\in\mathbb{N}}$ of strictly positive behavioral conjecture profiles is *proper* if for every $n \in \mathbb{N}$, every player i and all strategies $s_i, t_i \in S_i$ with $u_i(s_i, \sigma^n_{-i}) < u_i(t_i, \sigma^n_{-i})$ it holds that $\lim_{n\to\infty} \sigma^n_i(s_i)/\sigma^n_i(t_i) = 0$. The interpretation is that, whenever s_i performs worse that t_i against some conjecture in the approximating sequence, then player i's opponents should deem s_i "infinitely less likely" than t_i. The latter is formalized by the condition that $\lim_{n\to\infty} \sigma^n_i(s_i)/\sigma^n_i(t_i) = 0$.

Lemma 3.6.2 *A behavioral conjecture profile* σ *is a proper equilibrium if and only if there is a proper sequence* $(\sigma^n)_{n\in\mathbb{N}}$ *of strictly positive behavioral conjecture profiles converging to* σ.

Proof. Let μ be a proper equilibrium. Then, we can find sequences $(\epsilon^n)_{n\in\mathbb{N}}$ and $(\sigma^n)_{n\in\mathbb{N}}$ such that σ^n is ϵ^n-proper for every n, and $(\sigma^n)_{n\in\mathbb{N}}$ converges to σ. For every n and every player i, let $l^n_i : S_i \to \{1, 2, ..., |S_i|\}$ be a function such that $u_i(s_i, \sigma^n_{-i}) < u_i(t_i, \sigma^n_{-i})$ if and only if $l^n_i(s_i) < l^n_i(t_i)$. Since there are only finitely many possible functions l^n_i for every player, we can find a subsequence $(\sigma^{k(n)})_{n\in\mathbb{N}}$ such that $l^{k(n)}_i = l^{k(m)}_i$ for every player i and every $n, m \in \mathbb{N}$. We now show that the sequence $(\sigma^{k(n)})_{n\in\mathbb{N}}$ is proper. Suppose that $u_i(s_i, \sigma^{k(n)}_{-i}) < u_i(t_i, \sigma^{k(n)}_{-i})$ for some n, some player i and some $s_i, t_i \in S_i$. Then, by construction, $u_i(s_i, \sigma^{k(m)}_{-i}) < u_i(t_i, \sigma^{k(m)}_{-i})$ for all $m \in \mathbb{N}$, and hence $\sigma^{k(m)}_i(s_i) \leq \epsilon^{k(m)} \sigma^{k(m)}_i(t_i)$ for all m. It thus follows that $\lim_{n\to\infty} \sigma^{k(n)}_i(s_i)/\sigma^{k(n)}_i(t_i) = 0$, which implies that $(\sigma^{k(n)})_{n\in\mathbb{N}}$ is proper.

Now, let $(\sigma^n)_{n\in\mathbb{N}}$ be a proper sequence converging to σ. For every n, let

$$\epsilon^n = \max\{\sigma^n_i(s_i)/\sigma^n_i(t_i) \mid i \in I \text{ and } u_i(s_i, \sigma^n_{-i}) < u_i(t_i, \sigma^n_{-i})\}.$$

Then, by construction, σ^n is ϵ^n-proper for all n, and $(\epsilon^n)_{n\in\mathbb{N}}$ converges to zero. This completes the proof. ■

Blume, Brandenburger and Dekel (1991b) provide a characterization of proper equilibrium in which the requirement that inferior responses should be deemed infinitely less likely than superior responses is stated explicitly by means of *lexicographic probability systems.* The latter constitute an extension of conventional probability distributions by explicitly allowing a decision maker to deem one event infinitely less likely than some other event, while deeming both events "possible". For a precise definition and a discussion of the decision theoretic foundations of lexicographic probability systems, the reader is referred to Blume, Brandenburger and Dekel (1991a).

We next prove that a proper equilibrium always exists.

Theorem 3.6.3 *Every extensive form game has a proper equilibrium.*

Proof. Let Γ be an extensive form game with strategy spaces S_i. Choose some player i, a bijective function $\pi : S_i \to \{0, 1, ..., |S_i| - 1\}$ and some small strictly positive number $\epsilon > 0$. Let $\mu_i^\pi(\epsilon)$ be the mixed conjecture given by

$$\mu_i^\pi(\epsilon)(s_i) = \frac{\epsilon^{\pi(s_i)}}{\epsilon^0 + \epsilon^1 + ... + \epsilon^{|S_i|-1}}$$

for all $s_i \in S_i$. By $\Pi(S_i)$ we denote the collection of all such bijective functions π_i. Let $M_i(\epsilon) = \Delta(\{\mu_i^\pi(\epsilon)| \; \pi \in \Pi(S_i)\})$ be the set of mixed conjectures that can be written as convex combinations of such mixed conjectures $\mu_i^\pi(\epsilon)$. Let $M(\epsilon) = \times_{i \in I} M_i(\epsilon)$. For every player i and mixed conjecture profile $\mu \in M_i(\epsilon)$, let $u_i(\mu) = \sum_{s_i \in S_i} \mu_i(s_i)$ $u_i(s_i, \mu_{-i})$. Let $B_i^\epsilon : M(\epsilon) \twoheadrightarrow M_i(\epsilon)$ be the correspondence which assigns to every conjecture profile $\mu \in M(\epsilon)$ the set

$$B_i^\epsilon(\mu) = \{\bar{\mu}_i \in M_i(\epsilon)| \; u_i(\bar{\mu}_i, \mu_{-i}) = \max_{\mu_i' \in M_i(\epsilon)} u_i(\mu_i', \mu_{-i})\}.$$

Let $B^\epsilon : M(\epsilon) \twoheadrightarrow M(\epsilon)$ be the correspondence which assigns to every $\mu \in M(\epsilon)$ the set $B^\epsilon(\mu) = \times_{i \in I} B_i^\epsilon(\mu)$.

Since the set $M(\epsilon)$ is nonempty, convex and compact, the correspondence B^ϵ is upper-hemicontinuous, and the sets $B^\epsilon(\mu)$ are all nonempty, convex and compact, it follows with Kakutani's fixed point theorem that B^ϵ has some fixed point μ^* in $M(\epsilon)$. Hence, there is some $\mu^* \in M(\epsilon)$ such that $u_i(\mu_i^*, \mu_{-i}^*) = \max_{\mu_i \in M_i(\epsilon)} u_i(\mu_i, \mu_{-i}^*)$ for all $i \in I$.

We now show that this fixed point μ^* is an ϵ-proper mixed conjecture profile, that is, $u_i(s_i, \mu_{-i}^*) < u_i(t_i, \mu_{-i}^*)$ implies $\mu_i^*(s_i) \leq \epsilon \mu_i^*(t_i)$. Suppose that $u_i(s_i, \mu_{-i}^*) < u_i(t_i, \mu_{-i}^*)$ for some player i and some strategies $s_i, t_i \in S_i$. Since $u_i(\mu_i^*, \mu_{-i}^*) = \max_{\mu_i \in M_i(\epsilon)} u_i(\mu_i, \mu_{-i}^*)$, we have that μ_i^*, written as a convex combination over mixed conjectures $\mu_i^\pi(\epsilon)$, puts probability zero on all mixed conjectures $\mu_i^\pi(\epsilon)$ where $\pi(s_i) < \pi(t_i)$. But then, by construction, μ_i^* only puts positive probability on those mixed conjectures $\mu_i^\pi(\epsilon)$ where $\mu_i^\pi(\epsilon)(s_i) \leq \epsilon \mu_i^\pi(\epsilon)(t_i)$, and thus $\mu_i^*(s_i) \leq \epsilon \mu_i^*(t_i)$, which implies that μ^* is ϵ-proper.

Finally, choose some sequence $(\epsilon^n)_{n \in \mathbb{N}}$ of strictly positive numbers converging to zero and a corresponding sequence $(\mu^n)_{n \in \mathbb{N}}$ of fixed points of B^{ϵ^n}. Since μ^n is strictly positive, the local mixed conjectures λ_i^n induced by μ^n do not avoid any information set, and hence $\lambda^n = (\lambda_i^n)_{i \in I}$ corresponds to a behavioral conjecture profile σ^n. Above, we have seen that μ^n is ϵ^n-proper, and hence σ^n is ϵ^n-proper as well. Without loss of generality, we assume that $(\sigma^n)_{n \in \mathbb{N}}$ converges to some behavioral conjecture profile σ. But then, by construction, σ^n is a proper equilibrium. ■

We now show that every proper equilibrium is a quasi-perfect equilibrium. This result, as already mentioned above, is due to van Damme (1984).

Theorem 3.6.4 *Let Γ be an extensive form game and σ a proper equilibrium. Then, σ is a quasi-perfect equilibrium.*

Proof. Let σ be a proper equilibrium. Then, there is a proper sequence $(\sigma^n)_{n\in\mathbb{N}}$ of strictly positive behavioral conjecture profiles which converges to σ. We prove that for every player i information set h, and every $s_i \in S_i(h)$ it holds that $\sigma_{i|h}(s_i) > 0$ only if s_i is a sequential best response against σ^n_{-i} at h for every n. This would imply that σ is a quasi-perfect equilibrium. Let h be a player i information set, $s_i \in S_i(h)$ and suppose that s_i is not a sequential best response against σ^n_{-i} at h for some n. We show that $\sigma_{i|h}(s_i) = 0$.

Let β^n_{ih} be the belief vector at h induced by σ^n_{-i}. Since s_i is not a sequential best response against σ^n_{-i} at h, there is some $r_i \in S_i(h)$ with

$$u_i((s_i, \sigma^n_{-i})|\, h, \beta^n_{ih}) < u_i((r_i, \sigma^n_{-i})|\, h, \beta^n_{ih}). \tag{3.20}$$

Let $H^*_i(r_i)$ be the set of player i information sets in $H_i(r_i)$ that follow h, together with the information set h itself. Let t_i be the strategy which coincides with r_i at all information sets in $H^*_i(r_i)$, and coincides with s_i at all information sets in $H_i(s_i)\backslash H^*_i(r_i)$. We prove that $u_i(s_i, \sigma^n_{-i}) < u_i(t_i, \sigma^n_{-i})$.

Let Z_0 be the set of terminal nodes which are not preceded by the information set h. Then, we have that

$$u_i(s_i, \sigma^n_{-i}) = \sum_{x\in h} \mathbb{P}_{(s_i,\sigma^n_{-i})}(x)\, u_i((s_i, \sigma^n_{-i})|\, x) + \sum_{z\in Z_0} \mathbb{P}_{(s_i,\sigma^n_{-i})}(z)\, u_i(z).$$

Since $\mathbb{P}_{(s_i,\sigma^n_{-i})}(h) > 0$, it follows that

$$\begin{aligned}
u_i(s_i, \sigma^n_{-i}) &= \mathbb{P}_{(s_i,\sigma^n_{-i})}(h) \sum_{x\in h} \frac{\mathbb{P}_{(s_i,\sigma^n_{-i})}(x)}{\mathbb{P}_{(s_i,\sigma^n_{-i})}(h)}\, u_i((s_i, \sigma^n_{-i})|\, x) + \\
&\quad + \sum_{z\in Z_0} \mathbb{P}_{(s_i,\sigma^n_{-i})}(z)\, u_i(z) \\
&= \mathbb{P}_{(s_i,\sigma^n_{-i})}(h) \sum_{x\in h} \beta^n_{ih}(x)\, u_i((s_i, \sigma^n_{-i})|\, x) + \\
&\quad + \sum_{z\in Z_0} \mathbb{P}_{(s_i,\sigma^n_{-i})}(z)\, u_i(z) \\
&= \mathbb{P}_{(s_i,\sigma^n_{-i})}(h)\, u_i((s_i, \sigma^n_{-i})|\, h, \beta^n_{ih}) + \\
&\quad + \sum_{z\in Z_0} \mathbb{P}_{(s_i,\sigma^n_{-i})}(z)\, u_i(z).
\end{aligned} \tag{3.21}$$

Similarly,

$$u_i(t_i, \sigma^n_{-i}) = \mathbb{P}_{(t_i,\sigma^n_{-i})}(h)\, u_i((t_i, \sigma^n_{-i})|\, h, \beta^n_{ih}) + \sum_{z\in Z_0} \mathbb{P}_{(t_i,\sigma^n_{-i})}(z)\, u_i(z). \tag{3.22}$$

Since t_i coincides with r_i at all information sets in $H^*_i(r_i)$, it follows that $u_i((t_i, \sigma^n_{-i})|\, h, \beta^n_{ih}) = u_i((r_i, \sigma^n_{-i})|\, h, \beta^n_{ih})$. Since t_i coincides with s_i at all information sets in

$H_i(s_i)\backslash H_i^*(r_i)$, it holds that $\mathbb{P}_{(t_i,\sigma_{-i}^n)}(h) = \mathbb{P}_{(s_i,\sigma_{-i}^n)}(h)$ and $\mathbb{P}_{(t_i,\sigma_{-i}^n)}(z) = \mathbb{P}_{(s_i,\sigma_{-i}^n)}(z)$ for all $z \in Z_0$. With (3.22) it follows that

$$u_i(t_i, \sigma_{-i}^n) = \mathbb{P}_{(s_i,\sigma_{-i}^n)}(h)\; u_i((r_i, \sigma_{-i}^n)|\; h, \beta_{ih}^n) + \sum_{z \in Z_0} \mathbb{P}_{(s_i,\sigma_{-i}^n)}(z)\; u_i(z).$$

Since $\mathbb{P}_{(s_i,\sigma_{-i}^n)}(h) > 0$ and $u_i((r_i, \sigma_{-i}^n)|\; h, \beta_{ih}^n) > u_i((s_i, \sigma_{-i}^n)|\; h, \beta_{ih}^n)$, we have that

$$\begin{aligned} u_i(t_i, \sigma_{-i}^n) &> \mathbb{P}_{(s_i,\sigma_{-i}^n)}(h)\; u_i((s_i, \sigma_{-i}^n)|\; h, \beta_{ih}^n) + \sum_{z \in Z_0} \mathbb{P}_{(s_i,\sigma_{-i}^n)}(z)\; u_i(z) \\ &= u_i(s_i, \sigma_{-i}^n), \end{aligned}$$

which we wanted to show. Since $(\sigma^m)_{m\in\mathbb{N}}$ is a proper sequence, we may conclude that $\lim_{m\to\infty} \sigma_i^m(s_i)/\sigma_i^m(t_i) = 0$.

For every $m \in \mathbb{N}$, let $\sigma_{i|h}^m$ be the revised behavioral conjecture at h induced by σ_i^m. Let $h_1, ..., h_K$ be the unique sequence of player i information sets that lead to h, and for every h_k, let a_k be the unique action at h_k that leads to h. Let $H_i^* = \{h_1, ..., h_K\}$. Then, by definition of $\sigma_{i|h}^m$, $\sigma_{ih_k|h}^m(a_k) = 1$ for every $h_k \in H_i^*$, and $\sigma_{ih'|h}^m = \sigma_{ih'}^m$ for every $h' \in H_i\backslash H_i^*$. Here, $\sigma_{ih'|h}^m$ is the local probability distribution at h' specified by $\sigma_{i|h}^m$. By definition,

$$\frac{\sigma_{i|h}^m(s_i)}{\sigma_{i|h}^m(t_i)} = \frac{\prod_{h'\in H_i(s_i)} \sigma_{ih'|h}^m(s_i(h'))}{\prod_{h'\in H_i(t_i)} \sigma_{ih'|h}^m(t_i(h'))}.$$

Since $s_i, t_i \in S_i(h)$ we have that $H_i^* \subseteq H_i(s_i)$ and $H_i^* \subseteq H_i(t_i)$ and hence

$$\begin{aligned} \frac{\sigma_{i|h}^m(s_i)}{\sigma_{i|h}^m(t_i)} &= \frac{\prod_{h'\in H_i^*} \sigma_{ih'|h}^m(s_i(h')) \prod_{h'\in H_i(s_i)\backslash H_i^*} \sigma_{ih'|h}^m(s_i(h'))}{\prod_{h'\in H_i^*} \sigma_{ih'|h}^m(t_i(h')) \prod_{h'\in H_i(t_i)\backslash H_i^*} \sigma_{ih'|h}^m(t_i(h'))} \\ &= \frac{\prod_{h_k\in H_i^*} \sigma_{ih_k|h}^m(a_k) \prod_{h'\in H_i(s_i)\backslash H_i^*} \sigma_{ih'|h}^m(s_i(h'))}{\prod_{h_k\in H_i^*} \sigma_{ih_k|h}^m(a_k) \prod_{h'\in H_i(t_i)\backslash H_i^*} \sigma_{ih'|h}^m(t_i(h'))} \\ &= \frac{\prod_{h'\in H_i(s_i)\backslash H_i^*} \sigma_{ih'|h}^m(s_i(h'))}{\prod_{h'\in H_i(t_i)\backslash H_i^*} \sigma_{ih'|h}^m(t_i(h'))}. \end{aligned}$$

Since $\sigma_{ih'|h}^m = \sigma_{ih'}^m$ for all $h' \in H_i\backslash H_i^*$, we have

$$\begin{aligned} \frac{\sigma_{i|h}^m(s_i)}{\sigma_{i|h}^m(t_i)} &= \frac{\prod_{h'\in H_i(s_i)\backslash H_i^*} \sigma_{ih'}^m(s_i(h'))}{\prod_{h'\in H_i(t_i)\backslash H_i^*} \sigma_{ih'}^m(t_i(h'))} \\ &= \frac{\prod_{h_k\in H_i^*} \sigma_{ih_k}^m(a_k) \prod_{h'\in H_i(s_i)\backslash H_i^*} \sigma_{ih'}^m(s_i(h'))}{\prod_{h_k\in H_i^*} \sigma_{ih_k}^m(a_k) \prod_{h'\in H_i(t_i)\backslash H_i^*} \sigma_{ih'}^m(t_i(h'))} \\ &= \frac{\prod_{h'\in H_i^*} \sigma_{ih'}^m(s_i(h')) \prod_{h'\in H_i(s_i)\backslash H_i^*} \sigma_{ih'}^m(s_i(h'))}{\prod_{h'\in H_i^*} \sigma_{ih'}^m(t_i(h')) \prod_{h'\in H_i(t_i)\backslash H_i^*} \sigma_{ih'}^m(t_i(h'))} \\ &= \frac{\sigma_i^m(s_i)}{\sigma_i^m(t_i)}. \end{aligned}$$

Since we know that $\lim_{m\to\infty} \sigma_i^m(s_i)/\sigma_i^m(t_i) = 0$ it follows that

$$\lim_{m\to\infty} \sigma_{i|h}^m(s_i)/\sigma_{i|h}^m(t_i) = 0,$$

which implies that $\sigma_{i|h}(s_i) = 0$. This completes the proof. ■

Since we have seen that a proper equilibrium always exists, Theorem 3.6.4 implies that every game has a quasi-perfect equilibrium as well. We also know that every quasi-perfect equilibrium is a normal form perfect equilibrium, and hence every game has a normal form perfect equilibrium. We conclude this section by showing that a proper equilibrium need not be a perfect equilibrium. Consider again the game in Figure 3.9. It may be checked that $\sigma = (a, c)$ is a proper equilibrium. However, we have seen that σ is not a perfect equilibrium.

Chapter 4

Consistency and Sequential Rationality

In Chapter 3 we have discussed some restrictions on assessments which, roughly speaking, reflect the requirement that players should believe that opponents act optimally. In the concepts of Nash equilibrium and normal form perfect equilibrium, this requirement is only imposed at the beginning of the game, whereas in other concepts such as backward induction, subgame perfect equilibrium and quasi-perfect equilibrium, it is imposed at every stage of the game. The concepts discussed so far did not explicitly involve the belief system, however, since explicit restrictions were put on the behavioral conjecture profile only. In this chapter we discuss some rationality criteria which impose explicit conditions on the belief system as well. We consider two types of restrictions. The first type, which we refer to as *consistency conditions*, should guarantee that the behavioral conjectures and the beliefs in a given assessment should not contradict one another. For instance, the beliefs held by a player at an information set should be in accordance with his beliefs held at earlier information set, which, in turn, should be compatible with the behavioral conjecture held by this player about the opponents' behavior. Conditions of the second type are similar to those presented in Chapter 3, as they require players to attach positive probability only to those opponents' strategies that constitute sequential best responses. By imposing this requirement at every information set we obtain the notion of *sequential rationality*. If we impose it only at those information sets that are not avoided by the initial conjectures, the weaker notion of *weak sequential rationality* results. The chapter is organized as follows. In Section 4.1, we present two consistency requirements for assessments, called *updating consistency* and *consistency*. In Section 4.2, we introduce the notion of sequential rationality and define the concept of *sequential equilibrium* as the combination of consistency and sequential rationality. Section 4.3 discusses a "local" version of sequential rationality, extending the backward induction condition to the general class of games, and shows that it is equivalent to sequential rationality if the assessment is updating consistent. We show, moreover, that updating consistency is a necessary condition for this equivalence to hold. In Section 4.4 we discuss the notion of weak sequential rationality, and accordingly define the concept of weak sequential equilibrium as the combination of consistency and weak sequential rationality.

4.1 Consistency of Assessments

Let the players' conjectures about the opponents' behavior, and the revision of these conjectures during the game, be given by an assessment (σ, β). By definition, the behavioral conjecture σ_i specifies at each of player i's information sets $h \in H_i$ the common conjecture held by i's opponents about player i's behavior at h. On the other hand, the belief β_j specifies at each of player j's information sets $h \in H_j$ player j's personal conjecture about the opponents' past play that has led to h. In order for an assessment (σ, β) to be "acceptable", it seems to be a necessary requirement that the behavioral conjectures and beliefs be "consistent", that is, do not contradict one another. It is not clear, a-priori, what we mean by the phrase that "the behavioral conjectures and beliefs be consistent". In Section 2.6 we have seen a first such consistency-criterion, called *Bayesian consistency*, which states that the beliefs should be induced by the behavioral conjectures through Bayesian updating, whenever possible. More precisely, if some information set $h \in H_i$ is reached with positive probability under σ, then player i is able to compute, for every node $x \in h$, its probability under σ, conditional on the event that h is reached. Bayesian consistency requires the beliefs β_{ih} at h to coincide with these conditional probabilities. Bayesian consistency may be seen as a first, necessary requirement for an assessment to be "consistent". However, as the remainder of this section will demonstrate, this requirement is far from being sufficient. For this reason, we shall present two alternative conditions, called *updating consistency* and *consistency*. Other consistency conditions that have been proposed in the literature but which are not discussed here are structural consistency of beliefs (Kreps and Wilson, 1982a), structural consistency and convex structural consistency of assessments (Kreps and Ramey, 1987) and player consistency (Hendon, Jacobsen and Sloth, 1996).

In some games, Bayesian consistency is not enough to ensure that the behavioral conjectures and beliefs do not contradict one another. Consider, for instance, the game in Figure 4.1. Let (σ, β) be the assessment in which $\sigma = (a, (d, f))$ and player 2's belief vectors are $(1, 0)$ and $(0, 1)$, respectively. This assessment is Bayesian consistent since both of player 2's information sets are reached with probability zero under σ. Player 2's beliefs do not seem plausible, however, since player 2's beliefs at his second information set contradict his beliefs at his first information set. Indeed, player 2's uncertainty at both information sets uniquely concerns player 1's choice among $\{b, c\}$, and thus his beliefs at both information sets should be the same. We say that the assessment is not *updating consistent.*

The idea behind updating consistency is the following. Consider an assessment (σ, β), two information sets h^1 and h^2 which are controlled by the same player i, and assume that h^2 comes after h^1. Player i's conjecture about the opponents' past behavior at h^1 is reflected by the beliefs β_{ih^1}. On the other hand, player i's conjecture at h^1 about the opponents' future behavior is given by σ_{-i}. Updating consistency

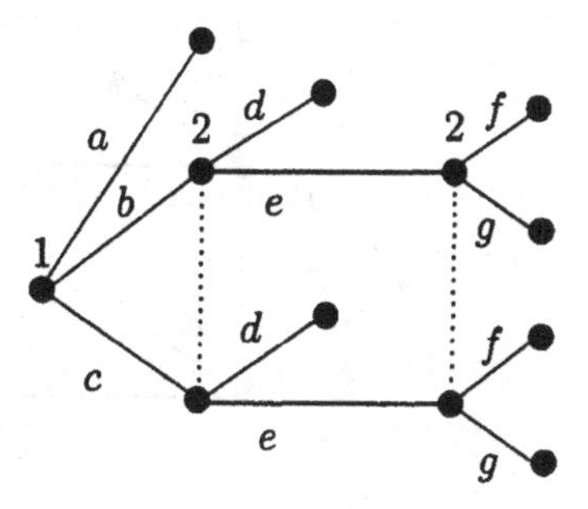

Figure 4.1

states that player i's conjecture about the opponents' past behavior at h^2 should be induced by his conjectures about past and future behavior at h^1 whenever the event of reaching h^2 is compatible with his conjectures at h^1.

Definition 4.1.1 *An assessment (σ, β) is called updating consistent if for every player i, every two information sets $h^1, h^2 \in H_i$ where h^2 comes after h^1, and every strategy $s_i \in S_i(h^2)$ with $\mathbb{P}_{(s_i,\sigma_{-i})}(h^2 \,|h^1, \beta_{h^1}) > 0$, it holds*

$$\beta_{ih^2}(x) = \frac{\mathbb{P}_{(s_i,\sigma_{-i})}(x \,|h^1, \beta_{ih^1})}{\mathbb{P}_{(s_i,\sigma_{-i})}(h^2 \,|h^1, \beta_{ih^1})}$$

for all $x \in h^2$.

Here, $\mathbb{P}_{(s_i,\sigma_{-i})}(x \,|h^1, \beta_{ih^1})$ is the probability that the node x is reached, conditional on h^1 being reached and given the beliefs β_{ih^1} at h^1. Formally,

$$\mathbb{P}_{(s_i,\sigma_{-i})}(x \,|h^1, \beta_{ih^1}) = \sum_{y \in h^1} \beta_{ih^1}(y) \; \mathbb{P}_{(s_i,\sigma_{-i})}(x \,|y).$$

By

$$\mathbb{P}_{(s_i,\sigma_{-i})}(h^2 |\, h^1, \beta^1_{ih}) = \sum_{x \in h^2} \mathbb{P}_{(s_i,\sigma_{-i})}(x \,|h^1, \beta_{ih^1})$$

we denote the probability that h^2 is reached, conditional on h^1 being reached and given the beliefs at h^1. By perfect recall, there is a unique sequence of player i actions that goes from h^1 and h^2, and hence the ratio in the definition of updating consistency does not depend on the choice of $s_i \in S_i(h^2)$. In Hendon, Jacobsen and Sloth (1996), assessments which are both Bayesian consistent and updating consistent are called *pre-consistent.* Fudenberg and Tirole (1991) introduce the notion of *reasonable*

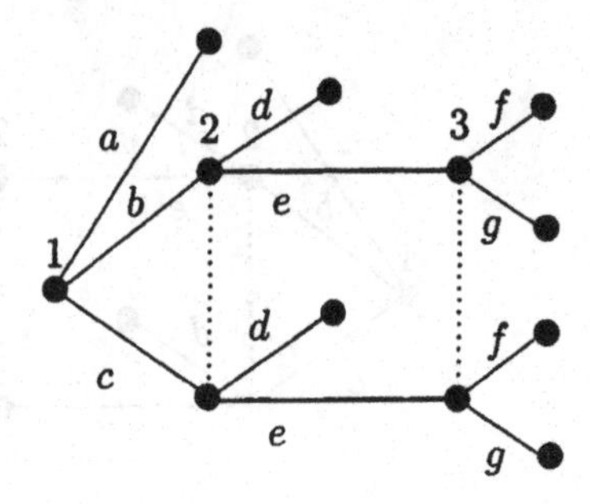

Figure 4.2

assessments for multistage games with observable actions. Their definition is essentially the same in spirit as updating consistency, as it requires a player to update his beliefs consistently.

Kreps and Wilson (1982a) proposed a condition on assessments, called *consistency,* that is stronger than both updating consistency and Bayesian consistency. The idea is the following. Suppose that instead of σ the players would hold a behavioral conjecture profile $\tilde{\sigma}$ in which all possible actions are taken into account, that is, every action is assigned a positive probability. In this case, $\tilde{\sigma}$ would induce a unique belief system $\tilde{\beta}$ since at every information set a unique belief vector is obtained by Bayesian updating. Moreover, the behavioral conjectures and the beliefs in $(\tilde{\sigma}, \tilde{\beta})$ would be "perfectly compatible" with one another. The idea in the concept of consistency is to take such "perfectly compatible" assessments $(\tilde{\sigma}, \tilde{\beta})$ in which all actions are given positive probability, as a reference point. More precisely, consistency states that for a given assessment (σ, β), it should be possible to approximate this assessment by a sequence of "perfectly compatible" assessments in which all actions are given positive probability.

Definition 4.1.2 *An assesment* (σ, β) *is called consistent if there is a sequence* $(\sigma^n, \beta^n)_{n \in \mathbb{N}}$ *of strictly positive Bayesian consistent assessments converging to* (σ, β).

The reader may easily verify that every consistent assessment is updating consistent and Bayesian consistent. In order to see that consistency can be a strictly stronger requirement than these two criteria, consider the example in Figure 4.2. Let (σ, β) be the assessment where $\sigma = (a, d, f)$, player 2 has beliefs $(1, 0)$ and player 3 has beliefs $(0, 1)$. This assessment is Bayesian consistent and updating consistent, but is not consistent. Namely, in every strictly positive Bayesian consistent assessment $(\tilde{\sigma}, \tilde{\beta})$, the beliefs for player 2 and 3 should be the same, and hence they should be the same in any consistent assessment.

The condition imposed by consistency in the example above is somewhat trivial. However, one can construct more complex examples in which the restrictions on the

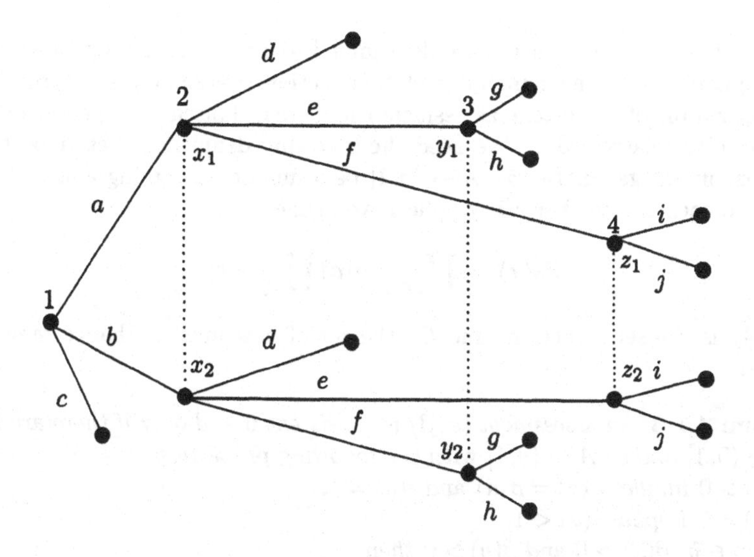

Figure 4.3

assessments imposed by consistency are not that obvious. Consider the game in Figure 4.3, which is taken from Kohlberg and Reny (1997) and Perea, Jansen and Peters (1997). Consider the assessment (σ^*, β^*) where $\sigma^* = (c, d, g, i)$ and the beliefs are $(0,1)$ for player 2, $(1/3, 2/3)$ for player 3 and $(2/3, 1/3)$ for player 4. The assessment is Bayesian consistent and updating consistent. It is not consistent, however. Consider an arbitrary strictly positive Bayesian consistent assessment (σ, β) in this game. Then,

$$\frac{\beta(y_1)}{\beta(y_2)} = \frac{\beta(x_1)\sigma_2(e)}{\beta(x_2)\sigma_2(f)} \text{ and } \frac{\beta(z_1)}{\beta(z_2)} = \frac{\beta(x_1)\sigma_2(f)}{\beta(x_2)\sigma_2(e)}$$

which implies that

$$\frac{\beta(y_1)}{\beta(y_2)} \frac{\beta(z_1)}{\beta(z_2)} = \frac{\beta(x_1)^2}{\beta(x_2)^2}. \tag{4.1}$$

Since a consistent assessment is a limit of strictly positive Bayesian consistent assessments, every consistent assessment should satisfy equation (4.1), if it is well defined. The assessment above does not satisfy this equation, and is therefore not consistent.

The example shows that consistency may put rather complex conditions on the assessment. Perea, Jansen and Peters (1997) and Kohlberg and Reny (1997) prove, however, that these conditions can always be reduced to a finite system of algebraic equations and inequalities on the behavioral conjectures and beliefs. Moreover, these papers explicitly *derive* the system of equations and inequalities that characterize the set of consistent assessments. Blume and Zame (1994) provide an indirect proof for the fact that consistent assessments are determined by finitely many algebraic equations and inequalities. In the remainder of this section, we shall present the algebraic characterization of consistent assessments in Perea, Jansen and Peters (1997) and compare it to related work. We need the following definition. Let A be the set of all actions in the game. Let $\bar{\sigma} : A \to (0,1]$ be a function assigning a strictly positive number to each action. For every node x we define

$$\mathbb{P}_{\bar{\sigma}}(x) := \prod_{a \in A_x} \bar{\sigma}(a) \prod_{c \in C_x} \tau(c),$$

where A_x is the set of actions and C_x the set of outcomes of chance moves on the path to x.

Theorem 4.1.3 *An assessment (σ, β) is consistent if and only if there are functions $\bar{\sigma} : A \to (0,1]$ and $\epsilon : A \to (0,1]$ with the following properties:*
(a) $\sigma(a) > 0$ implies $\bar{\sigma}(a) = \sigma(a)$ and $\epsilon(a) = 1$,
(b) $\sigma(a) = 0$ implies $\epsilon(a) < 1$,
(c) if $x, y \in h$, $\beta(x) > 0$ and $\beta(y) > 0$ then

$$\prod_{a \in A_x} \epsilon(a) = \prod_{a \in A_y} \epsilon(a) \text{ and } \frac{\beta(x)}{\beta(y)} = \frac{\mathbb{P}_{\bar{\sigma}}(x)}{\mathbb{P}_{\bar{\sigma}}(y)},$$

(d) if $x, y \in h, \beta(x) = 0$ and $\beta(y) > 0$ then

$$\prod_{a \in A_x} \epsilon(a) < \prod_{a \in A_y} \epsilon(a).$$

Whenever we write $\sigma(a)$, we mean $\sigma_{ih}(a)$, where $h \in H_i$ is the information set at which action a is available. Whenever we write $\beta(x)$, we mean $\beta_{ih}(x)$ where $h \in H_i$ is the information set at which x is present. A possible way to interpret the algebraic equations and inequalities in the theorem is the following. For every action a, the number $\epsilon(a)$ may be interpreted as the "order of likelihood" of a. Intuitively, if actions a and b are available at the same information set, $\epsilon(a) < \epsilon(b)$ means that a is deemed "infinitely less likely" than b by the opponents. This notion of "order of likelihood" may be mimicked by taking the limit ratio of the probabilities of a and b in some sequence of behavioral conjecture profiles. For instance, if $(\sigma^n)_{n \in \mathbb{N}}$ is a sequence of strictly positive behavioral conjecture profiles, the limit ratio $\lim_{n \to \infty} \sigma^n(a)/\sigma^n(b)$ reflects the order of likelihood of a compared to that of b in the sequence $(\sigma^n)_{n \in \mathbb{N}}$. If $\lim_{n \to \infty} \sigma^n(a)/\sigma^n(b) = 0$, then the action a is deemed "infinitely less likely" than b in the limit. If, on the other hand, $0 < \lim_{n \to \infty} \sigma^n(a)/\sigma^n(b) < \infty$, then a and b have the same likelihood order in the limit. In this case, we would write $\epsilon(a) = \epsilon(b)$.

Now, the pair $(\bar{\sigma}, \epsilon)$ can be interpreted as an *extended* behavioral conjecture profile, in which to every action a, in addition to a probability, some likelihood order $\epsilon(a)$ is assigned. Among the actions that are assigned the same likelihood order, $\bar{\sigma}(a)$ denotes the probability assigned to this action, conditional on the event that some action with likelihood order $\epsilon(a)$ is played. Conditions (a) and (b) simply state that the "conventional" behavioral conjecture profile σ is obtained by restricting $\bar{\sigma}$ to those actions that occur with maximum likelihood order, that is, with likelihood order 1. All actions with likelihood order less than 1 are assigned probability zero in σ. As such, the extended behavioral conjecture profile $(\bar{\sigma}, \epsilon)$ is an extension of σ. For a given node x, the number $\prod_{a \in A_x} \epsilon(a)$ can be interpreted as the "likelihood order" of reaching x. Conditions (c) and (d) state that the beliefs should be induced by $(\bar{\sigma}, \epsilon)$ by applying Bayes' rule to every information set, taking into account the likelihood order of nodes. More specifically, a node can only have positive belief if its likelihood order is maximal within this information set. Otherwise, this node should have belief zero. Hence, the theorem above could be restated as follows: An assessment (σ, β) is consistent if and only if σ can be extended to some extended behavioral conjecture profile $(\bar{\sigma}, \epsilon)$ which induces β.

The key ingredient in this particular characterization of consistent assessments is thus the introduction of a function ϵ, which allows us to compare the order of likelihood of two actions at the same information set that both have probability zero in σ. A similar approach can be found in McLennan (1989a,b), Battigalli (1996b) and Kohlberg and Reny (1997) in which consistent assessments are characterized by means of *conditional probability systems* (see Section 7.4) and the closely related notion of *relative probability systems*. The latter two notions are comparable to the concept of *lexicographic probability systems* (see Section 3.6) in that they explicitly describe situations in which one event is deemed infinitely less likely than some other event, while both events are deemed possible. In fact, Hammond (1994) shows that all three notions are "equivalent". The functions $\bar{\sigma}$ and ϵ used in the characterization above may be viewed as a refinement of these three notions, since $(\bar{\sigma}, \epsilon)$ induces a conditional probability system on the set of actions, as well as a relative probability system and a lexicographic probability system. However, in contrast to these three notions, the function ϵ not only states whether some action is infinitely less likely than some other action, it also *quantifies* their respective likelihood orders. Another difference is that the functions $\bar{\sigma}$ and ϵ above compare the likelihood of *actions,* whereas the conditional probability system used in Battigalli (1996b) and the relative probability system used in Kohlberg and Reny (1997) compare the likelihood of *strategies.* Note, however, that the function ϵ induces for every strategy s_i a likelihood order $\epsilon(s_i)$ by the formula

$$\epsilon(s_i) = \prod_{h \in H_i(s_i)} \epsilon(s_i(h)),$$

and as such, $\bar{\sigma}$ and ϵ induce a conditional probability system and a relative probability system on the set of strategies as well. In this sense, the pair of functions $(\bar{\sigma}, \epsilon)$ used in the characterization above may be viewed as a refinement of the notion of *strategic extended assessment* applied in Battigalli (1996b). We should also mention the work by Fudenberg and Tirole (1991) who provide, for a special class of so-called multi-stage games, a characterization of consistent assessments that does not make use of sequences of assessments.

In Kohlberg and Reny (1997) and Perea, Jansen and Peters (1997), the algebraic characterization of consistent assessments is used to develop an algorithm which computes the set of consistent assessments. Moreover, Perea, Jansen and Peters (1997) exploits the characterization to analyze the topological properties of the set of consistent assessments. See also McLennan (1989a,b) and Blume and Zame (1994) for a topological analysis of the set of consistent assessments.

Proof of Theorem 4.1.3. Let (σ, β) be an assessment. For every information set h, let $r(h)$ be the "first" node in h with positive belief. For convenience, we assume that the nodes in an information set are ordered. For a node $x \in h$, let $r(x) := r(h)$. Let A^0 be the set of actions assigned probability zero in σ and A^+ the set of actions assigned positive probability. Let X^0 be the set of nodes with belief zero in β and X^+ the set of nodes with positive belief. We construct a matrix M whose rows correspond with the set X of all nodes and whose columns correspond with the set A^0 of actions with probability zero, so $M = (m_{x,a})_{x \in X, a \in A^0}$. The elements $m_{x,a}$ are given by

$$m_{x,a} := \begin{cases} 1 & \text{if } a \in A_x \text{ and } a \notin A_{r(x)} \\ -1 & \text{if } a \notin A_x \text{ and } a \in A_{r(x)} \\ 0 & \text{otherwise.} \end{cases}$$

Furthermore, we define the vector $s = (s_x)_{x \in X}$ by

$$s_x := \sum_{c \in C_x} \log \tau(c) + \sum_{a \in A^+ \cap A_x} \log \sigma(a) - \sum_{c \in C_{r(x)}} \log \tau(c) - \sum_{a \in A^+ \cap A_{r(x)}} \log \sigma(a).$$

Let the vector $b = (b_x)_{x \in X}$ be given by

$$b_x := \log \beta(x) - \log \beta(r(x)),$$

where $\log 0 := -\infty$. Note that b_x can be $-\infty$ since $\beta(x)$ can be 0.

Claim 1. The assessment (σ, β) is consistent if and only if there is a sequence $(w^n)_{n \in \mathbb{N}}$ in $\mathbb{R}^{A^0}$ converging coordinatewise to $-\infty$ such that $b = s + \lim_{n \to \infty} M w^n$.

Proof of claim 1. Suppose that (σ, β) is a consistent assessment with supporting sequence $(\sigma^n, \beta^n)_{n \in \mathbb{N}}$. For every node x we have

$$\frac{\beta(x)}{\beta(r(x))} = \lim_{n \to \infty} \frac{\mathbb{P}_{\sigma^n}(x)}{\mathbb{P}_{\sigma^n}(r(x))}.$$

By definition,

$$\mathbb{P}_{\sigma^n}(x) = \prod_{c \in C_x} \tau(c) \cdot \prod_{a \in A_x \cap A^+} \sigma^n(a) \cdot \prod_{a \in A_x \cap A^0} \sigma^n(a).$$

Since $\lim_{n \to \infty} \sigma^n(a) = \sigma(a) > 0$ for every $a \in A^+$, it follows that

$$\frac{\beta(x)}{\beta(r(x))} = \frac{\prod_{c \in C_x} \tau(c) \cdot \prod_{a \in A^+ \cap A_x} \sigma(a)}{\prod_{c \in C_{r(x)}} \tau(c) \cdot \prod_{a \in A^+ \cap A_{r(x)}} \sigma(a)} \cdot \lim_{n \to \infty} \frac{\prod_{a \in A^0 \cap A_x} \sigma^n(a)}{\prod_{a \in A^0 \cap A_{r(x)}} \sigma^n(a)}.$$

If we take the logarithm on both sides, we obtain

$$b_x = s_x + \lim_{n \to \infty} \sum_{a \in A^0} m_{x,a} \log \sigma^n(a).$$

If we define the sequence $(w^n)_{n\in\mathbb{N}}$ in $\mathbb{R}^{A^0}$ by $w_a^n := \log \sigma^n(a)$ for every $a \in A^0$, it follows that $(w^n)_{n\in\mathbb{N}}$ converges coordinatewise to $-\infty$ and $b = s + \lim_{n\to\infty} M\, w^n$.

Now, let the node x be fixed and let $(w^n)_{n\in\mathbb{N}}$ be a sequence as described in claim 1. Then,

$$b_x = s_x + \lim_{n\to\infty} \sum_{a\in A^0} m_{x,a}\, w_a^n.$$

By definition of the elements $m_{x,a}$ it holds that

$$b_x = s_x + \lim_{n\to\infty} \Big(\sum_{a\in A^0\cap A_x} w_a^n - \sum_{a\in A^0\cap A_{r(x)}} w_a^n \Big).$$

If we take the exponential function on both sides, we obtain

$$\exp b_x = \exp s_x \cdot \lim_{n\to\infty} \frac{\prod_{a\in A^0\cap A_x} \exp w_a^n}{\prod_{a\in A^0\cap A_{r(x)}} \exp w_a^n},$$

where $\exp(-\infty) := 0$. Using the definition of b_x and s_x yields

$$\frac{\beta(x)}{\beta(r(x))} = \frac{\prod_{c\in C_x} \tau(c) \cdot \prod_{a\in A^+\cap A_x} \sigma(a)}{\prod_{c\in C_{r(x)}} \tau(c) \cdot \prod_{a\in A^+\cap A_{r(x)}} \sigma(a)} \cdot \lim_{n\to\infty} \frac{\prod_{a\in A^0\cap A_x} \exp w_a^n}{\prod_{a\in A^0\cap A_{r(x)}} \exp w_a^n}. \tag{4.2}$$

Define the sequence $(\hat{\sigma}^n)_{n\in\mathbb{N}}$ by

$$\hat{\sigma}^n(a) := \begin{cases} \exp w_a^n & \text{if } a \in A^0 \\ \sigma(a) & \text{if } a \in A^+. \end{cases}$$

Obviously, the sequence $(\hat{\sigma}^n)_{n\in\mathbb{N}}$ converges to σ. For every n, let σ^n be given by

$$\sigma^n(a) := R^n(a) \cdot \hat{\sigma}^n(a)$$

where $R^n(a) = 1/(\sum_{a'\in A(h)} \hat{\sigma}^n(a'))$ and h is the information set at which the action a is available. Since $R^n(a)$ converges to 1 for every a and $\hat{\sigma}^n(a) > 0$ for every a it follows that $(\sigma^n)_{n\in\mathbb{N}}$ is a sequence of strictly positive behavioral conjecture profiles converging to σ. With equation (4.2) it follows that

$$\frac{\beta(x)}{\beta(r(x))} = \lim_{n\to\infty} \frac{\prod_{c\in C_x} \tau(c) \cdot \prod_{a\in A^+\cap A_x} \sigma^n(a)}{\prod_{c\in C_{r(x)}} \tau(c) \cdot \prod_{a\in A^+\cap A_{r(x)}} \sigma^n(a)} \cdot \frac{\prod_{a\in A^0\cap A_x} \sigma^n(a)}{\prod_{a\in A^0\cap A_{r(x)}} \sigma^n(a)}.$$

This equality implies that

$$\frac{\beta(x)}{\beta(r(x))} = \lim_{n\to\infty} \frac{\mathbb{P}_{\sigma^n}(x)}{\mathbb{P}_{\sigma^n}(r(x))} \tag{4.3}$$

for every $x \in X$. Define the sequence β^n of beliefs by $\beta^n(x) = \mathbb{P}_{\sigma^n}(x)/\mathbb{P}_{\sigma^n}(h)$ where h is the information set at which x is present. Then, with equation (4.3) we have

$$\frac{\beta(x)}{\beta(r(x))} = \lim_{n\to\infty} \frac{\beta^n(x)}{\beta^n(r(x))}$$

for every x, which implies that β^n converges to β. Hence, the assessment (σ, β) is consistent. This completes the proof of claim 1.

Let M^+ and M^0 be the restrictions of M to the rows corresponding to nodes in X^+ and X^0 respectively and let s^+, b^+ be the restrictions of the vectors s, b to nodes in X^+.

Claim 2. There is a sequence $(w^n)_{n\in\mathbb{N}}$ in $\mathbb{R}^{A^0}$ converging coordinatewise to $-\infty$ such that $b = s + \lim_{n\to\infty} M\, w^n$ if and only if
(1) there is a vector $w \in \mathbb{R}^{A^0}, w < 0$ with $M^+ w = 0$ and $M^0 w < 0$ and
(2) $b^+ \in s^+ + \operatorname{Im}(M^+)$, where $\operatorname{Im}(M^+)$ denotes the image of the linear operator M^+.

Proof of claim 2. Assume, that $(w^n)_{n\in\mathbb{N}}$ is a sequence as described in claim 1. Let the vector v be given by $v := b - s$ and let v^+ be the restriction of v to the nodes in X^+. By construction, $v_x \in \mathbb{R}$ for all $x \in X^+$ and $v_x = -\infty$ for all $x \notin X^+$. Since $\lim_{n\to\infty} M^+ w^n = v^+$ and $\operatorname{Im}(M^+)$ is closed, it follows that $v^+ \in \operatorname{Im}(M^+)$, so $b^+ \in s^+ + \operatorname{Im}(M^+)$.

Now, let $z \in \mathbb{R}^{A^0}$ with $M^+ z = v^+$. Let $B := \{w \in \mathbb{R}^{A^0} \mid w \le -1, M^0 w \le -1\}$ and $C := \{M^+ w \mid w \in B\}$. Here, the inequality $w \le -1$ should be read coordinatewise. Obviously, B is a closed and convex set. Moreover, B is non-empty since $w^n \in B$ for large n. It follows that C is a non-empty closed convex set. Suppose that $0 \notin C$. By Corollary 11.4.2 in Rockafellar (1970), there exists a hyperplane which separates the sets C and $\{0\}$ strongly. In other words, we can find a non-zero vector p and a number $\alpha \in \mathbb{R}$ such that $p \cdot c > \alpha$ for all $c \in C$ and $p \cdot 0 < \alpha$. The last inequality implies that $\alpha > 0$. From the first inequality, it follows that $p \cdot M^+ w > \alpha$ for every $w \in B$. Since $w^n - z \in B$ for large n, it follows that $\lim_{n\to\infty} p \cdot M^+(w^n - z) \ge \alpha$. We know that $\lim_{n\to\infty} M^+ w^n = v^+ = M^+ z$. Therefore $0 \ge \alpha$, which is a contradiction. So $0 \in C$, which implies that there is a $w \in \mathbb{R}^{A^0}$, $w \le -1$ with $M^0 w \le -1$ and $M^+ w = 0$. Hence, (1) and (2) in claim 2 are satisfied.

Now, let (1) and (2) in claim 2 be satisfied and let $z \in \mathbb{R}^{A^0}$ with $b^+ = s^+ + M^+ z$. Define the sequence $(w^n)_{n\in\mathbb{N}}$ by $w^n := z + n\, w$. It is easy to check that w^n has the desired properties. This completes the proof of claim 2.

Combining claims 1 and 2 leads to the following result.

Claim 3. The assessment (σ, β) is consistent if and only if (1) there is a vector $w < 0$ with $M^+ w = 0$ and $M^0 w < 0$ and (2) $b^+ \in s^+ + \operatorname{Im}(M^+)$.

Claim 3 now allows us to prove the theorem. Let (σ, β) be a consistent assessment. By (2) in claim 3 there is a vector $z = (z_a)_{a\in A^0}$ such that $b^+ = s^+ + M^+ z$. Moreover, z can be chosen such that $z < 0$ since there is a vector $w < 0$ with $M^+ w = 0$. We define the function $\bar{\sigma} : A \to (0,1]$ by

$$\bar{\sigma}(a) := \begin{cases} \sigma(a) & \text{if } a \in A^+ \\ \exp(z_a) & \text{if } a \in A^0. \end{cases}$$

Since

$$b_x = s_x + \sum_{a \in A^0} m_{x,a}\, z_a = s_x + \sum_{a \in A^0 \cap A_x} z_a - \sum_{a \in A^0 \cap A_{r(x)}} z_a$$

for every $x \in X^+$, we obtain by taking the exponential function on both sides and using the definitions of b_x, v_x and $m_{x,a}$ that

$$\begin{aligned} \frac{\beta(x)}{\beta(r(x))} &= \frac{\prod_{c \in C_x} \tau(c) \prod_{a \in A^+ \cap A_x} \sigma(a)}{\prod_{c \in C_{r(x)}} \tau(c) \prod_{a \in A^+ \cap A_{r(x)}} \sigma(a)} \frac{\prod_{a \in A^0 \cap A_x} \bar{\sigma}(a)}{\prod_{a \in A^0 \cap A_{r(x)}} \bar{\sigma}(a)} \\ &= \frac{\mathbb{P}_{\bar{\sigma}}(x)}{\mathbb{P}_{\bar{\sigma}}(r(x))} \end{aligned}$$

for every $x \in X^+$. Hence, for all x, y in the same information set with $\beta(x) > 0, \beta(y) > 0$ we have that $\beta(x)/\beta(y) = \mathbb{P}_{\bar{\sigma}}(x)/\mathbb{P}_{\bar{\sigma}}(y)$.

By (1) of claim 3 there is some $w < 0$ with $M^+ w = 0$ and $M^0 w < 0$. Since $M^+ w = 0$ it follows that

$$\sum_{a \in A^0 \cap A_x} w_a - \sum_{a \in A^0 \cap A_{r(x)}} w_a = 0$$

for every $x \in X^+$. Taking the exponential function on both sides, we obtain

$$\frac{\prod_{a \in A^0 \cap A_x} \exp w_a}{\prod_{a \in A^0 \cap A_{r(x)}} \exp w_a} = 1$$

for every $x \in X^+$. Since $M^0 w < 0$, we can show in a similar way that

$$\frac{\prod_{a \in A^0 \cap A_x} \exp w_a}{\prod_{a \in A^0 \cap A_{r(x)}} \exp w_a} < 1$$

for every $x \in X^0$. If we define the function $\epsilon : A \to (0,1]$ by $\epsilon(a) := \exp(w_a)$ for every $a \in A^0$ and $\epsilon(a) := 1$ for every $a \in A^+$ then we have

$$\prod_{a \in A_x} \epsilon(a) = \prod_{a \in A_{r(x)}} \epsilon(a)$$

whenever $\beta(x) > 0$ and

$$\prod_{a \in A_x} \epsilon(a) < \prod_{a \in A_{r(x)}} \epsilon(a)$$

if $\beta(x) = 0$ which implies that ϵ has the desired properties. We have thus shown that for every consistent assessment there exist functions $\bar{\sigma}, \epsilon$ that satisfy the properties (1) - (4) in the theorem.

Suppose now that there exist functions $\bar{\sigma} : A \to (0,1]$ and $\epsilon : A \to (0,1]$ as described in the theorem. We define the sequence $(\sigma^n)_{n\in\mathbb{N}}$ of strictly positive behavioral conjecture profiles by

$$\sigma^n(a) := R_h^n \cdot \bar{\sigma}(a) \cdot \epsilon(a)^n$$

where $R_h^n = 1/(\sum_{a'\in A(h)} \bar{\sigma}(a') \cdot \epsilon(a')^n)$ and h is the information set at which a is available. Let β^n be the unique belief system such that (σ^n, β^n) is Bayesian consistent. We show that (σ^n, β^n) converges to (σ, β). By construction, $(\bar{\sigma}(a)\cdot\epsilon(a)^n)_{n\in\mathbb{N}}$ converges to $\sigma(a)$ for every a. Consequently, $(R_h^n)_{n\in\mathbb{N}}$ converges to 1 and $(\sigma^n(a))_{n\in\mathbb{N}}$ converges to $\sigma(a)$ for every a. Let x, y be nodes in h. Then,

$$\begin{aligned}
\frac{\beta^n(x)}{\beta^n(y)} &= \frac{\mathbb{P}_{\sigma^n}(x)}{\mathbb{P}_{\sigma^n}(y)} = \frac{\prod_{c\in C_x} \tau(c) \cdot \prod_{a\in A_x} \sigma^n(a)}{\prod_{c\in C_y} \tau(c) \cdot \prod_{a\in A_y} \sigma^n(a)} \\
&= \frac{\prod_{c\in C_x} \tau(c) \cdot \prod_{a\in A_x} R_h^n \cdot \bar{\sigma}(a) \cdot \epsilon(a)^n}{\prod_{c\in C_y} \tau(c) \cdot \prod_{a\in A_y} R_h^n \cdot \bar{\sigma}(a) \cdot \epsilon(a)^n} \\
&= \left[\frac{\prod_{a\in A_x} \epsilon(a)}{\prod_{a\in A_y} \epsilon(a)}\right]^n \cdot \frac{\prod_{a\in A_x} R_{h(a)}^n}{\prod_{a\in A_y} R_{h(a)}^n} \cdot \frac{\prod_{c\in C_x} \tau(c) \cdot \prod_{a\in A_x} \bar{\sigma}(a)}{\prod_{c\in C_y} \tau(c) \cdot \prod_{a\in A_y} \bar{\sigma}(a)} \\
&= \left[\frac{\prod_{a\in A_x} \epsilon(a)}{\prod_{a\in A_y} \epsilon(a)}\right]^n \cdot \frac{\prod_{a\in A_x} R_{h(a)}^n}{\prod_{a\in A_y} R_{h(a)}^n} \cdot \frac{\mathbb{P}_{\bar{\sigma}}(x)}{\mathbb{P}_{\bar{\sigma}}(y)}.
\end{aligned}$$

Here, $h(a)$ denotes the information set at which the action a is available. If $\beta(x) > 0$ and $\beta(y) > 0$ then, by assumption,

$$\prod\nolimits_{a\in A_x} \epsilon(a) = \prod\nolimits_{a\in A_y} \epsilon(a) \text{ and } \frac{\beta(x)}{\beta(y)} = \frac{\mathbb{P}_{\bar{\sigma}}(x)}{\mathbb{P}_{\bar{\sigma}}(y)}$$

which implies that

$$\lim_{n\to\infty} \frac{\beta^n(x)}{\beta^n(y)} = \frac{\mathbb{P}_{\bar{\sigma}}(x)}{\mathbb{P}_{\bar{\sigma}}(y)} = \frac{\beta(x)}{\beta(y)}.$$

If $\beta(x) = 0$ and $\beta(y) > 0$ then, by assumption,

$$\prod\nolimits_{a\in A_x} \epsilon(a) < \prod\nolimits_{a\in A_y} \epsilon(a)$$

which implies that

$$\lim_{n\to\infty} \frac{\beta^n(x)}{\beta^n(y)} = 0 = \frac{\beta(x)}{\beta(y)}.$$

Hence we have that

$$\lim_{n\to\infty} \frac{\beta^n(x)}{\beta^n(y)} = \frac{\beta(x)}{\beta(y)}$$

for every $x, y \in h$ with $\beta(y) > 0$. This implies that β^n converges to β. Hence, the assessment (σ, β) is consistent. This completes the proof of the theorem. ■

In order to illustrate the characterization, consider again the example in Figure 4.3. Consider the assessment (σ, β) with $\sigma = (c, d, g, i)$, $\beta(x_1) = 1/3$, $\beta(y_1) = 1/5$ and $\beta(z_1) = 1/2$. We apply Theorem 4.1.3 to verify whether (σ, β) is consistent. Let the functions $\bar{\sigma}$ and ϵ be given by the following table.

	a	b	c	d	e	f	g	h	i	j
$\bar{\sigma}$	1/3	2/3	1	1	3/10	6/10	1	1	1	1
ϵ	1/2	1/2	1	1	1/2	1/2	1	1/2	1	1/2

It can be checked easily that $\bar{\sigma}$ and ϵ satisfy the conditions of the characterization. As a consequence, (σ, β) is consistent.

Now, consider the assessment (σ, β) with $\sigma = (c, d, g, i)$, $\beta(x_1) = 0, \beta(y_1) = 1/3$ and $\beta(z_1) = 2/3$. We use the characterization to verify whether (σ, β) is consistent. Assume that there are functions $\bar{\sigma}$ and ϵ with the properties of Theorem 4.1.3. Since $\beta(x_1) = 0$ and $\beta(x_2) > 0$ it follows that $\epsilon(a) < \epsilon(b)$. Since $\beta(y_1) > 0$ and $\beta(y_2) > 0$ it must hold that $\epsilon(a)\ \epsilon(e) = \epsilon(b)\ \epsilon(f)$ which implies that $\epsilon(e) > \epsilon(f)$. However, this would mean that $\epsilon(a)\ \epsilon(f) < \epsilon(b)\ \epsilon(e)$ which is a contradiction since $\beta(z_1) > 0$ and $\beta(z_2) > 0$. Hence, (σ, β) is not a consistent assessment.

4.2 Sequential Rationality

In addition to consistency requirements, a "reasonable" assessments should satisfy conditions which reflects the players' belief that opponents act rationally in the game. Note that the conditions of updating consistency and consistency discussed in the previous section are "non-strategic" conditions, in the sense that they are independent of the players' utility functions u_i, and thus are independent of the players' strategic considerations in the game. They simply state that the behavioral conjectures and beliefs should be compatible, given the extensive form structure at hand. Some conditions on the assessment (σ, β) that *do* involve strategic considerations have already been discussed in the previous chapter. The concepts of Nash equilibrium, subgame perfect equilibrium, perfect equilibrium, quasi-perfect equilibrium and proper equilibrium all put "strategic" conditions on the behavioral conjecture profile σ. However, these concepts do not use *explicitly* the belief system β which is part of the assessment. Some of the concepts, like perfect equilibrium and quasi-perfect equilibrium, *implicitly* involve the belief system β by considering sequences of strictly positive Bayesian consistent assessments converging to the original assessment, and imposing optimality conditions against assessments in this sequence. In this section, we shall present a "strategic" condition on assessments, called *sequential rationality,* that *explicitly* uses the belief system.

For a given assessment (σ, β) and information $h \in H_i$, say that a strategy $s_i \in S_i(h)$ is a *sequential best response against* (σ, β) *at* h if

$$u_i((s_i, \sigma_{-i})|\, h, \beta_{ih}) = \max_{s_i' \in S_i(h)} u_i((s_i', \sigma_{-i})|\, h, \beta_{ih}).$$

Definition 4.2.1 *An assessment is called sequentially rational if for every player i, every $h \in H_i$ and every $s_i \in S_i(h)$, we have that $\sigma_{i|h}(s_i) > 0$ only if s_i is a sequential best response against (σ, β) at h.*

Recall that $\sigma_{i|h}$ is the revised conjecture about player i at h, induced by σ_i. Sequential rationality thus requires the players to believe, at every information set h, that the player at h is playing a sequential best response against (σ, β) at h. By simultaneously imposing consistency and sequential rationality, we obtain the concept of sequential equilibrium (Kreps and Wilson, 1982a).

Definition 4.2.2 *An assessment (σ, β) is called a sequential equilibrium if (σ, β) is consistent and sequentially rational.*

Note that the concept of quasi-perfect equilibrium also involves a form of sequential rationality. A quasi-perfect equilibrium σ can be approximated by a sequence $(\sigma^n, \beta^n)_{n \in \mathbb{N}}$ of strictly positive Bayesian consistent assessments such that $\sigma_{i|h}(s_i) > 0$ only if s_i is a sequential best response against (σ^n, β^n) at h for every n. Assume, without loss of generality, that $(\beta^n)_{n \in \mathbb{N}}$ converges to some belief system β. Then it immediately follows that (σ, β) is sequentially rational. Moreover, by construction, (σ, β) is consistent. We have thus shown the following result.

Lemma 4.2.3 *Let σ be a quasi-perfect equilibrium. Then, there is a belief system β such that (σ, β) is a sequential equilibrium.*

Since we already know that a quasi-perfect equilibrium always exists, every game contains at least one sequential equilibrium.

Theorem 4.2.4 *Every extensive form game contains a sequential equilibrium.*

Lemma 4.2.3 also holds if we replace "quasi-perfect equilibrium" by "perfect equilibrium". By definition, a perfect equilibrium σ can be approximated by a sequence $(\sigma^n, \beta^n)_{n \in \mathbb{N}}$ of strictly positive Bayesian consistent assessments such that $\sigma_{ih}(a) > 0$ only if a is a local best response against (σ^n, β^n) for all n. Without loss of generality, assume that (β^n) converges to some belief system β. Then, by construction, (σ, β) is consistent. In the proof of Theorem 3.4.3 we have shown that $\sigma_i(s_i) > 0$ only if s_i is a sequential best response against (σ, β). This result can be sharpened; one can show, by the same techniques, that $\sigma_{i|h}(s_i) > 0$ only if s_i is a sequential best response against (σ, β) at h. Hence, (σ, β) is sequentially rational. We have shown the following result.

Lemma 4.2.5 *Let σ be a perfect equilibrium. Then, there is a belief system β such that (σ, β) is a sequential equilibrium.*

A normal form perfect equilibrium σ need not always be extendable to a sequential equilibrium. Consider again the game in Figure 3.11. The unique sequential equilibrium in this game is $\sigma = (b_1, b_2, b_3)$, together with the trivial belief system β. However, as we have seen, (a_1, a_2, a_3) is a normal form perfect equilibrium, and is therefore not extendable to a sequential equilibrium. We finally prove that every sequential equilibrium induces a subgame perfect equilibrium, and therefore satisfies backward induction in every game with perfect information.

Lemma 4.2.6 *Let (σ, β) be a sequential equilibrium. Then, σ is a subgame perfect equilibrium.*

Proof. Let (σ, β) be a sequential equilibrium. We first show that σ is a Nash equilibrium. Let $s_i \in S_i$ such that $\sigma_i(s_i) > 0$. We prove that s_i is a best response against σ. Since (σ, β) is Bayesian consistent, it suffices to show that s_i is a sequential best response against (σ, β). Let $h \in H_i(s_i)$. Then, $\sigma_{i|h}(s_i) = \sigma_i(s_i) > 0$. Since (σ, β) is sequentially rational and $\sigma_{i|h}(s_i) > 0$, we have that s_i is a sequential best response against (σ, β) at h. Since this holds for every $h \in H_i(s_i)$, we have that s_i is a sequential best response against (σ, β), and therefore a best response against σ. We may thus conclude that σ is a Nash equilibrium.

We next show that σ is a subgame perfect equilibrium. Let Γ_x be a subgame, and let (σ^x, β^x) be the restriction of (σ, β) on this subgame. We prove that (σ^x, β^x) is a sequential equilibrium in Γ_x. Since (σ, β) is consistent, there is a sequence $(\sigma^n, \beta^n)_{n \in \mathbb{N}}$ of strictly positive Bayesian consistent assessments converging to (σ, β). Let $(\sigma^{x,n}, \beta^{x,n})$ be the restriction of (σ^n, β^n) to Γ_x. Then, $(\sigma^{x,n}, \beta^{x,n})_{n \in \mathbb{N}}$ is a sequence of strictly positive Bayesian consistent assessments in Γ_x converging to (σ^x, β^x), and hence (σ^x, β^x) is consistent. In order to show that (σ^x, β^x) is sequentially rational, let $h \in H_i$ be a player i information set in Γ_x, and let $s_i^x \in S_i^x(h)$ be a strategy in Γ_x which does not avoid h. Suppose that $\sigma^x_{i|h}(s_i^x) > 0$. Here, $\sigma^x_{i|h}$ is the revised conjecture at h induced by σ_i^x. Let s_i be the strategy in the original game Γ which chooses all player i actions leading to x, and which coincides with s_i^x in Γ_x. Then, we have that $s_i \in S_i(h)$ and $\sigma_{i|h}(s_i) > 0$. The latter follows from the fact that $\sigma^x_{i|h}(s_i^x) > 0$. Since (σ, β) is sequentially rational and $\sigma_{i|h}(s_i) > 0$, we have that s_i is a sequential best response against (σ, β) at h, and hence s_i^x is a sequential best response against (σ^x, β^x) at h. Since this holds for every h in Γ_x, and every s_i^x with $\sigma^x_{i|h}(s_i^x) > 0$, it follows that (σ^x, β^x) is sequentially rational in Γ_x, and hence (σ^x, β^x) is a sequential equilibrium in Γ_x. By the first part of the proof, we know that σ^x is then a Nash equilibrium in Γ_x. Since this holds for every subgame Γ_x, we may conclude that σ is a subgame perfect equilibrium in Γ. ■

4.3 One-Deviation Property

Recall that in games with perfect information, the concept of backward induction imposes the requirement that at every node of the game, positive probability should only be assigned to those actions at that node which are a local best response against σ. This restriction may be viewed as a weak version of sequential rationality, to which we shall refer as *local sequential rationality.* The notion of local sequential rationality may be extended to the general class of games as follows.

Definition 4.3.1 *We say that an assessment (σ, β) is locally sequentially rational if for every player i and every information set $h \in H_i$ we have that $\sigma_{ih}(a) > 0$ only if a is a local best response against (σ, β).*

In contrast to sequential rationality, the notion of local sequential rationality requires players to believe, at every stage of the game, that the opponents choose *optimal actions*, instead of *optimal strategies.* We first show that local sequential rationality is implied by sequential rationality.

Lemma 4.3.2 *Every sequentially rational assessment is locally sequentially rational.*

Proof. Let (σ, β) be sequentially rational. Let $h \in H_i$ and let $a^* \in A(h)$ with $\sigma_{ih}(a^*) > 0$. We show that a^* is a local best response against (σ, β). Let $s_i^* \in S_i(h)$ be such that $\sigma_{i|h}(s_i^*) > 0$ and $s_i^*(h) = a^*$. Such a strategy s_i^* exists since $\sigma_{ih}(a^*) > 0$. Since (σ, β) is sequentially rational, we have that s_i^* is a sequential best response against (σ, β) at h. Moreover, $\sigma_{i|h}(s_i) > 0$ only if s_i is a sequential best response against (σ, β) at h. Define

$$u_i(\sigma|\ h, \beta_{ih}) := \sum_{s_i \in S_i(h)} \sigma_{i|h}(s_i) u_i((s_i, \sigma_{-i})|\ h, \beta_{ih}).$$

Then, we may conclude that

$$u_i(\sigma|\ h, \beta_{ih}) = \max_{s_i' \in S_i(h)} u_i((s_i', \sigma_{-i})|\ h, \beta_{ih})$$

and hence

$$u_i((s_i^*, \sigma_{-i})|\ h, \beta_{ih}) = u_i(\sigma|\ h, \beta_{ih}) = \max_{s_i' \in S_i(h)} u_i((s_i', \sigma_{-i})|\ h, \beta_{ih}).$$

In fact, it may be verified easily that

$$\max_{s_i' \in S_i(h)} u_i((s_i', \sigma_{-i})|\ h, \beta_{ih}) = \max_{\sigma_i'} u_i((\sigma_i', \sigma_{-i})|\ h, \beta_{ih}).$$

Then, we obtain

$$u_i((s_i^*, \sigma_{-i})|\ h, \beta_{ih}) = u_i(\sigma|\ h, \beta_{ih}) = \max_{\sigma_i'} u_i((\sigma_i', \sigma_{-i})|\ h, \beta_{ih}). \tag{4.4}$$

Now, suppose that a^* is not a local best best response against (σ, β). Then, we have that $u_i((a^*, \sigma_{-h})|\ h, \beta_{ih}) < \max_{a \in A(h)} u_i((a, \sigma_{-h})|\ h, \beta_{ih})$. By definition,

$$u_i(\sigma|\ h, \beta_h) = \sum_{a \in A(h)} u_i((a, \sigma_{-h})|\ h, \beta_{ih}).$$

Since $\sigma_{ih}(a^*) > 0$, it follows that

$$u_i(\sigma|\ h, \beta_{ih}) < \max_{a \in A(h)} u_i((a, \sigma_{-h})|h, \beta_{ih}).$$

Similarly, it may be verified that

$$\max_{a \in A(h)} u_i((a, \sigma_{-h})|\ h, \beta_{ih}) = \max_{\sigma'_{ih}} u_i((\sigma'_{ih}, \sigma_{-h})|\ h, \beta_{ih}),$$

and hence

$$u_i(\sigma|\ h, \beta_{ih}) < \max_{\sigma'_{ih}} u_i((\sigma'_{ih}, \sigma_{-h})|h, \beta_{ih}). \quad (4.5)$$

However, from (4.4) and (4.5) it follows that

$$\max_{\sigma'_{ih}} u_i((\sigma'_{ih}, \sigma_{-h})|\ h, \beta_{ih}) > \max_{\sigma'_i} u_i((\sigma'_i, \sigma_{-i})|\ h, \beta_{ih}),$$

which is clearly a contradiction. Hence, we may conclude that a^* is a local best response against (σ, β). This completes the proof. ■

For games with perfect information, it may be shown that every locally sequentially rational assessment is sequentially rational. Namely, consider a game with perfect information, and an assessment (σ, β) that is locally sequentially rational, that is, satisfies backward induction. In the proof of Theorem 3.1.4 we have shown that $\sigma_i(s_i) > 0$ only if s_i is a sequential best response against (σ, β). In fact, by the same proof we can show that for every $h \in H_i$ and every $s_i \in S_i(h)$, $\sigma_{i|h}(s_i) > 0$ only if s_i is a sequential best response against (σ, β) at h. Therefore, (σ, β) is sequentially rational. The following result is therefore obtained.

Lemma 4.3.3 *In a game with perfect information, an assessment is sequentially rational if and only if it is locally sequentially rational.*

This equivalence result need no longer hold if the game does not have perfect information. Consider, for instance, the game in Figure 4.4. Consider the assessment (σ, β) where $\sigma = (a, (d, g))$ and player 2 has beliefs $(1, 0)$ and $(0, 1)$ respectively. The assessment is locally sequentially rational. Let h be the first information set for player 2. Then, $\sigma_{2|h}$ assigns probability one to strategy (d, g). However, (d, g) is not a sequential best response against (σ, β) at h since (e, f) yields a higher expected utility there. Hence, (σ, β) is not sequentially rational. Even stronger, it may be verified that the beliefs $(1, 0)$ and $(0, 1)$ cannot be part of a sequentially rational assessment.

The reason for the failure of Lemma 4.3.3 in the game above is the fact that player 2's beliefs at his first and second information set contradict one another. Note that the assessment in this example is not updating consistent, as updating consistency implies that player 2's beliefs should be the same at his first and second information set. This example thus raises the following question: which kind of consistency condition on the assessment would assure that local sequential rationality implies sequential rationality? In the remainder of this section we shall prove that under updating consistency,

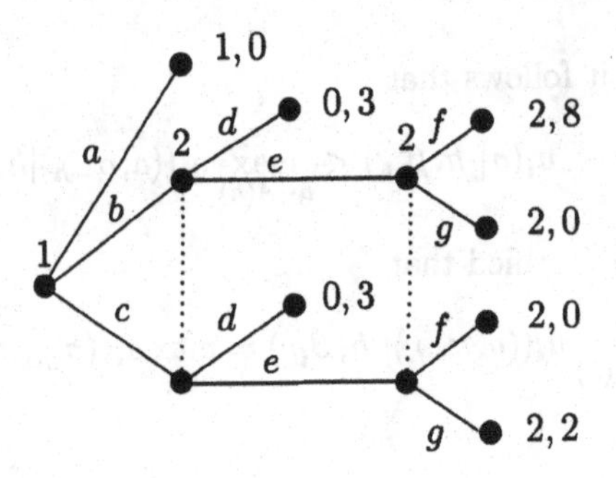

Figure 4.4

local sequential rationality implies sequential rationality. This result is basically due to Hendon, Jacobsen and Sloth (1996), who have shown that under *pre-consistency* of assessments, every locally sequentially rational assessment is sequentially rational. Here, an assessment is called *pre-consistent* if it is updating consistent and Bayesian consistent. However, Bayesian consistency is not needed for their proof, and hence the same property holds under updating consistency as well. In Perea (2001) it is shown that updating consistency is not only a *sufficient* condition for the equivalence between local sequential rationality and sequential rationality, but is also *necessary* for this equivalence to hold.

We should formalize what we mean by the phrase that "updating consistency is a necessary condition for the equivalence between local sequential rationality and sequential rationality." To this purpose, we define the "one-deviation property" for assessments.

Definition 4.3.4 *Let S be an extensive form structure and (σ, β) an assessment in S. We say that (σ, β) satisfies the one-deviation property if for every profile $u = (u_i)_{i \in I}$ of utility functions the following holds: (σ, β) is sequentially rational in the game $\Gamma = (S, u)$ if and only if (σ, β) is locally sequentially rational in Γ.*

Note that the space of possible assessments only depends upon the extensive form structure of the game, and not on the particular utility functions u_i at the terminal nodes. As such, the phrase "assessment in S" is well-understood. By definition, if (σ, β) satisfies the one-deviation property, then for every game with this extensive form structure, the requirements of sequential rationality and local sequential rationality are equivalent. On the other hand, if (σ, β) does not satisfy the one-deviaition property, then we can find utility functions for the players such that in the resulting game, (σ, β) is locally sequentially rational but not sequentially rational.

Note that if the utilities for all players would be constant, then in the resulting game local sequential rationality and sequential rationality would be trivially equivalent, since every assessment would be sequentially rational in this game. The one-deviation property, however, requires equivalence between local sequential rationality and sequential rationality *for all possible utility functions*, and is thus a condition which only depends upon the extensive form structure, and not on the particular utility functions.

The following theorem shows that updating consistency of assessments is both a necessary and sufficient condition for the one-deviation property to hold. The first part of the proof, which shows that updating consistency implies the one-deviation property, is due to Hendon, Jacobsen and Sloth (1996). The second part, which proves that every assessment that is not updating consistent fails to satisfy the one-deviation property, is due to Perea (2001).

Theorem 4.3.5 *Let S be an extensive form structure and (σ, β) an assessment in S. Then, (σ, β) satisfies the one-deviation property if and only if (σ, β) is updating consistent.*

Proof. (a) We first prove that every updating consistent assessment satisfies the one-deviation property. This proof is very similar to the proof of Theorem 3.4.3, and we therefore omit some steps that are identical to the one in Theorem 3.4.3. Let Γ be an extensive form game and (σ, β) an updating consistent assessment in Γ. Assume that (σ, β) is locally sequentially rational. We prove that (σ, β) is sequentially rational.

Let $h \in H_i$ and let $s_i^* \in S_i(h)$ be a strategy with $\sigma_{i|h}(s_i^*) > 0$. We must prove that s_i^* is a sequential best response against (σ, β) at h. In other words, we must show

$$u_i((s_i^*, \sigma_{-i})|\, h, \beta_{ih}) = \max_{s_i \in S_i(h)} u_i((s_i, \sigma_{-i})|\, h, \beta_{ih}). \tag{4.6}$$

Let $u_i(\sigma|\, h, \beta_{ih}) = \sum_{s_i \in S_i(h)} \sigma_{i|h}(s_i)\, u_i((s_i, \sigma_{-i})|\, h, \beta_{ih})$ be the expected utility at h, generated by σ and β_{ih}. We shall prove that

$$u_i((s_i^*, \sigma_{-i})|\, h, \beta_{ih}) = u_i(\sigma|\, h, \beta_h) = \max_{s_i \in S_i(h)} u_i((s_i, \sigma_{-i})|\, h, \beta_{ih}). \tag{4.7}$$

We prove (4.7) by induction on the number of player i information sets that follow h. Suppose first that h is not followed by any other information set of player i. Let a be some action at h with $\sigma_{ih}(a) > 0$. Since (σ, β) is locally sequentially rational, we have that

$$u_i((a, \sigma_{-h})|\, h, \beta_{ih}) = \max_{a' \in A(h)} u_i((a', \sigma_{-h})|\, h, \beta_{ih}). \tag{4.8}$$

Since h is not followed by any other player i information set, the expression $u_i((s_i, \sigma_{-i})|$ $h, \beta_{ih})$ does not depend on the actions prescribed by s_i at information sets other than h. As such, (4.8) implies that $u_i((a, \sigma_{-h})|\, h, \beta_{ih}) = \max_{s_i \in S_i(h)} u_i((s_i, \sigma_{-i})|\, h, \beta_{ih})$. Since this holds for all actions a with $\sigma_{ih}(a) > 0$, it follows that

$$\begin{aligned} u_i(\sigma|\, h, \beta_h) &= \sum_{a \in A(h)} \sigma_{ih}(a)\, u_i((a, \sigma_{-h})|\, h, \beta_{ih}) \\ &= \max_{s_i \in S_i(h)} u_i((s_i, \sigma_{-i})|h, \beta_{ih}). \end{aligned}$$

Now, let a^* be the action prescribed by s_i^* at h. Since $\sigma_{i|h}(s_i^*) > 0$, we know that $\sigma_{ih}(a^*) > 0$ and hence, by the argument above,

$$u_i((a^*, \sigma_{-h})|\, h, \beta_{ih}) = \max_{s_i \in S_i(h)} u_i((s_i, \sigma_{-i})|\, h, \beta_{ih}).$$

Since h is not followed by any other player i information set, we have that $u_i((s_i^*, \sigma_{-i})| h, \beta_{ih}) = u_i((a^*, \sigma_{-h})| h, \beta_{ih})$. We have thus shown (4.7).

Now, let $k \geq 1$ and assume that for every player i information set $h' \in H_i$ followed by at most $k-1$ other player i information sets, (4.7) holds. Let $h \in H_i(s_i^*)$ be such that h is followed by at most k player i information sets. Let $s_i \in S_i(h)$ and $x \in h$. Similar to the proof of Theorem 3.4.3, we can show that

$$\begin{aligned} u_i((s_i, \sigma_{-i})| x) &= \sum_{h' \in H_i^*(s_i)} \sum_{y \in Y(x,h')} \mathbb{P}_{(s_i,\sigma_{-i})}(y| x)\, u_i((s_i, \sigma_{-i})| y) + \\ &\quad + \sum_{z \in Z_0(x)} \mathbb{P}_{(s_i,\sigma_{-i})}(z| x)\, u_i(z). \end{aligned} \tag{4.9}$$

Here, $H_i^*(s_i)$ contains those player i information sets h' such that (1) h' follows h, (2) there is no player i information set between h' and h, and (3) h' is not avoided by s_i. By $Y(x, h')$ we denote those nodes in h' that follow x. By $Z_0(x)$ we denote the terminal nodes z following x such that there is no player i information set on the path between x and z.

Since (σ, β) is updating consistent, for every $h' \in H_i^*(s_i)$ with $\mathbb{P}_{(s_i,\sigma_{-i})}(h'|h, \beta_{ih}) > 0$ and every $y \in h'$

$$\beta_{ih'}(y) = \frac{\mathbb{P}_{(s_i,\sigma_{-i})}(y\,|h, \beta_{ih})}{\mathbb{P}_{(s_i,\sigma_{-i})}(h'\,|h, \beta_{ih})}.$$

If $y \in Y(x, h')$, then $\mathbb{P}_{(s_i,\sigma_{-i})}(y\,|h, \beta_{ih}) = \beta_{ih}(x)\, \mathbb{P}_{(s_i,\sigma_{-i})}(y \mid x)$ and hence

$$\beta_{ih'}(y) = \frac{\beta_{ih}(x)\, \mathbb{P}_{(s_i,\sigma_{-i})}(y \mid x)}{\mathbb{P}_{(s_i,\sigma_{-i})}(h'\,|h, \beta_{ih})}.$$

It therefore follows for every $h' \in H_i^*(s_i)$ and every $y \in Y(x, h')$ that

$$\beta_{ih}(x)\, \mathbb{P}_{(s_i,\sigma_{-i})}(y| x) = \beta_{ih'}(y)\, \mathbb{P}_{(s_i,\sigma_{-i})}(h'| h, \beta_{ih}). \tag{4.10}$$

Note that (4.10) is identical to the property in Claim 1 in the proof of Theorem 3.4.3. Similar to the proof in Theorem 3.4.3, we may show, using (4.9) and (4.10), that

$$\begin{aligned} u_i((s_i, \sigma_{-i})| h, \beta_{ih}) &= \sum_{h' \in H_i^*(s_i)} \mathbb{P}_{(s_i,\sigma_{-i})}(h'| h, \beta_{ih})\, u_i((s_i, \sigma_{-i})| h', \beta_{ih'}) + \\ &\quad + \sum_{x \in h} \beta_{ih}(x) \sum_{z \in Z_0(x)} \mathbb{P}_{(s_i,\sigma_{-i})}(z| x)\, u_i(z). \end{aligned} \tag{4.11}$$

Similarly, we have for every action $a \in A(h)$,

$$\begin{aligned} u_i((a, \sigma_{-h})| h, \beta_{ih}) &= \sum_{h' \in H_i^*(a)} \mathbb{P}_{(a,\sigma_{-h})}(h'|h, \beta_{ih})\, u_i((a, \sigma_{-h})| h', \beta_{ih'}) + \\ &\quad + \sum_{x \in h} \beta_{ih}(x) \sum_{z \in Z_0(x)} \mathbb{P}_{(a,\sigma_{-h})}(z| x)\, u_i(z), \end{aligned} \tag{4.12}$$

where $\mathbb{P}_{(a,\sigma_{-h})}(h'|h,\beta_{ih})$ is the probability that h' is reached under (a,σ_{-h}), given the beliefs β_{ih}. By $H_i^*(a)$ we denote those player i information sets in H_i^* that are not avoided if player i chooses a at h.

Let a be the action chosen by s_i at h. Since every $h' \in H_i^*(s_i)$ is followed by at most $k-1$ player i information sets, we know by induction assumption that $u_i((s_i,\sigma_{-i})| h',\beta_{ih'}) \leq u_i(\sigma| h',\beta_{ih'})$ for every $h' \in H_i^*(s_i)$. Since $u_i(\sigma| h',\beta_{ih'})$ does not depend on player i's behavior at h, we have that $u_i(\sigma| h',\beta_{ih'}) = u_i((a,\sigma_{-h})| h',\beta_{ih'})$. For every terminal node $z \in Z_0(x)$, the probability $\mathbb{P}_{(s_i,\sigma_{-i})}(z| x)$ does not depend upon player i's choices at information sets other than h. Therefore, $\mathbb{P}_{(s_i,\sigma_{-i})}(z| x) = \mathbb{P}_{(a,\sigma_{-h})}(z| x)$ for every $x \in h$ and every $z \in Z_0(x)$. For every $h' \in H_i^*(s_i)$ there is no player i information set between h and h', and therefore $\mathbb{P}_{(s_i,\sigma_{-i})}(h'|h,\beta_{ih})$ does not depend upon player i's choices at information sets other than h. It thus follows that $\mathbb{P}_{(s_i,\sigma_{-i})}(h'|h,\beta_{ih}) = \mathbb{P}_{(a,\sigma_{-h})}(h'|h,\beta_{ih})$. Combining these insights with (4.11) leads to

$$\begin{aligned} u_i((s_i,\sigma_{-i})| h,\beta_{ih}) &\leq \sum_{h'\in H_i^*(a)} \mathbb{P}_{(a,\sigma_{-h})}(h'|h,\beta_{ih})\, u_i((a,\sigma_{-h})| h',\beta_{ih'}) + \\ &\quad + \sum_{x\in h} \beta_{ih}(x) \sum_{z\in Z_0(x)} \mathbb{P}_{(a,\sigma_{-h})}(z| x)\, u_i(z) \\ &= u_i((a,\sigma_{-h})| h,\beta_{ih}). \end{aligned} \tag{4.13}$$

Now, let a' be some action at h with $\sigma_{ih}(a') > 0$. Since (σ,β) is locally sequentially rational, we know that a' is a local best response against (σ,β) and hence $u_i((a',\sigma_{-h})| h,\beta_{ih}) = \max_{a''\in A(h)} u_i((a'',\sigma_{-h})| h,\beta_{ih})$. Since this holds for every $a' \in A(h)$ with $\sigma_{ih}(a') > 0$, it follows that $u_i(\sigma| h,\beta_{ih}) = \max_{a''\in A(h)} u_i((a'',\sigma_{-h})| h,\beta_{ih})$. By (4.13) we may conclude that $u_i((s_i,\sigma_{-i})| h,\beta_{ih}) \leq u_i(\sigma| h,\beta_{ih})$ for all $s_i \in S_i(h)$.

We next show that $u_i((s_i^*,\sigma_{-i})| h,\beta_{ih}) = u_i(\sigma| h,\beta_{ih})$, which would imply (4.7). By (4.11) we know that

$$\begin{aligned} u_i((s_i^*,\sigma_{-i})| h,\beta_{ih}) &= \sum_{h'\in H_i^*(s_i^*)} \mathbb{P}_{(s_i^*,\sigma_{-i})}(h'|h,\beta_{ih})\, u_i((s_i^*,\sigma_{-i})| h',\beta_{ih'}) + \\ &\quad + \sum_{x\in h} \beta_{ih}(x) \sum_{z\in Z_0(x)} \mathbb{P}_{(s_i^*,\sigma_{-i})}(z| x)\, u_i(z). \end{aligned} \tag{4.14}$$

Let $h' \in H_i^*(s_i^*)$. Since by assumption $\sigma_{i|h}(s_i^*) > 0$, h' follows h and h' is not avoided by s_i^*, it follows that $\sigma_{i|h'}(s_i^*) > 0$. Since h' is followed by at most $k-1$ player i information sets, we know by induction assumption that $u_i((s_i^*,\sigma_{-i})| h',\beta_{ih'}) = u_i(\sigma| h',\beta_{ih'})$. Let a^* be the action chosen by s_i^* at h. Since $u_i(\sigma| h',\beta_{ih'})$ does not depend on the conjecture at h, we have that $u_i(\sigma| h',\beta_{ih'}) = u_i((a^*,\sigma_{-h})| h',\beta_{ih'})$ and hence $u_i((s_i^*,\sigma_{-i})| h',\beta_{ih'}) = u_i((a^*,\sigma_{-h})| h',\beta_{ih'})$ for every $h' \in H_i^*(s_i^*)$. By the same argument as used above for (s_i,σ_{-i}), we may conclude that $\mathbb{P}_{(s_i^*,\sigma_{-i})}(h'|h,\beta_{ih}) = \mathbb{P}_{(a^*,\sigma_{-h})}(h'|h,\beta_{ih})$ for all $h' \in H_i^*(s_i^*)$ and $\mathbb{P}_{(s_i^*,\sigma_{-i})}(z| x) = \mathbb{P}_{(a^*,\sigma_{-h})}(z| x)$ for all

$x \in h$ and $z \in Z_0(x)$. Combining these insights with (4.14) leads to

$$\begin{aligned} u_i((s_i^*, \sigma_{-i})|\, h, \beta_{ih}) &= \sum_{h' \in H_i^*(s_i^*)} \mathbb{P}_{(a^*, \sigma_{-h})}(h'|h, \beta_{ih})\, u_i((a^*, \sigma_{-h})|\, h', \beta_{ih'}) + \\ &\quad + \sum_{x \in h} \beta_{ih}(x) \sum_{z \in Z_0(x)} \mathbb{P}_{(a^*, \sigma_{-h})}(z|\, x)\, u_i(z) \\ &= u_i((a^*, \sigma_{-h})|\, h, \beta_{ih}). \end{aligned} \tag{4.15}$$

By assumption, $\sigma_{i|h}(s_i^*) > 0$ and hence $\sigma_{ih}(a^*) > 0$. We know, from above, that then $u_i((a^*, \sigma_{-h})|\, h, \beta_{ih}) = \max_{a \in A(h)} u_i((a, \sigma_{-h})|\, h, \beta_{ih}) = u_i(\sigma|\, h, \beta_{ih})$. By (4.15) it follows that $u_i((s_i^*, \sigma_{-i})|\, h, \beta_{ih}) = u_i(\sigma|\, h, \beta_{ih})$, which was to show. By induction, it follows that s_i^* is a sequential best response against (σ, β) at h. Since this holds for every $h \in H_i$ and every $s_i^* \in S_i(h)$ with $\sigma_{i|h}(s_i^*) > 0$, it follows that (σ, β) is sequentially rational. We have thus shown that every updating consistent assessment satisfies the one-deviation property.

(b) Next, we prove that updating consistency is also a necessary condition for the one-deviation property. Let (σ, β) be an assessment in $\mathcal{S}$ which is not updating consistent. We show that there is a profile $u = (u_i)_{i \in I}$ of utility functions such that in the extensive form game $\Gamma = (\mathcal{S}, u)$ the assessment (σ, β) is locally sequentially rational but not sequentially rational.

Since (σ, β) is not updating consistent, there is some player i, two information sets $h^1, h^2 \in H_i$ where h^2 follows h^1, and some strategy $s_i^* \in S_i(h^2)$ such that $\mathbb{P}_{(s_i^*, \sigma_{-i})}(h^2|h^1, \beta_{ih^1}) > 0$ but

$$\beta_{ih^2}(x^*) \neq \frac{\mathbb{P}_{(s_i^*, \sigma_{-i})}(x^*|h^1, \beta_{ih^1})}{\mathbb{P}_{(s_i^*, \sigma_{-i})}(h^2|h^1, \beta_{ih^1})}$$

for some node $x^* \in h^2$. Since $\beta_{ih^2}(\cdot)$ and $\mathbb{P}_{(s_i^*, \sigma_{-i})}(\cdot|h^1, \beta_{ih^1}) / \mathbb{P}_{(s_i^*, \sigma_{-i})}(h^2|h^1, \beta_{ih^1})$ are both probability distributions on the set of nodes at h^2, we can choose x^* such that

$$\beta_{ih^2}(x^*) < \frac{\mathbb{P}_{(s_i^*, \sigma_{-i})}(x^*|h^1, \beta_{ih^1})}{\mathbb{P}_{(s_i^*, \sigma_{-i})}(h^2|h^1, \beta_{ih^1})}. \tag{4.16}$$

The reader may verify that h^1 and h^2 can always be chosen in such a way that (4.16) holds and there is no further player i information set between h^1 and h^2.

By perfect recall, there is a unique sequence $h_1, ..., h_K$ of player i information sets with the following properties: (1) h_k follows h_{k-1} for all k, (2) there is no player i information set between h_{k-1} and h_k for all k, (3) there is no player i information set before h_1 and (4) $h_{K-1} = h^1$ and $h_K = h^2$. We define the player i utilities following h_k by induction on k.

We first define the player i utilities following $h_K = h^2$. Let a_K be some action at h^2 with $\sigma_{ih^2}(a_K) < 1$. Such an action exists since, by definition of an extensive form structure, there are at least two actions at h^2. For every terminal node z following node x^* (see (4.16)) and action a_K, set $u_i(z) = 1$. For all terminal nodes z following action a_K but not following node x^*, set $u_i(z) = 0$. For every terminal node z following h_K but not following action a_K, set $u_i(z) = \beta_{ih^2}(x^*)$.

Now, suppose that $k < K$ and that the player i utilities $u_i(z)$ have been defined for all terminal nodes z following h_{k+1}. We define the player i utilities following h_k but not following h_{k+1} in the following way. Let a_k be the unique action at h_k that leads to h_{k+1}. For every terminal node z following a_k but not following h_{k+1}, set $u_i(z) = 0$. Let $u_i((a_k, \sigma_{-h_k})|h_k, \beta_{ih_k})$ be the expected utility induced by (a_k, σ_{-h_k}) at h_k, given the beliefs β_{ih_k} and the utilities following a_k, which have already been defined above. For all terminal nodes z following h_k but not action a_k, we set $u_i(z) = u_i((a_k, \sigma_{-h_k})|h_k, \beta_{ih_k})$.

Finally, for all terminal nodes not covered by the procedure above, we set $u_i(z) = 0$. For all players $j \neq i$, we set $u_j(z) = 0$ for all terminal nodes z.

It can be verified easily that the assessment (σ, β) is locally sequentially rational, given the profile $u = (u_i)_{i \in I}$ of utility functions. Note that the player i utilities are constructed in such a way that at every information set h_k, for $k = 1, ..., K$, player i is indifferent between action a_k and all other actions available at h_k, given his beliefs β_{ih_k}, and given σ_{-h_k}. If a player i information set h does not belong to $\{h_1, ..., h_K\}$, then, by construction of the utilities, for every node $x \in h$ all utilities following x are equal, and hence local sequential rationality follows trivially.

We finally show that (σ, β) is not sequentially rational at $h^1 = h_{K-1}$. Let $\tilde{s}_i \in S_i(h^1)$ be such that $\tilde{s}_i(h^1) = a_{K-1}$ and $\tilde{s}_i(h^2) = a_K$. Here, a_{K-1} is the unique action at h^1 which leads to h^2. Then, by construction of the utilities u_i, it follows that $u_i((\tilde{s}_i, \sigma_{-i})|\ h^1, \beta_{ih^1})$ equals $\mathbb{P}_{(a_{K-1}, \sigma_{-h^1})}(x^*|\ h^1, \beta_{ih^1})$ and hence

$$\max_{s_i' \in S_i(h^1)} u_i((s_i', \sigma_{-i})|\ h^1, \beta_{ih^1}) \geq \mathbb{P}_{(a_{K-1}, \sigma_{-h^1})}(x^*|\ h^1, \beta_{ih^1}). \tag{4.17}$$

We shall now prove that there is some $s_i \in S_i(h^1)$ with $\sigma_{i|h^1}(s_i) > 0$ but $u_i((s_i, \sigma_{-i})|\ h^1, \beta_{ih^1}) < \mathbb{P}_{(a_{K-1}, \sigma_{-h^1})}(x^*|\ h^1, \beta_{ih^1})$. By (4.17), this would imply that (σ, β) is not sequentially rational. We will distinguish two cases.

Case 1. If there is no $s_i \in S_i(h^2)$ with $\sigma_{i|h^1}(s_i) > 0$. In this case, choose an arbitrary $s_i \in S_i(h^1)$ with $\sigma_{i|h^1}(s_i) > 0$. Then, $s_i \notin S_i(h^2)$, and hence s_i does not choose a_{K-1} at h^1. By construction of the utilities, it follows that $u_i((s_i, \sigma_{-i})|\ h^1, \beta_{ih^1}) = u_i((a_{K-1}, \sigma_{-h^1})|h^1, \beta_{ih^1})$. After choosing a_{K-1} at h^1, the only feasible player i utilities different from zero are the ones following h^2. By definition of the utilities following $h^2 = h_K$, we have that

$$\begin{aligned} u_i((a_{K-1}, \sigma_{-h^1})|h^1, \beta_{ih^1}) &= \mathbb{P}_{(a_{K-1}, \sigma_{-h^1})}(x^*|h^1, \beta_{ih^1})\ \sigma_{ih^2}(a_K)1 + \\ &\quad + \mathbb{P}_{(a_{K-1}, \sigma_{-h^1})}(h^2|h^1, \beta_{ih^1})(1 - \sigma_{ih^2}(a_K))\beta_{ih^2}(x^*). \end{aligned} \tag{4.18}$$

By (4.16), there exists a strategy $s_i^* \in S_i(h^2)$ with $\mathbb{P}_{(s_i^*, \sigma_{-i})}(h^2|h^1, \beta_{ih^1}) > 0$ such that

$$\beta_{ih^2}(x^*) < \frac{\mathbb{P}_{(s_i^*, \sigma_{-i})}(x^*|h^1, \beta_{ih^1})}{\mathbb{P}_{(s_i^*, \sigma_{-i})}(h^2|h^1, \beta_{ih^1})}.$$

Since there is no player i information set between h^1 and h^2, and a_{K-1} is the unique action that leads from h^1 to h^2, it follows that $\mathbb{P}_{(a_{K-1}, \sigma_{-h^1})}(h^2|h^1, \beta_{ih^1}) > 0$. Moreover,

$$\frac{\mathbb{P}_{(s_i^*,\sigma_{-i})}(x^*|h^1,\beta_{ih^1})}{\mathbb{P}_{(s_i^*,\sigma_{-i})}(h^2|h^1,\beta_{ih^1})} = \frac{\mathbb{P}_{(a_{K-1},\sigma_{-h^1})}(x^*|h^1,\beta_{ih^1})}{\mathbb{P}_{(a_{K-1},\sigma_{-h^1})}(h^2|h^1,\beta_{ih^1})}$$

which implies that

$$\beta_{ih^2}(x^*) < \frac{\mathbb{P}_{(a_{K-1},\sigma_{-h^1})}(x^*|h^1,\beta_{ih^1})}{\mathbb{P}_{(a_{K-1},\sigma_{-h^1})}(h^2|h^1,\beta_{ih^1})}.$$

Hence,

$$\beta_{ih^2}(x^*)\ \mathbb{P}_{(a_{K-1},\sigma_{-h^1})}(h^2|h^1,\beta_{ih^1}) < \mathbb{P}_{(a_{K-1},\sigma_{-h^1})}(x^*|h^1,\beta_{ih^1}). \tag{4.19}$$

Since, by definition, $\sigma_{ih^2}(a_K) < 1$, it follows from (??) and (4.19) that

$$u_i((a_{K-1},\sigma_{-h^1})|h^1,\beta_{ih^1}) < \mathbb{P}_{(a_{K-1},\sigma_{-h^1})}(x^*|h^1,\beta_{ih^1}).$$

Since $u_i((s_i,\sigma_{-i})|\ h^1,\beta_{ih^1}) = u_i((a_{K-1},\sigma_{-h^1})|h^1,\beta_{ih^1})$ it follows that $u_i((s_i,\sigma_{-i})|\ h^1,\beta_{ih^1}) < \mathbb{P}_{(a_{K-1},\sigma_{-h^1})}(x^*|h^1,\beta_{ih^1})$, which was to show.

Case 2. If there is some $s_i \in S_i(h^2)$ with $\sigma_{i|h^1}(s_i) > 0$. Since $\sigma_{ih^2}(a_K) < 1$, we can choose an $s_i \in S_i(h^2)$ with $\sigma_{i|h^1}(s_i) > 0$ and $s_i(h^2) \neq a_K$. Since $s_i \in S_i(h^2)$, we have that $s_i(h^1) = a_{K-1}$. Since $s_i(h^2) \neq a_K$ we have, by definition of the utilities,

$$\begin{aligned} u_i((s_i,\sigma_{-i})|h^1,\beta_{ih^1}) &= \mathbb{P}_{(a_{K-1},\sigma_{-h^1})}(h^2|h^1,\beta_{ih^1})\ \beta_{ih^2}(x^*) \\ &< \mathbb{P}_{(a_{K-1},\sigma_{-h^1})}(x^*|h^1,\beta_{ih^1}) \end{aligned}$$

by (4.19). This was to show.

Hence, we may conclude that there is some $s_i \in S_i(h^1)$ with $\sigma_{i|h^1}(s_i) > 0$ and

$$u_i((s_i,\sigma_{-i})|h^1,\beta_{ih^1}) < u_i((\tilde{s}_i,\sigma_{-i})|h^1,\beta_{ih^1}).$$

However, this implies that (σ,β) is not sequentially rational. This completes the proof. ■

Since every consistent assessment is updating consistent, it follows from the theorem above that a consistent assessment is sequentially rational if and only if it is locally sequentially rational. The following characterization of sequential equilibrium immediately follows.

Lemma 4.3.6 *An assessment (σ,β) is a sequential equilibrium if and only if it is consistent and locally sequentially rational.*

Sometimes, this characterization is convenient for checking whether a given assessment is a sequential equilibrium or not. The reason is that for games where players have many consecutive information sets, checking for local sequential rationality is easier than checking for sequential rationality. McLennan (1989b) uses the lemma above to provide a direct proof for the existence of sequential equilibria. In his paper, a correspondence is defined which to every consistent assessment (σ, β) assigns the set of consistent assessments (σ', β') with the following property: $\sigma'_{ih}(a) > 0$ only if a is a local best response against (σ, β). Let us call this correspondence B. By exploiting the topological properties of the set of consistent assessments derived in McLennan (1989a,b) it can be shown that this correspondence B satisfies the conditions of the Eilenberg-Montgomery fixed point theorem (Eilenberg and Montogomery, 1946). As such, the correspondence B has a fixed point (σ^*, β^*), which then, by definition, is a consistent and locally sequentially rational assessment. By the lemma above, it follows that (σ^*, β^*) is a sequential equilibrium.

4.4 Weak Sequential Rationality

The notion of sequential rationality requires players, at every information set of the game, to assign positive probability only to those opponents' strategies which are a sequential best response at this information set. In particular, they should do so at information sets which can only be reached if one of the opponents, say player i, has played a strategy that is *not* a sequential best response. Even in this case, i's opponents should believe that player i will choose a sequential best response in the future. Similar to backward induction and subgame perfect equilibrium, the reasonableness of this property may be questioned in some examples. Consider again the game in Figure 3.11. Let (σ, β) be the assessment with $\sigma = ((a_1, a_3), a_2)$. Then, (σ, β) is not a sequential equilibrium since player 2, after observing that player 1 has chosen b_1, believes that player 1 will choose (b_1, a_3), which is not a sequential best response at player 1 's second information set. However, if player 2 observes b_1, then he knows that player 1 is not choosing a sequential best response against (σ, β), since a_1 is the unique sequential best response for player 1 against (σ, β). As an alternative approach, one could allow player 2 to believe that player 1 is not rational after observing the "irrational" move b_1. In that case, player 2 could be allowed to believe that player 1 chooses a_3 at his second information set. Given this belief revision process for player 2, it would be optimal for player 2 to choose a_2. Player 1, anticipating on this belief revision process by player 2, may therefore believe that player 2 chooses a_2. Then, a_1 would be the unique sequential best response for him. As such, the assessment (σ, β) with $\sigma = ((a_1, a_3), a_2)$ may be defended on the ground that, *as long as the past play does not contradict the players' initial conjectures*, positive probability is assigned only to strategies that are sequential best responses. We say that the assessment is *weakly sequentially rational* (Reny, 1992b). The difference with sequential rationality is that, once the initial conjecture about player i's behavior has been contradicted, i's opponents are no longer required to believe that player i will choose rationally in the future.

Formally, the definition of weak sequential rationality is as follows. For a given behavioral conjecture profile σ, let $H_i(\sigma_i)$ be the set of those player i information sets that are not avoided by σ_i. In other words, $H_i(\sigma_i)$ contains those information sets $h \in H_i$ at which the initial conjecture about player i is not contradicted by player i's behavior in the past.

Definition 4.4.1 *An assessment (σ, β) is called weakly sequentially rational if for every player i, every $h \in H_i(\sigma_i)$ and every $s_i \in S_i(h)$, it holds that $\sigma_{i|h}(s_i) > 0$ only if s_i is a sequential best response against (σ, β) at h.*

Obviously, sequential rationality implies weak sequential rationality, but not vice versa. By simultaneously imposing consistency and weak sequential rationality, we obtain the concept of *weak sequential equilibrium.*

Definition 4.4.2 *An assessment (σ, β) is called a weak sequential equilibrium if it is consistent and weakly sequentially rational.*

The example above shows that a weak sequential equilibrium need not satisfy backward induction, and hence need not be a subgame perfect equilibrium. Consider the assessment (σ, β) with $\sigma = ((a_1, a_3), a_2)$. We have already verified that (σ, β) is weakly sequentially rational, and hence (σ, β) is a weak sequential equilibrium. However, σ does not satisfy backward induction.

It may be easily verified that a weak sequential equilibrium always constitutes a Nash equilibrium.

Lemma 4.4.3 *Let (σ, β) be a weak sequential equilibrium. Then, σ is a Nash equilibrium.*

Proof. Let (σ, β) be a weak sequential equilibrium. Let $\sigma_i(s_i) > 0$. We prove that s_i is a best response against σ. Since (σ, β) is consistent and thus Bayesian consistent, it suffices to show that s_i is a sequential best response against (σ, β). Let $h \in H_i(s_i)$. Since $\sigma_i(s_i) > 0$, it follows that $h \in H_i(\sigma_i)$. Moreover, $\sigma_{i|h}(s_i) > 0$ since $\sigma_i(s_i) > 0$ and $s_i \in S_i(h)$. Since (σ, β) is weakly sequentially rational, $h \in H_i(\sigma_i)$ and $\sigma_{i|h}(s_i) > 0$, it follows that s_i is a sequential best response against (σ, β) at h. Since this holds for every $h \in H_i(s_i)$, we have that s_i is a sequential best response against (σ, β), and thus a best response against σ. Consequently, σ is a Nash equilibrium. ■

It may be shown that every normal form perfect equilibrium is extendable to a weak sequential equilibrium. This result is taken from Reny (1992b).

Theorem 4.4.4 *Let σ be a normal form perfect equilibrium. Then, there is a belief system β such that (σ, β) is a weak sequential equilibrium.*

Proof. Let σ be a normal form perfect equilibrium. Then, there is a sequence $(\sigma^n)_{n \in \mathbb{N}}$ of strictly positive behavioral conjecture profiles converging to σ such that $\sigma_i(s_i) > 0$ only if s_i is a best response against σ^n for all n. For every n, let β^n be the belief system induced by σ^n. Without loss of generality, we assume that $(\beta^n)_{n \in \mathbb{N}}$ converges to some belief system β. We prove that (σ, β) is a weak sequential equilibrium. By construction, (σ, β) is consistent, and it therefore only remains to prove weak sequential rationality. Let $h \in H_i(\sigma_i)$, and suppose that $\sigma_{i|h}(s_i) > 0$ for some $s_i \in S_i(h)$. We have to show that s_i is a sequential best response against (σ, β) at h.

Since $h \in H_i(\sigma_i)$ and $\sigma_{i|h}(s_i) > 0$, it holds that $\sigma_i(s_i) > 0$. Then, by construction, s_i is a best response against σ^n for all n. Now, suppose that s_i is not a sequential best response against (σ, β) at h. Then, there is some n such that s_i is not a sequential best response against (σ^n, β^n) at n. Let $r_i \in S_i(h)$ be a sequential best response against (σ^n, β^n) at h. Hence,

$$u_i((s_i, \sigma^n_{-i})| \; h, \beta^n_{ih}) < u_i((r_i, \sigma^n_{-i})| \; h, \beta^n_{ih}). \tag{4.20}$$

Let Z_0 be the set of terminal nodes that are not preceded by h. Then, it may be verified that

$$u_i(s_i, \sigma^n_{-i}) = \mathbb{P}_{(s_i, \sigma^n_{-i})}(h) \; u_i((s_i, \sigma^n_{-i})| \; h, \beta^n_{ih}) + \sum_{z \in Z_0} \mathbb{P}_{(s_i, \sigma^n_{-i})}(z) \; u_i(z). \tag{4.21}$$

Let H_i^* be the set of those player i information sets that follow h, together with h itself. Let t_i be the strategy which coincides with r_i at all information sets in H_i^*, and coincides with s_i at all other information sets. Then, it holds that

$$\begin{aligned}
u_i(t_i, \sigma^n_{-i}) &= \mathbb{P}_{(t_i, \sigma^n_{-i})}(h) \; u_i((t_i, \sigma^n_{-i})| \; h, \beta^n_{ih}) + \sum_{z \in Z_0} \mathbb{P}_{(t_i, \sigma^n_{-i})}(z) \; u_i(z) \\
&= \mathbb{P}_{(s_i, \sigma^n_{-i})}(h) \; u_i((r_i, \sigma^n_{-i})| \; h, \beta^n_{ih}) + \sum_{z \in Z_0} \mathbb{P}_{(s_i, \sigma^n_{-i})}(z) \; u_i(z) \\
&> \mathbb{P}_{(s_i, \sigma^n_{-i})}(h) \; u_i((s_i, \sigma^n_{-i})| \; h, \beta^n_{ih}) + \sum_{z \in Z_0} \mathbb{P}_{(s_i, \sigma^n_{-i})}(z) \; u_i(z) \\
&= u_i(s_i, \sigma^n_{-i}),
\end{aligned}$$

where the inequality follows from the fact that $\mathbb{P}_{(s_i, \sigma^n_{-i})}(h) > 0$ and $u_i((s_i, \sigma^n_{-i})| \; h, \beta^n_{ih}) < u_i((r_i, \sigma^n_{-i})| \; h, \beta^n_{ih})$. Hence, s_i is not a best response against σ^n, which is a contradiction. We may therefore conclude that s_i is a sequential best response against (σ, β) at h, which completes the proof. ■

Note that a normal form perfect equilibrium need not be extendable to a sequential equilibrium. Consider again the game in Figure 3.11. We have seen, in Section 3.5, that $\sigma = ((a_1, a_3), a_2)$ is a normal form perfect equilibrium. However, σ is not extendable to a sequential equilibrium, since $((b_1, b_3), b_2)$ is the unique sequential equilibrium in this game. To conclude this chapter, Figure 4.5 depicts the logical relationships that exist between the various rationality concepts discussed in Chapters 3 and 4. An arrow without circle between two concepts indicates that the former always implies the latter in *every* extensive form game. An arrow with circle means that this implication holds only for games with perfect information. If there is no arrow between two concepts, then this implies that there is no logical relationship between them.

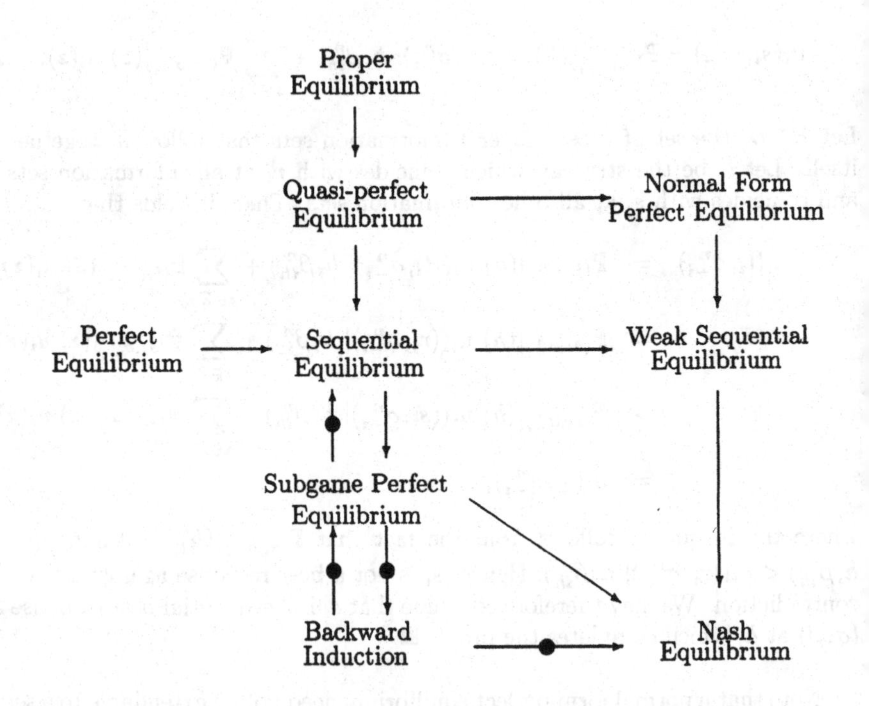
Proper
Equilibrium
Quasi-perfect
Equilibrium
Normal Form
Perfect Equilibrium
Perfect
Equilibrium
Sequential
Equilibrium
Weak Sequential
Equilibrium
Subgame Perfect
Equilibrium
Backward
Induction
Nash
Equilibrium

Figure 4.5

Chapter 5

Forward Induction

In this chapter we present different *forward induction criteria* which restrict the way in which players should revise their conjectures during the game. Although there does not exist a unique definition of forward induction in the literature, its main idea can be described as follows. Suppose that a player's initial conjecture about the opponents' behavior has been contradicted by the play of the game. In this case, this player is required to form an alternative conjecture compatible with the current situation in the game. In other words, the player should study all possible scenarios that could have led to this situation, and form some subjective belief about these scenarios. Roughly speaking, a *forward induction criterion* requires the player to make a distinction between "more plausible" and "less plausible" scenarios, and to assign positive probability only to the "more plausible" scenarios. The various criteria presented in this chapter differ in the formalization of "more plausible" and "less plausible" scenarios. The forward induction criteria discussed in this chapter are iterated weak dominance, stable sets of equilibria, forward induction equilibrium, justifiable sequential equilibrium and stable sets of beliefs. The idea of forward induction is particularly succesful in eliminating "implausible" sequential equilibria in *signaling games*. For this class of games, which is relevant due to many applications to information economics, several forward induction criteria have been developed that exploit the specific structure of signaling games. Among these, we mention the intuitive criterion and related criteria (Cho and Kreps, 1987), divine equilibrium (Banks and Sobel, 1987), perfect sequential equilibrium (Grossman and Perry, 1986) and undefeated equilibrium (Mailath, Okuno-Fujiwara and Postlewaite, 1993). In this chapter, however, we confine our attention to forward induction concepts applicable to the general class of extensive form games. For an overview of forward induction criteria for signaling games the reader may consult Kreps and Sobel (1994), van Damme (1991) or Kohlberg (1990), among others.

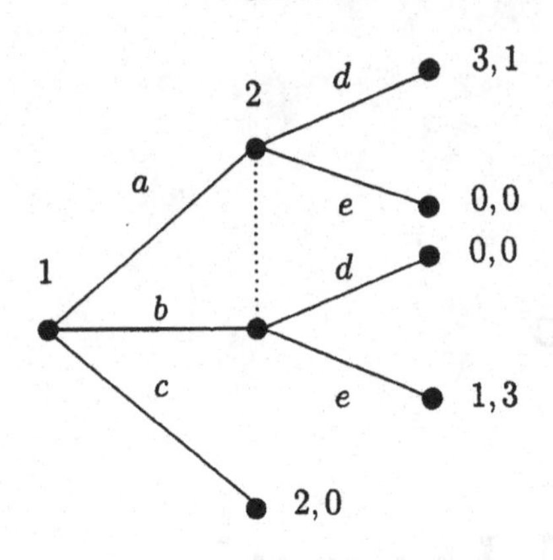

Figure 5.1

5.1 Examples

Before turning to a formal analysis of forward induction, we illustrate its basic ideas by means of some examples. The examples show that forward induction may be formalized in different directions, and that the effectiveness of a specific forward induction criterion depends heavily on the type of game considered. In the remainder of this chapter, these examples will be cited repeatedly in order to illustrate the different concepts.

Example 1. *Battle of the Sexes with Outside Option.*
Consider the game in Figure 5.1, which is taken from Kohlberg (1981) and Kohlberg and Mertens (1986). Player 1 has the choice between the outside option (action c) which leads to utility 2 for sure, and entering the "risky" part of the game in which his utility depends on player 2's action. Both (a, d) and (c, e) can be extended to a sequential equilibrium, where the latter equilibrium requires that player 2 attaches at least belief $1/4$ to his lower node. If player 2 is called upon to play, he knows that player 1 has either chosen a or b. Among these two possibilities, action b seems the less plausible one since it would give at most utility 1 to player 1, whereas he could guarantee 2 by choosing his outside option. Action a could possibly give him more than 2, if he believes that player 2 will respond with d. According to this kind of reasoning, action a seems the only plausible secenario which could have led to player 2's information set, and therefore player 2's beliefs should attach probability zero to his lower node. Consequently, player 2 should choose d and player 1, anticipating on this reasoning, should play a. The sequential equilibrium (c, e) can thus be eliminated by this argument.

Suppose now that player 1's utility at his outside option would be 4 instead of 2. Then the forward induction argument used above can no longer be applied to restrict player 2's beliefs, since both a and b will give player 1 always less than his outside option. Player 2, when called upon to play, should therefore conclude that player 1 has definitely made an irrational choice as there is no plausible scenario which could

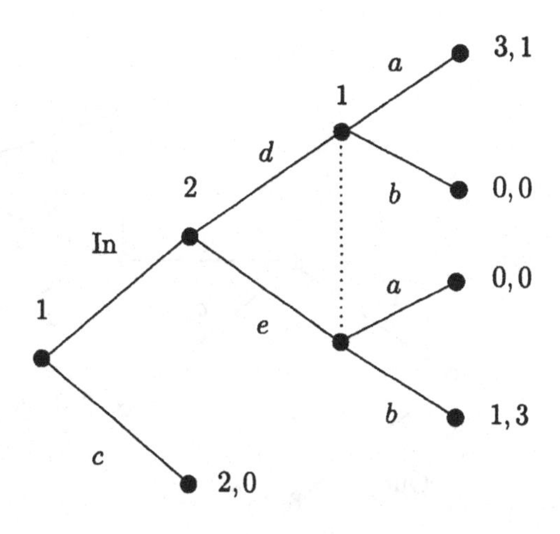

Figure 5.2

have led to player 2's information set.

Consider the game in Figure 5.2, which is a variation on the game in Figure 5.1: the order of player 1 and player 2 has been exchanged after player 1 has decided not to take the outside option. The assessment (σ, β) with $\sigma = ((c, b), e)$ and $\beta = (0, 1)$ is a sequential equilibrium. Suppose, however, that player 2 is called upon to play. He therefore knows with certainty that player 1 has not chosen his outside option. Player 2 may now reason that for player 1 choosing b in the future is less plausible than choosing a, since b would give him at most utility 1, which is less than he can guarantee in the outside option, whereas playing a possibly gives him more than 2. Comparing both continuation strategies, player 2 should conclude that player 1 will continue with a, and therefore player 2 should play d. Player 1, anticipating on this kind of reasoning, will play a at his second information set and expects utility 3 when choosing "In" at the beginning. Consequently, player 1 should not take his outside option, ruling out the sequential equilibrium (σ, β). Observe that this sequential equilibrium can not be excluded by restricting player 2's beliefs, as above, since player 2's information set is now a singleton. Instead, the restrictions are put on player 2's conjecture about how the game will be played after his information set is reached.

Example 2. *Backward induction versus forward induction.*
The forward induction principle applied in the previous example can be expressed as a procedure in which "implausible" strategies are eliminated iteratively. In the example of Figure 5.1, the implausible strategy b can be eliminated since c does always better than b, after which strategy e can be excluded for player 2. In Figure 5.2, one first eliminates the strategy (In, b) for player 1, after which e becomes implausible for player 2. Such iterative elimination procedures are studied in Section 5.2.

The example in Figure 5.3, which is taken from Kohlberg (1990), shows that in some games it is not clear in which order strategies have to be eliminated. Moreover,

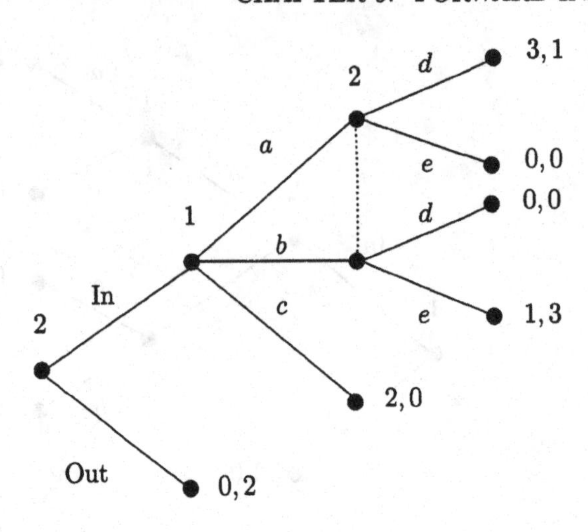

Figure 5.3

these different orders of elimination may yield different restrictions on the assessment. At the beginning, player 2 decides whether or not to take the outside option, which would guarantee him utility 2. If he does not take the outside option, he faces the game of Figure 5.1. Suppose that the game reaches player 1. Inspecting player 2's utilities in the game, and knowing that player 2 has chosen In, player 1 should conclude that player 2 will choose e at his second information set, since the strategy (In, d) for player 2 does worse than the ouside option, whereas playing e could give player 2 possibly more than his outside option. Anticipating on this conjecture, player 1 will choose his outside option c. Player 2, expecting player 1 to reason in this way, should therefore take his outside option. This argument uniquely leaves the behavioral conjecture profile $(c, (Out, e))$. The unique sequential best responses are c and Out, respectively.

However, one could also reason as follows. Suppose that the game reaches player 1. In this subgame, strategy b for player 1 does always worse than c, whereas a could possibly do better. Hence, player 2 should conclude that player 1 has chosen a whenever his second information set is reached. Player 2 will therefore play d at this information set. Player 1, anticipating on this conjecture, will prefer a, inducing player 2 to choose his outside option at the beginning of the game. This argument uniquely leaves the behavioral conjecture profile $(a, (Out, d))$. The unique sequential best responses are a and Out, respectively.

Both forms of reasoning thus induce different assessments, and a different optimal behavior for player 1 at his information set. The difference between the two criteria is that the second criterion first applies forward induction to the subgame starting with player 1, yielding a unique assessment at this subgame, and afterwards applies backward induction to player 2's decision at the beginning of the game. The first criterion first applies forward induction to player 2's decision at the beginning of the game, leaving action e for player 2 in the subgame, and afterwards applies backward induction to the subgame. There does not seem to be an obvious reason to prefer either of these criteria, and therefore both $(c, (Out, e))$ and $(a, (Out, d))$ seem plausible conjecture profiles. Note that both criteria end up with player 1 believing that player 2 chooses his outside option. Therefore, the conjectures about the players' behavior only differ once the initial conjecture about player 2's behavior has been contradicted. The fact that both forward induction criteria seem equally reasonable but lead to different assessments may indicate that one needs some ambiguity on the players' conjectures in certain parts of the game in order to support forward induction.

This ambiguity problem of forward induction has been one of the reasons in Kohlberg and Mertens (1986) to study *set-valued* solution concepts, i.e. concepts which propose a *set* of conjecture profiles instead of *one particular* conjecture profile as a solution to the game. According to Kohlberg and Mertens' approach, a solution to the game of Figure 5.3 should contain both conjecture profiles $(c, (Out, e))$ and $(a, (Out, d))$. In Section 5.3 we present the set-valued solution concept proposed by Kohlberg and Mertens and analyze its relation with forward induction.

Example 3. *Burning money.*
The example depicted in Figure 5.4 is due to van Damme (1989) and Ben-Porath and Dekel (1992). At the beginning of the game, player 1 decides whether or not to publicly burn one dollar (or one utility unit). After this decision, player 1 and 2 play the Battle of the Sexes game of Example 1, without outside option. Suppose that player 2 finds himself at the upper information set, knowing that player 1 has burned a dollar. The only sensible action for player 1 after burning a dollar seems a: by playing b player 1 can obtain at most 0, whereas he could guarantee 0.75 by not burning the dollar and playing a' and b' with probabilities 0.25 and 0.75 respectively. The action a could possibly give him 2, which is more than he would get after not burning, given that player 2 would respond with d' there. If player 2 believes that player 1 will choose a after burning money, he will respond with c at his upper information set, and player 1 will anticipate on this reasoning by choosing a. As such, player 1 expects to get 2 if he would decide to burn a dollar. Suppose now that player 2 finds himself at the lower information set, knowing that payer 1 has decided not to burn money. Since player 2 can reproduce player 1's reasoning above, he knows that player 1 expects utility 2 after burning a dollar. Given that player 1 has decided not to burn money, the only reasonable action seems a', since b' will always give him less than 2, whereas a' leaves open the possibility of receiving 3. Player 2 should therefore think that player 1 has chosen a', and should respond with c'. Player 1, believing that player 2 will reason in this way, should choose a'. Hence, player 1 expects utility 3 after not burning the dollar. According to this reasoning, player 1 is believed not to burn money and to choose a', whereas player 2 is believed to choose c'. The unique sequential best responses are $(Not\ burn, a')$ and (c, c'), respectively, rewarding player 1 with his most preferred outcome in the game.

Note that in the unique sequential best response, player 1 eventually does not use the opportunity of burning a dollar, but the mere presence of this opportunity helps him to obtain his most preferred outcome, given that it is common knowledge that both players reason in the spirit of forward induction. This role of the "potential for sacrifice" is studied in detail in Ben-Porath and Dekel (1992) and van Damme (1989).

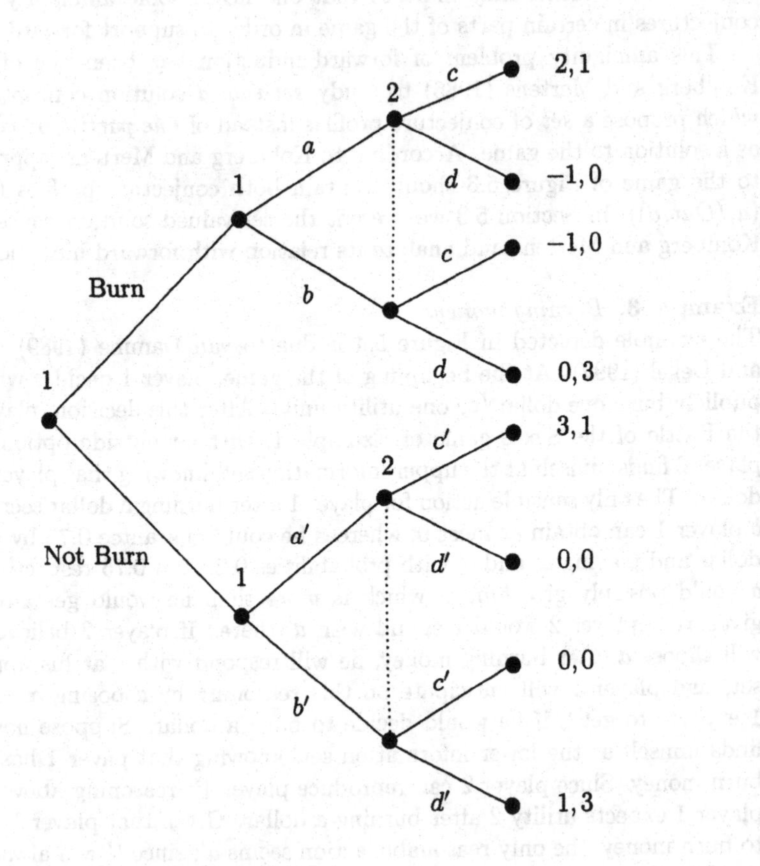

Figure 5.4

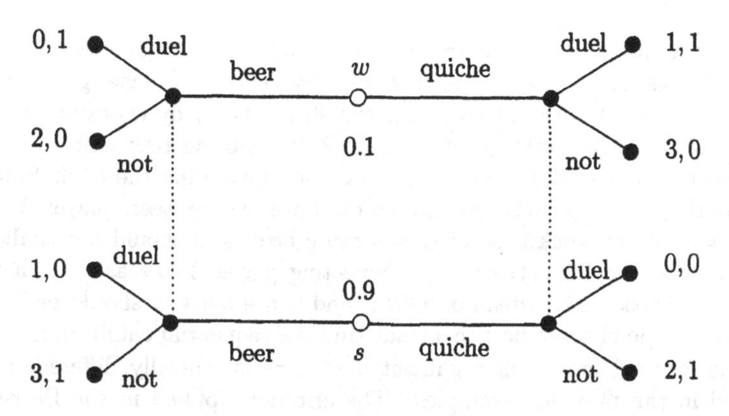

Figure 5.5

Example 4. *Beer Quiche Game.*
Consider the game in Figure 5.5, which is due to Cho and Kreps (1987). Player 1 can be of two possible types: weak (w) or strong (s), depicted by open circles in the figure. At the beginning, a chance move selects player 1's type according to the probabilities specified in the figure. Player 1, after learning his type, can choose beer or quiche for breakfast. Player 2 observes the breakfast chosen by player 1, but does not know his type. Finally, player 2 decides whether or not to challenge player 1 by means of a duel.

The utilities in the game are generated as follows. Player 1 extracts a utility of 1 from a breakfast that fits best his type (quiche if weak and beer if strong) and receives an additional utility of 2 if he is not challenged. Player 2 will win the duel if he is faced with a weak opponent, whereas he will loose against the strong type. In case he challenges the weak type or avoids the strong type, player 2 receives 1.

In this game there are two classes of sequential equilibria. In equilibria of the first class, player 1 is believed to choose beer, irrespective of his type, player 2 is believed to avoid a duel if he observes beer and is believed to challenge with probability greater than 0.5 if he observes quiche. In order to support these equilibria, player 2 after observing quiche should believe that player 1 is weak with probability greater than 0.5.

In the second class, player 1 is believed to choose quiche, irrespective of his type, player 2 is believed to challenge with probability greater than 0.5 if he observes beer and is believed to avoid a duel if he observes quiche. These equilibria can be supported by beliefs that attach a probability of at least 0.5 to the weak type if player 2 observes beer.

Consider a sequential equilibrium (σ, β) of the second class in which player 2 is believed to challenge player 1 with probability 1 after observing beer for breakfast. Suppose that player 2 would be faced with the event that player 1 has chosen beer, thus contradicting his initial conjecture in the assessment. The weak type has no incentive at all to choose beer, since it would give him at most 2, irrespective of player 2's response, whereas he receives 3 by choosing quiche, given the assessment (σ, β) at hand. On the other hand, the strong type, by choosing quiche at (σ, β), gets 2, whereas he might get more than 2 by choosing beer instead of quiche, if he believes that player 2 will respond to beer with avoiding the duel. Since the strong type is the only type who could credibly have chosen beer, player 2 should attach belief zero to the weak type after observing beer, and should not challenge player 1 on this occasion. The strong type, expecting player 2 to reason in this way, should therefore choose beer instead of quiche, and hence player 2 should believe that player 1's strong type chooses beer, contradicting the sequential equilibrium (σ, β).

The forward induction argument used here is crucially different from the ones applied in the previous examples. The criterion applied in the Beer-Quiche game compares, for a given assessment (σ, β), the expected utility that a player gets by choosing a sequential best response against (σ, β) with the expected utilities that this player can possibly achieve by choosing some other strategy, against any possible assessment. The criterion therefore crucially depends on the specific assessment under consideration. On the other hand, the criteria used in the previous examples only use the publicly available data of the game and do not depend on the specific assessment to be tested.

5.2 Iterated Weak Dominance

In order to catch the intuition behind the criterion of iterated weak dominance, let us return to the Battle of the Sexes game of Figure 5.1. If player 2 is called upon to play, he knows that payer 1 has chosen a or b. Between these two strategies b seems the less plausible one, since b is never a sequential best response against any assessment, whereas strategy a is. Player 2, arguing in this way, should thus conclude that player 1 has chosen a. Given this conjecture, player 2 should choose d. The only sequential equilibrium which is compatible with this principle is (a, d), together with the beliefs $(1, 0)$.

The criterion applied in the example may be generalized as follows. Suppose that, at a given stage of the game, the initial conjecture about player i has been falsified, and let S_i^* be the set of player i strategies which are still "possible" at this stage. If there is some strategy in S_i^* which is a "potential sequential best response", that is, constitutes a sequential best response against *some* assessment (not necessarily the assessment at hand), then i's opponents, at this stage of the game, should attach positive probability only to player i strategies in S_i^* that are potential sequential best responses. In other words, at every stage of the game player i's opponents partition the set S_i^* of possible player i strategies into "more rational" and "less rational" strategies, where "more rational" is to be read as "potential sequential best response", and assign positive probability only to "more rational" strategies. On the other hand, if at a given stage of the game the set S_i^* of possible strategies contains no potential sequential best responses, then player i's opponents regard all strategies in S_i^* as "equally rational", or "equally irrational", and no further restrictions are put on the conjecture about player i's behavior at this stage of the game.

In two-player games, the criterion in iterated weak dominance is in fact a bit stronger than descibed above. Instead of requiring that player i's opponents, if possible, attach positive probability only to strategies in S_i^* that are potential sequential best responses, iterated weak dominance in two-player requires the players to attach positive probability only to strategies that are sequential best responses against some *strictly positive* Bayesian consistent assessment. Recall that an assessment is called strictly positive if it assigns positive probability to all actions in the game. The following lemma characterizes, for two-player games, the latter sequential best responses by means of *weakly dominated strategies.*

In order to define weakly dominated strategies we need the following notation. For a given profile of strategies (s_i, s_{-i}), let $u_i(s_i, s_{-i})$ be the induced expected utility for player i. For a given $\mu_i \in \Delta(S_i)$, let $u_i(\mu_i, s_{-i}) = \sum_{s_i \in S_i} \mu_i(s_i)\ u_i(s_i, s_{-i})$. We say that a strategy s_i is *weakly dominated* if there is some $\mu_i \in \Delta(S_i)$ with $u_i(s_i, s_{-i}) \leq u_i(\mu_i, s_{-i})$ for all $s_{-i} \in S_{-i}$, and $u_i(s_i, \tilde{s}_{-i}) < u_i(\mu_i, \tilde{s}_{-i})$ for some $\tilde{s}_{-i}$. In this case, we say that s_i is weakly dominated by the randomization $\mu_i \in \Delta(S_i)$.

Lemma 5.2.1 *Let Γ be a two-player game and s_i a strategy for player i. Then, the following two statements are equivalent: (1) there is a strictly positive Bayesian consistent assessment (σ, β) such that s_i is a sequential best response against (σ, β), (2) s_i is not weakly dominated.*

Proof. Suppose that $s_i^* \in S_i$ is not weakly dominated. We prove that there is some strictly positive Bayesian consistent assessment (σ, β) such that s_i is a sequential best response against (σ, β). In the proof of Theorem 4.4.4 we have shown the following: if s_i is a best response against some strictly positive behavioral conjecture profile σ, then s_i is a sequential best response against (σ, β), where β is the belief system induced by σ. Together with Lemma 2.6.4 we may conclude, for a given strictly positive Bayesian consistent assessment (σ, β), that s_i is a sequential best response against (σ, β) if and only if s_i is a best response against σ. It is thus sufficient to show that there is some strictly positive behavioral conjecture profile σ such that s_i is a best response against σ. Assume that this were not the case.

We construct a zero-sum game Γ^* with utility functions u_i^*, u_j^* where $u_i^*(s_i, s_j) = u_i(s_i, s_j) - u_i(s_i^*, s_j)$ for all $(s_i, s_j) \in S_i \times S_j$, and $u_j^*(s_i, s_j) = -u_i^*(s_i, s_j)$ for all $(s_i, s_j) \in S_i \times S_j$. Let (σ_i^*, σ_j^*) be a normal form perfect equilibrium for Γ^*. Let μ_i^*, μ_j^* be the mixed conjectures induced by σ_i^*, σ_j^*. We prove that μ_i^* weakly dominates s_i^*. Let $\mu_j \in \Delta(S_j)$ be arbitrary. Since (μ_i^*, μ_j^*) is a Nash equilibrium, we have that $u_j^*(\mu_i^*, \mu_j) \leq u_j^*(\mu_i^*, \mu_j^*)$ and hence $u_i^*(\mu_i^*, \mu_j) \geq u_i(\mu_i^*, \mu_j^*)$. On the other hand, we have that $u_i^*(\mu_i^*, \mu_j^*) \geq u_i^*(s_i^*, \mu_j^*) = 0$. We may thus conclude that $u_i^*(\mu_i^*, \mu_j) \geq 0$ for all $\mu_j \in \Delta(S_j)$, which implies that $u_i(\mu_i^*, \mu_j) \geq u_i(s_i^*, \mu_j)$ for all $\mu_j \in \Delta(S_j)$.

Since (μ_i^*, μ_j^*) is a normal form perfect equilibrium in Γ^*, there exists some sequence $(\mu_j^n)_{n \in \mathbb{N}}$ of strictly positive probability measures converging to μ_j^* such that $\mu_i^*(s_i) > 0$ only if $u_i^*(s_i, \mu_j^n) = \max_{s_i' \in S_i} u_i^*(s_i', \mu_j^n)$ for every n. Hence, $u_i(\mu_i^*, \mu_j^n) = \max_{s_i \in S_i} u_i(s_i, \mu_j^n)$ for every n. Take some arbitrary μ_j^n from this sequence. Since μ_j^n is strictly positive, and s_i^* is, by assumption, never a best response against a strictly positive probability measure, we have that $u_i(\mu_i^*, \mu_j^n) > u_i(s_i^*, \mu_j^n)$. We thus have that $u_i(\mu_i^*, \mu_j) \geq u_i(s_i^*, \mu_j)$ for all $\mu_j \in \Delta(S_j)$ and $u_i(\mu_i^*, \mu_j) > u_i(s_i^*, \mu_j)$ for at least one

μ_j. Consequently, $u_i(\mu_i^*, s_j) \geq u_i(s_i^*, s_j)$ for all $s_j \in S_j$ and $u_i(\mu_i^*, \tilde{s}_j) > u_i(s_i^*, \tilde{s}_j)$ for at least one $\tilde{s}_j$. However, this contradicts the assumption that s_i^* is not weakly dominated.

Suppose now that s_i^* is weakly dominated. Then, there is some $\mu_i \in \Delta(S_i)$ such that $u_i(\mu_i, s_j) \geq u_i(s_i^*, s_j)$ for all $s_j \in S_j$, with strict inequality for some $\tilde{s}_j \in S_j$. Then, $u_i(\mu_i, \mu_j) > u_i(s_i^*, \mu_j)$ for all strictly positive probability measures $\mu_j \in \Delta(S_j)$. This implies that s_i^* can never be a best response against a strictly positive behavioral conjecture σ. Consequently, s_i^* can never be a sequential best response against a strictly positive Bayesian consistent assessment, which completes the proof of this lemma. ■

For games with more than two players, the above equivalence result may no longer hold, since statement (2) need no longer imply statement (1). The implication from (1) to (2) is true for all games, however. In the remainder of this section we shall assume that we are dealing uniquely with two-player games.

In view of Lemma 5.2.1, the requirement that players, whenever possible, should assign probability zero to strategies that are never a sequential best response against any strictly positive Bayesian consistent assessment, is equivalent to the requirement that players should assign probability zero to weakly dominated strategies, whenever possible. The latter may be mimicked by "eliminating", for a given player i, his strategies that are weakly dominated, and considering the "reduced" game that is left after removing these strategies. Formally, this elimination procedure is carried out within the so-called *normal form* of the game Γ.

Definition 5.2.2 *Let Γ be an extensive form game. Let $N(\Gamma) = ((S_i)_{i\in I}, (v_i)_{i\in I})$ be the pair where S_i is the set of strategies for every player i and $v_i : \times_{j\in I} S_j \to \mathbb{R}$ is the function which assigns to every strategy profile $s = (s_j)_{j\in I}$ the induced expected utility $v_i(s) := u_i(s)$. We refer to $N(\Gamma)$ as the normal form of Γ.*

The normal form $((S_i)_{i\in I}, (v_i)_{i\in I})$ of an extensive form game $\Gamma = (S, u)$ is sufficient to compute, for every strategy s_i and every mixed conjecture profile μ_{-i} about the opponents, the corresponding expected utility $u_i(s_i, \mu_{-i})$. Namely, by construction, $u_i(s_i, \mu_{-i}) = \sum_{s_{-i}\in S_{-i}} \mu_{-i}(s_{-i})\; v_i(s_i, s_{-i})$ where $\mu_{-i}(s_{-i}) = \prod_{j\neq i} \mu_j(s_j)$. Similarly, we have that $u_i(\mu_i, s_{-i}) = \sum_{s_i\in S_i} \mu_i(s_i)\; v_i(s_i, s_{-i})$ for all $\mu_i \in \Delta(S_i)$ and $s_{-i} \in S_{-i}$, and hence, in order to find the weakly dominated strategies in an extensive form game, it is sufficient to know the normal form of that game. Pairs $((S_i)_{i\in I}, (v_i)_{i\in I})$ of the form above are called *normal form games*.

Definition 5.2.3 *A normal form game is a pair $\Gamma = ((S_i)_{i\in I}, (v_i)_{i\in I})$ where $v_i : \times_{j\in I} S_j \to \mathbb{R}$ for all players i.*

If we say that we *eliminate* the strategy s_i from an extensive form game Γ, then we mean that the normal form $N(\Gamma)$ is transformed into a reduced normal form game $N^*(\Gamma)$ where $S_i^* = S_i \backslash \{s_i\}$, and the new utility functions v_j^* are simply the restrictions of the original utility functions to the reduced strategy spaces.

Suppose now that, according to the forward induction criterion above, we have eliminated some weakly dominated strategy s_i from the game, and that this fact is common knowledge among the players. In other words, it is common knowledge that player i's opponents, whenever possible, assign probability zero to s_i. Within this reduced game, we may again apply the forward induction criterion above. That is, within the reduced game, players should, whenever possible, assign probability zero to strategies that are never a sequential best response against some "strictly positive Bayesain consistent assessment within the reduced game". By the latter, we formally mean an assessment that assigns positive probability to all strategies not yet deleted in the game. However, by Lemma 5.2.1, this equivalent to deleting those strategies that are weakly dominated "within the reduced game". Recall that we are concentrating our attention on two-player games for the moment. By repeating this elimation process iteratively, we obtain a procedure that is called *iterated weak dominance.*

Before formally presenting this procedure, let us illustrate it by the example of Figure 5.1 The normal form of this game is represented by the table below.

	d	e
a	3,1	0,0
b	0,0	1,3
c	2,0	2,0

Strategy b is weakly dominated, since c always performs strictly better than b. After removing b from the game, e performs strictly worse for player 2 than d if player 1 chooses a, and performs equally well if he chooses c. Hence, e is weakly dominated by d in the reduced game. If we eliminate e from the game, c is weakly dominated by a and hence (a, d) is the only behavioral conjecture profile that survives iterated weak dominance.

Formally, iterated weak dominance is defined as follows. Let S_i be the set of strategies for every player i. Consider a subset $T = \times_{i\in I}T_i$ of strategy profiles, where $T_i \subseteq S_i$ for all i. We say that a strategy $s_i \in T_i$ is *weakly dominated* on T if there is some $\mu_i \in \Delta(T_i)$ such that $u_i(s_i, s_{-i}) \leq u_i(\mu_i, s_{-i})$ for all $s_{-i} \in T_{-i}$ and $u_i(s_i, \tilde{s}_{-i}) < u_i(\mu_i, \tilde{s}_{-i})$ for at least one $\tilde{s}_{-i} \in T_{-i}$.

Definition 5.2.4 *An iterated weak dominance procedure is a sequence $T^1, T^2, ..., T^K$ of subsets of strategy profiles such that: (1) $T^1 = \times_{i\in I}S_i$, (2) T^{k+1} is obtained from T^k by eliminating, for one player i, exactly one strategy in T_i^k which is weakly dominated on T^k, and (3) there is no strategy in T^K which is weakly dominated on T^K. We say that strategies in T^K survive this particular iterated weak dominance procedure.*

Accordingly, we say that a behavioral conjecture profile σ *respects* an iterated weak dominance procedure $T^1, T^2, ..., T^K$ if the initial mixed conjectures induced by σ assign positive probability only to strategies in T^K. A problem with iterated weak dominance is that the set of strategies surviving this process will depend, in general, on the particular procedure chosen, that is, on the order in which strategies are eliminated. Consider, for instance, the game in Figure 5.3. The strategy b is weakly dominated by c. After eliminating b, player 2's strategy (In, e) is weakly dominated by *Out*. After eliminating (In, e), the strategy c is weakly dominated by a, and after eliminating c the strategy (In, d) is weakly dominated by *Out*. This order of elimination only leaves strategy a for player 1 and strategy *Out* for player 2.

One could also eliminate in the following order. First, (In, d) can be eliminated for player 2 since it is weakly dominated by *Out*. Then, both a and b are weakly dominated by c. After eliminating both strategies, (In, e) is weakly dominated for player 2. This order of elimination only leaves strategy c for player 1 and strategy *Out* for player 2. Both orders of elimination thus provide different predictions on player 2's conjecture about player 1. However, in both cases the unique sequential best response for player 2 is to choose *Out*, and hence for both orders of elimination the game is expected to end at the same terminal node.

The first order of elimination coincides with the reasoning in which first forward induction is applied to the subgame, yielding the conjecture profile (a, d), and afterwards backward induction is applied to the whole game, anticipating on the conjecture profile (a, d) in the subgame. The second order of elimination corresponds to first applying forward induction to player 2's decision, excluding (In, d) for player 2, and afterwards applying backward induction.

The fact that iterated weak dominance may lead to contradictory predictions on the players' conjectures in certain parts of the game can be viewed in two different ways. On the one hand, one could say that in general there is no single conjecture profile which is robust against every possible elimination of weakly dominated strategies, and hence there may be no assessment compatible with forward induction. On the other hand, one could argue that perhaps both mixed conjecture profiles (a, Out) and (c, Out) should be viewed as reasonable. After all, both lead to the same outcome in the game, as we have seen above, and differ only at parts of the game which are not expected to be reached anyhow. This is exactly the viewpoint taken by Kohlberg and Mertens, and has been one of various reasons for introducing set-valued solution concepts such as stable sets of equilibria. In the next section we introduce stable sets of equilibria and discuss the extent to which this concept reflects the idea of forward induction.

In Marx and Swinkels (1997) it is shown that for games satisfying the so-called *transference of decisionmaker indifference condition,* different orders of iterative elimination of weakly dominated strategies lead to "equivalent" sets of strategies that survive it. By the latter, we mean that one set of strategies can be obtained from the other after adding or removing duplicate strategies, and relabeling of strategies, within the final reduced game. In the example above, for instance, the two different procedures lead to the strategy profiles (a, Out) and (c, Out), respectively. After player 2's strategies (In, d) and (In, e) have been eliminated, the strategies a and c may be viewed identical within the reduced game, since they always yield the same expected utility. Note that in both procedures, only the strategy *Out* survives for player 2, and hence player 1's subgame is not expected to be reached if player 1 holds a conjecture that respects either of both elimination procedures.

For the Burning Money game of Figure 5.4, the reader may verify that every iterated weak dominance procedure yields the same unique strategy profile, namely $((not\ burn, a'), (c, c'))$. Iterated weak dominance thus selects the most preferred outcome for player 1, and the option of burning a dollar is not used. Note, however, that

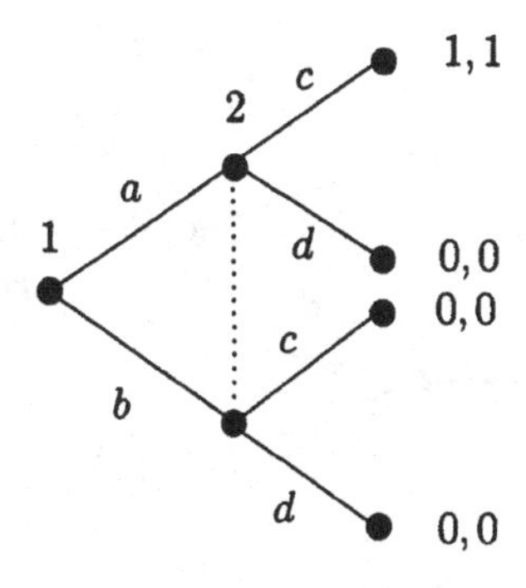

Figure 5.6

the presence of this option is crucial for the elimination procedure. This particular property of iterated weak dominance holds for a larger class of burning money games, as is shown by Ben-Porath and Dekel (1992). In their paper, situations are studied in which player 1 and 2 face some simultaneous move game but before playing this game player 1 has the opportunity to publicly burn a non-negative integer multiple of some positive amount ϵ. They prove that, if the simultaneous move game contains a pure behavioral conjecture profile which is a strict Nash equilibrium and which leads to the unique best outcome for player 1, then iterated weak dominance selects a unique outcome where player 1 does not burn any money and receives his best outcome, provided that ϵ is sufficiently small. Here, a Nash equilibrium $\sigma = (\sigma_i)_{i \in I}$ in pure behavioral conjectures is called *strict* if every σ_i assigns probability 1 to a strategy s_i with $u_i(s_i, \sigma_{-i}) > u_i(s_i', \sigma_{-i})$ for all $s_i' \neq s_i$.

There are some other situations, however, where iterated weak dominance does not really reflect the idea of forward induction. Consider the game in Figure 5.6. Iterated weak dominance uniquely selects the strategy profile (a, c), since both b and d are weakly dominated. There is no forward induction story, however, which supports the elimination of the sequential equilibrium (b, d). We may thus conclude that iterated weak dominance is too restrictive in this case.

The example in Figure 5.7 shows that iterated weak dominance may also be too weak in certain games. This game is obtained by the Battle of the Sexes game of Figure 5.1 after replacing the outcome $(0, 0)$ corresponding to (b, d) by some zero-sum game in which the unique Nash equilibrium is $\sigma' = (0.5f + 0.5g, 0.5h + 0.5i)$. If the players' conjectures in this subgame are assumed to constitute a Nash equilibrium, both players expect to receive a utility of zero in this subgame, since against σ' every strategy yields expected utility zero. As such, if we assume that the players' conjectures form a subgame perfect equilibrium, this game is equivalent to Figure 5.1. We would therefore expect that forward induction leads to (a, d) also in this new game. However, iterated weak dominance in not able to eliminate the choice b in this game, and hence is not able to exclude the sequential equilibrium in which player 1 is believed to choose c and player 2 is believed to choose e. The reason is that in this new game no strategy is weakly dominated. In particular, no strategy in which player 1 chooses b is weakly dominated by c as player 1 might conceivably

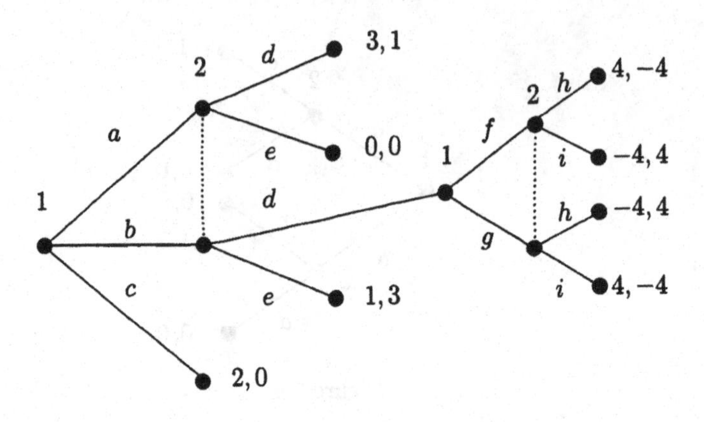

Figure 5.7

get more than 2 in the zero-sum game. The basic problem here is that iterated weak dominance is not able to filter the unique Nash equilibrium in the zero-sum game. Once this Nash equilibrium is filtered, b becomes weakly dominated, after which e could be eliminated.

The paper by Glazer and Rubinstein (1996) highlights an interesting relationship between iterated weak dominance and backward induction. It is shown that for normal form games in which an iterated weak dominance procedure yields a unique strategy profile, the elimination process is equivalent to backward induction in some appropriately chosen extensive form game.

5.3 Stable Sets of Equilibria

5.3.1 Motivation

In their paper, Kohlberg and Mertens (1986) present a list of requirements that a "reasonable" solution concept for extensive form games should satisfy, and show by means of examples that single-valued solution concepts can not satisfy all of them. They use this "shortcoming" as a motivation for the introduction of set-valued solution concepts. Here, by a *single-valued solution concept* we mean a correspondence which assigns to every extensive form game a set of assessments. Each assessment in this set is interpreted as a "solution" to the game. This is in contrast to *set-valued solution concepts*, which to every game assign a collection of sets of assesments. Within such concepts, a solution to the game is a *set of assessments*, rather than one particular assessment. In this section we discuss two requirements proposed by Kohlberg and Mertens that are related to forward induction: "iterated weak dominance" and the "never a best response property".

In addition to these two properties, Kohlberg and Mertens also impose that the solution be invariant against certain transformations of the extensive form game, to which we shall refer as *Kohlberg-Mertens transformations*. For a detailed discussion of these tranformations, the reader is referred to Chapter 6, and in particular Section 6.6. In that section it is shown that invariance against such transformations implies that the solution should only depend on the so-called *mixed reduced normal form* of the game, which is some reduced presentation of the normal form. In particular, it follows that the solution should only depend on the information available in the normal form of the game. For this reason, Kohlberg and Mertens (1986) assume that a solution is defined on the class of normal form games, instead of on the class of extensive form games. Within a normal form game $\Gamma = ((S_i)_{i\in I}, (u_i)_{i\in I})$ the conjectures held by the players about the opponents' strategy choices may be represented by a mixed conjecture profile $\mu = (\mu_i)_{i\in I}$ where $\mu_i \in \Delta(S_i)$ for all players i. Accordingly, a *single-valued solution concept* is a correspondence φ which to every normal form game Γ assigns a nonempty set $\varphi(\Gamma)$ of mixed conjecture profiles. A *set-valued solution concept* φ assigns to every normal form game Γ a collection $\varphi(\Gamma)$ of sets of mixed conjecture profiles. Within a set-valued solution concept φ, a solution to the game Γ is a set $S \in \varphi(\Gamma)$ of mixed conjecture profiles, instead of one particular mixed conjecture profile. The intuition is that it is common knowledge that some conjecture profile $\mu \in S$ is held by the players, but players are not necessarily informed about the exact conjecture profile in S held by the opponents.

We now discuss the requirement of "iterated weak dominance", as proposed by Kohlberg and Mertens. As we have seen in the previous section, the procedure of iterated weak dominance may single out different strategy profiles, depending on the order in which weakly dominated strategies are eliminated. Consequently, there is no single mixed conjecture profile, and hence no single-valued solution concept, which is robust against every possible elimination of weakly dominated strategies. Instead, one could focus on set-valued solution concepts. One would be tempted to define iterated weak dominance for set-valued solutions as follows: if a set S of mixed conjecture profiles is a solution in Γ and s_i is a weakly dominated strategy, then S should be a solution in the reduced game $\Gamma\backslash s_i$ where the strategy s_i is eliminated. However, this would lead to nonexistence for the same reasons as above; in the game of Figure 5.3, different orders of elimination single out (a, Out) and (c, Out) respectively. By the definition of iterated weak dominance above, a solution to the game should on the one hand be equal to the single point (a, Out) (since b and c are eliminated by the first elimination order) but should on the other hand be equal to the single point (c, Out), (since a and b are eliminated by the second elimination order), which is impossible. Instead, we apply the following definition.

Definition 5.3.1 *A set-valued solution concept φ is said to satisfy iterated weak dominance if for every game Γ and every weakly dominated strategy s_i the following holds: if $S \in \varphi(\Gamma)$, then there is some $T \in \varphi(\Gamma\backslash s_i)$ with $T \subseteq S$.*

In words, a solution to the game *contains* a solution of the reduced game after eliminating a weakly dominated strategy. The phrase "$T \subseteq S$" needs some clarification here, since the games Γ and $\Gamma\backslash s_i$ have different strategy spaces, and therefore S and T do not belong to the same space. The set T of mixed conjecture profiles in $\Gamma\backslash s_i$ may be identified with the set T' of mixed conjecture profiles in Γ where (1) every $\mu \in S'$ is such that $\mu_i(s_i) = 0$ and (2) the projection of T' on $\Gamma\backslash s_i$ coincides with T. When we say that "$T \subseteq S$" we formally mean that $T' \subseteq S$.

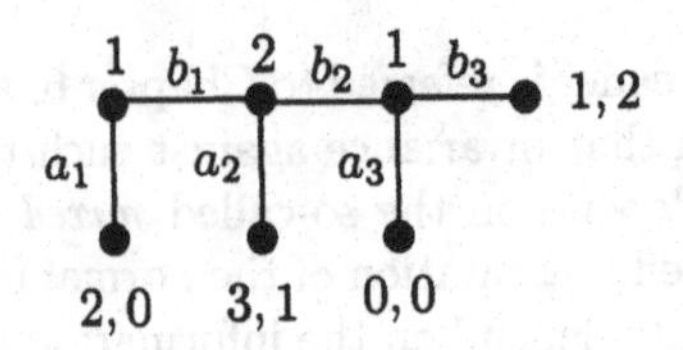

Figure 5.8

In the game of Figure 5.3, the property of iterated weak dominance implies that a solution to the original game Γ should include both mixed conjecture profiles (a, Out) and (c, Out). Consequently, in a solution satisfying iterated weak dominance, player 1 should be uncertain about player 2's conjecture about player 1, since player 1 should deem both conjectures a and c possible.

The second requirement, the never a best response property, states that a solution should be robust against elimination of strategies which are inferior responses to (all conjecture profiles in) the solution. For a mixed conjecture profile $\mu = (\mu_i)_{i \in I}$, a strategy s_i is called an inferior response to μ if there is some s_i' with $u_i(s_i, \mu_{-i}) < u_i(s_i', \mu_{-i})$.

Definition 5.3.2 *A set-valued solution concept φ is said to satisfy the never a best response property if for every game Γ and every $S \in \varphi(\Gamma)$ the following holds: if s_i is an inferior response to all conjecture profiles in S, then there is some $T \in \varphi(\Gamma \backslash s_i)$ with $T \subseteq S$.*

As for iterated weak dominance, the requirement for set-valued solution concepts is that a solution of the original game should *contain* a solution of the reduced game after elimination of an inferior response to this solution. The phrase "$T \subseteq S$" should be read as above. Instead of "inferior response to all conjecture profiles in S" we often write "inferior response to S". The intuition of the never a best ressponse property is the following. Suppose that it is common knowledge that the players hold some conjecture profile in the set $S \in \varphi(\Gamma)$, but that players are not necessarily informed about the exact conjecture profile in S held by the opponents. If all players believe that their opponents choose best responses against their conjectures, then all players should assign probability zero to opponents' strategies that are inferior responses to all conjecture profiles in S. If there is common knowledge about this reasoning by the players, then the solution S should induce a solution in the reduced game obtained after deleting such inferior responses from Γ.

The example in Figure 5.8 shows that a single-valued solution concept cannot be compatible with both backward induction and the never a best response property. A single-valued solution concept compatible with backward induction should prescribe the conjecture profile $((a_1, b_3), b_2)$ in this game. The strategy (b_1, b_3) is an inferior response to this solution. In the reduced game after eliminating (b_1, b_3), strategy b_2 leads to $(0, 0)$. The unique conjecture profile satisfying backward induction in this reduced game is therefore $((b_1, a_3), a_2)$, contradicting the conjecture profile selected in the original game. Hence, if one accepts backward induction and the never a best response property as necessary requirements, one is led to set-valued solution concepts.

5.3.2 Definitions and Properties

The basic idea behind stable sets of equilibria is the following. Take a normal form game and a closed set of Nash equilibria in this game. This set of Nash equilibria is considered stable if for any small perturbation of the utility functions in the game, there is a Nash equilibrium of the perturbed game close to this set. Moreover, the set of equilibria should be minimal with respect to this property, i.e. there should be no smaller set that also satisfies this property.

Kohlberg and Mertens present three different stability concepts; hyperstability, full stability and stability, which differ in the way utilities may be perturbed. Hyperstability allows for arbitrary perturbation of the utilities, whereas full stability and stability only allow for specific utility perturbations. In turn, full stability allows for a larger class of utility perturbations than stability. By minimality, it follows that every stable set is contained in a fully stable set and every fully stable set is contained in a hyperstable set. The reason for analyzing different stability concepts is the trade-off between the "stability" and the "size" of sets of Nash equilibria. On the one hand, it is desirable to have sets of equilibria that are robust against many possible utility perturbations (an extreme case being hyperstability) but this leads in general to large sets that may contain some "undesirable" equilibria. On the other hand, small sets of equilibria are, by definition, less robust against utility perturbations than large sets. The following two facts demonstrate that this trade-off is important. Hyperstable and fully stable sets may contain Nash equilibria that assign positive probability to weakly dominated strategies. If one regards the use of weakly dominated strategies as a fundamental drawback, hyperstable and fully stable sets should be considered too large. Stable sets, on the other hand, never contain Nash equilibria which assign positive probability to weakly dominated strategies, but may fail to contain sequential equilibria as opposed to hyperstable and fully stable sets. If one uses the concept of sequential equilibrium as a reference point, stable sets may be considered too small. In this section, we concentrate solely on stable sets and, more specifically, on its relation to forward induction.

We need the following notation. Let $\Gamma = ((S_i)_{i \in I}, (u_i)_{i \in I})$ be a normal form game. Choose a mixed conjecture profile $\mu = (\mu_i)_{i \in I}$ and a vector $\delta = (\delta_i)_{i \in I}$ of numbers with $0 \leq \delta_i < 1$ for all i. For a strategy s_i, let $s_i^{\delta,\mu} \in \Delta(S_i)$ be given by

$$s_i^{\delta,\mu} = (1 - \delta_i)s_i + \delta_i \mu_i.$$

For a given strategy profile $s = (s_i)_{i \in I}$, let $s^{\delta,\mu} = (s_i^{\delta,\mu})_{i \in I} \in \times_{i \in I} \Delta(S_i)$. Let the perturbed utility functions $u_i^{\delta,\mu} : \times_{j \in I} S_j \to \mathbb{R}$ be given by

$$u_i^{\delta,\mu}(s) = u_i(s^{\delta,\mu})$$

for every strategy profile s. The perturbed normal form game $((S_i)_{i\in I}, (u_i^{\delta,\mu})_{i\in I})$ is denoted by $\Gamma^{\delta,\mu}$. A possible interpretation is that whenever a player chooses some strategy s_i, then for player i's opponents it seems as if player i randomizes according to $s_i^{\delta,\mu}$. For every set E of mixed conjecture profiles and $\epsilon > 0$, let $B_\epsilon(E)$ be the set of mixed conjecture profiles μ for which there is some $\mu' \in E$ with $d(\mu, \mu') < \epsilon$. Here, $d(\mu, \mu')$ is the Euclidean distance between μ and μ'.

Definition 5.3.3 *Let Γ be a normal form game and E a nonempty and closed set of Nash equilibria in Γ. We say that E satisfies property (S) if for every $\epsilon > 0$ there exists some $\delta_0 > 0$ such that for every mixed conjecture profile μ and every $\delta = (\delta_i)_{i\in I}$ with $0 < \delta_i < \delta_0$ for all i, the perturbed game $\Gamma^{\delta,\mu}$ has a Nash equilibrium $\tilde{\mu}^{\delta,\mu} \in B_\epsilon(E)$. The set E is called a stable set of equilibria if it is a minimal set with property (S).*

In the original definition by Kohlberg and Mertens, it is required that the property above holds for every *strictly positive* mixed conjecture profile μ. However, both definitions are equivalent. In the sequel, when we say that E is a stable set for an extensive form game Γ, we mean that E is a stable set for its normal form $N(\Gamma)$. Stable sets may be seen as the set-valued analogue of *strictly perfect equilibrium* (Okada, 1981) since every stable set containing exactly one mixed conjecture profile is a strictly perfect equilibrium.

In order to show that stable sets always exist, we need the following result for which the proof is omitted. It states that there is always a connected component of Nash equilibria which is robust against arbitrary small perturbations of the utilities. For an exact statement of this result, we need the following definitions, which are discussed in more detail in Section 6.6. The *mixed reduced normal form* of a normal form game is obtained by deleting from the game all strategies s_i that are equivalent to some randomization over player i strategies other than s_i. Two normal form games are called *equivalent* if they have the same mixed reduced normal form. In Section 6.6 it is shown that for two equivalent games Γ_1 and Γ_2 there is a mapping which transforms mixed conjecture profiles in Γ_1 to "equivalent" mixed conjecture profiles in Γ_2. We call this mapping τ.

Definition 5.3.4 *Let E be a nonempty and closed set of Nash equilibria of some normal form game Γ. We say that E satisfies property (H) if for every equivalent game $\Gamma' = ((S_i)_{i\in I}, u')$ the following holds: for every $\epsilon > 0$ there exists some $\delta > 0$ such that every game $\Gamma'' = ((S_i)_{i\in I}, u'')$ with $d(u'', u') < \delta$ has a Nash equilibrium $\mu'' \in B_\epsilon(\tau(E))$.*

Here, $\tau(E)$ is the set of Nash equilibria in Γ' which is obtained from E through the above mentioned mapping τ. By u' and u'' we denote profiles of utility functions, and $d(u'', u')$ is the Euclidean distance between them.

It is exactly this property that leads to the concept of hyperstable sets: a hyperstable set is simply a minimal set of equilibria with property (H).

Theorem 5.3.5 *(Kohlberg and Mertens, 1986). Let Γ be a normal form game. Then, the set of Nash equilibria consists of finitely many connected components and at least one connected component satisfies property (H).*

The proof can be found in Kohlberg and Mertens (1986). With this result at hand, it is not difficult to prove the following property.

Theorem 5.3.6 *Every normal form game has a stable set of equilibria which is contained in a single connected component of Nash equilibria.*

Proof. Let Γ be a normal form game. By Theorem 5.3.5, there is a connected component E of Nash equilibria that satisfies property (H). Since property (H) implies property (S), E satisfies property (S). Denote by $\mathcal{F}$ the family of nonempty and closed subsets of E that satisfy property (S). We show that $\mathcal{F}$ has a minimal element, which would imply the statement in the theorem. Since $E \in \mathcal{F}$, the family $\mathcal{F}$ is nonempty. By the countable version of Zorn's Lemma, it is sufficient to show that every decreasing chain of elements in $\mathcal{F}$ has a lower bound. Let $E_1 \supseteq E_2 \supseteq ...$ be a decreasing chain of sets in $\mathcal{F}$. Since all sets E_i are nonempty and compact, the intersection $E_\infty = \bigcap_{n \in \mathbb{N}} E_n$ is nonempty and compact, and hence closed. It can easily be verified that E_∞ satisfies property (S). Hence, $E_\infty \in \mathcal{F}$ and E_∞ is a lower bound in $\mathcal{F}$. By Zorn's Lemma, the statement follows. ■

Every mixed conjecture profile μ induces a probability distribution $\mathbb{P}_\mu \in \Delta(Z)$ on the set of terminal nodes. A probability distribution on the terminal nodes is called an *outcome*. Say that an outcome is *stable* if there is a stable set of equilibria such that all Nash equilibria in this set induce this particular outcome. We prove that for a given extensive form structure and "generic utilities" at the terminal nodes, there exists a stable outcome. By "generic utilities at the terminal nodes" we mean the following. For a given extensive form structure, the utilities at the terminal nodes can be written as a vector $u \in \mathbb{R}^{Z \times I}$ where Z is the set of terminal nodes. Fix an extensive form structure S. Say that a property holds for *generic utilities at the terminal nodes* if the closure of the set of utility vectors for which the property does not hold has Lebesgue measure zero within $\mathbb{R}^{Z \times I}$. In order to prove the above mentioned result, we use Theorem 2 in Kreps and Wilson (1982a) which is stated here without proof. A proof is given in Appendix C of their paper.

Theorem 5.3.7 *Let S be an extensive form structure. Then, for generic utilities at the terminal nodes, the set of Nash equilibria induces only finitely many different outcomes.*

Say that an extensive form game is *regular* if the set of Nash equilibrium outcomes is finite. The theorem above thus states that for a given extensive form structure, the game is regular for generic utilities at the terminal nodes.

Lemma 5.3.8 *Every regular extensive form game has a stable outcome.*

Proof. Let Γ be a regular extensive form game. Since the outcome of a mixed conjecture profile depends continuously on the mixed conjecture profile, and the number of Nash equilibrium outcomes is finite, it follows that every connected component of Nash equilibria should induce a unique outcome. By Theorem 5.3.6, there is stable set contained within a single connected component of Nash equilibria. This stable set thus induces a unique outcome, guaranteeing the existence of a stable outcome. ■

Let us return to the example of Figure 5.3. This game has a unique Nash equilibrium outcome, namely reaching the terminal node after *Out* with probability one, which trivially implies that this outcome is stable. We show, however, that every stable set should contain the Nash equilibria in mixed conjectures (a, Out) and (c, Out), prescribing different conjectures about player 1. This illustrates the fact that a stable outcome may have to be supported by different Nash equilibria. The normal form Γ of the game is given by the following table.

	(In, d)	(In, e)	Out
a	3,1	0,0	0,2
b	0,0	1,3	0,2
c	2,0	2,0	0,2

Consider a perturbed game $\Gamma^{\delta,\mu}$ as used in the definition of stable sets, where

$$\delta = (\delta_1, \delta_2), \delta_1 > 0, \delta_2 > 0$$

and $\mu = (\mu_1, \mu_2)$ with

$$\begin{aligned} \mu_1 &= p_1 a + q_1 b + (1 - p_1 - q_1)c \text{ and} \\ \mu_2 &= p_2(In, d) + q_2(In, e) + (1 - p_2 - q_2)Out. \end{aligned}$$

It may be verified easily that strategies b, (In, d) and (In, e) are assigned probability zero in any Nash equilibrium of the perturbed game $\Gamma^{\delta,\mu}$. If player 2 chooses *Out* in the perturbed game, player 1's expected utility by playing a is

$$3(1 - \delta_1 + \delta_1 p_1)\delta_2 p_2 + \delta_1 \delta_2 q_1 q_2 + 2\delta_1(1 - p_1 - q_1)\delta_2(p_2 + q_2)$$

whereas player 1's expected utility by playing c is

$$3\delta_1\delta_2 p_1 p_2 + \delta_1 \delta_2 q_1 q_2 + 2(1 - \delta_1 + \delta_1(1 - p_1 - q_1))\delta_2(p_2 + q_2).$$

If one chooses $p_2 = 3/5$ and $q_2 = 1/5$, then, for δ_1, δ_2 small enough, player 1's utility by playing a is strictly larger than his utility by playing c, and hence (a, Out) is the unique Nash equilibrium in $\Gamma^{\delta,\mu}$. By definition of stable sets, every stable set should contain (a, Out).

If one chooses $p_2 = 1/5$ and $q_2 = 3/5$, then, for δ_1, δ_2 small enough, player 1's utility by playing c is strictly larger than by playing a, which induces that (c, Out) is the unique Nash equilibrium in $\Gamma^{\delta,\mu}$. Consequently, every stable set should contain (c, Out) also.

In order to show that in a stable set, all Nash equilibria always assign probability zero to weakly dominated strategies, we prove that every Nash equilibrium in a stable set is a normal form perfect equilibrium. Here, a mixed conjecture profile μ is called a normal form perfect equilibrium if there is some sequence $(\mu^n)_{n\in\mathbb{N}}$ of strictly positive mixed conjecture profiles converging to μ such that $\mu_i(s_i) > 0$ only if s_i is a best response against μ^n for all n.

Lemma 5.3.9 *Every Nash equilibrium in some stable set of equilibria is a normal form perfect equilibrium.*

Proof. Let E be a stable set of equilibria and $\mu^* \in E$. Since E is a *minimal* set of Nash equilibria with property (S), there should be a sequence $(\Gamma^{\delta^n,\mu^n})_{n\in\mathbf{N}}$ of perturbed games with $\delta^n > 0$, μ^n strictly positive and $\lim_{n\to\infty} \delta^n = 0$, and a sequence $(\mu^{*n})_{n\in\mathbf{N}}$ of Nash equilibria in Γ^{δ^n,μ^n} with $\lim_{n\to\infty} \mu^{*n} = \mu^*$. For every n and every player i, let $\tilde{\mu}_i^n = (1-\delta_i^n)\mu_i^{*n} + \delta_i^n \mu_i^n$. Since $\delta_i^n > 0$ and μ_i^n strictly positive, it follows that $\tilde{\mu}_i^n$ is strictly positive for all n. Moreover, $(\tilde{\mu}_i^n)_{n\in\mathbf{N}}$ converges to μ_i^* since $(\delta_i^n)_{n\in\mathbf{N}}$ converges to 0 and μ_i^{*n} converges to μ_i^*. Let $\tilde{\mu}^n = (\tilde{\mu}_i^n)_{i\in I}$. Then, $(\tilde{\mu}^n)_{n\in\mathbf{N}}$ is a sequence of strictly positive mixed conjecture profiles in Γ converging to μ^*. For every n, let $B_i(\tilde{\mu}^n) \subseteq S_i$ be the set of best responses against $\tilde{\mu}^n$ in the original game Γ. Since S_i is finite, there is a subsequence of $(\tilde{\mu}^n)_{n\in\mathbf{N}}$ along which $B_i(\tilde{\mu}^n)$ is constant for all i. Without loss of generality, we may assume that $B_i(\tilde{\mu}^n)_{n\in\mathbf{N}}$ is constant along the sequence $(\tilde{\mu}^n)_{n\in\mathbf{N}}$. We show that μ^* is a normal form perfect equilibrium "justified" by the sequence $(\tilde{\mu}^n)_{n\in\mathbf{N}}$. Assume that s_i is not a best response against $\tilde{\mu}^n$ for some n. We prove that $\mu_i^*(s_i) = 0$. Let r_i be a best response against $\tilde{\mu}^n$ for this n. Then, by construction, s_i is not a best response against $\tilde{\mu}^n$ for all n, and r_i is a best response against $\tilde{\mu}^n$ for all n. Hence, $u_i(s_i, \tilde{\mu}_{-i}^n) < u_i(r_i, \tilde{\mu}_{-i}^n)$ for all n. By definition of the perturbed games Γ^{δ^n,μ^n} and by construction of $\tilde{\mu}^n$, it holds: $u_i(s_i, \tilde{\mu}_{-i}^n) < u_i(r_i, \tilde{\mu}_{-i}^n)$ if and only if $u_i^{\delta^n,\mu^n}(s_i, \mu_{-i}^{*n}) < u_i^{\delta^n,\mu^n}(r_i, \mu_{-i}^{*n})$. We may thus conclude that $u_i^{\delta^n,\mu^n}(s_i, \mu_{-i}^{*n}) < u_i^{\delta^n,\mu^n}(r_i, \mu_{-i}^{*n})$ for all n. Since μ^{*n} is a Nash equilibrium in Γ^{δ^n,μ^n}, it should hold that $\mu_i^{*n}(s_i) = 0$ for all n. However, this implies that $\mu_i^*(s_i) = 0$. We may thus conclude that μ^* is a normal form perfect equilibrium justified by the sequence $(\tilde{\mu}^n)_{n\in\mathbf{N}}$. ■

It may be easily verified that a normal form perfect equilibrium always assigns probability zero to weakly dominated strategies.

Lemma 5.3.10 *Let μ be a normal form perfect equilibrium and s_i a weakly dominated strategy. Then, $\mu_i(s_i) = 0$.*

Proof. Let μ be a normal form perfect equilibrium. Then, there is some sequence $(\mu^n)_{n\in\mathbf{N}}$ of strictly positive mixed conjecture profiles converging to μ such that $\mu_i(s_i) > 0$ only if s_i is a best response against μ^n for all n. Let s_i be a weakly dominated strategy. Then, there is some $\tilde{\mu}_i \in \Delta(S_i)$ such that $u_i(\tilde{\mu}_i, s_{-i}) \geq u_i(s_i, s_{-i})$ for all s_{-i}, with strict inequality for at least one s_{-i}. But then, $u_i(\tilde{\mu}_i, \mu_{-i}) > u_i(s_i, \mu_{-i})$ for all strictly positive probability measures $\mu_{-i} \in \Delta(S_{-i})$. This implies that s_i is never a best response against any strictly positive mixed conjecture profile $\tilde{\mu}$, and therefore $\mu_i(s_i) = 0$. ■

It therefore follows that every Nash equilibrium in a stable set always assigns probability zero to weakly dominated strategies. In the following theorem, it is shown that stable sets are compatible with two forward induction principles: iterated weak dominance and the never a best response property. Let the stable set correspondence be the set-valued solution which to every normal form game assigns the collection of stable sets.

Theorem 5.3.11 *The stable set correspondence satisfy iterated weak dominance and the never a best response property.*

Proof. First, we show that the stable set correspondence satisfy iterated weak dominance. Let Γ be a normal form game and E a stable set in Γ. Let s_i^* be a weakly dominated strategy and let $\tilde{\Gamma}$ be the game in which s_i^* is eliminated. We know, by the two lemmas above, that in every Nash equilibrium of E the strategy s_i^* is assigned probability zero. By $\tilde{E}$ we denote the projection of E on the strategy spaces of $\tilde{\Gamma}$. In order to show that E contains a stable set of $\tilde{\Gamma}$, it suffices to show that $\tilde{E}$ satisfies property (S). Let $\tilde{\Gamma}^{\delta,\mu}$ be a perturbation of $\tilde{\Gamma}$ with $\delta > 0$. Define a perturbation Γ^{δ,μ^*} of the original game Γ as follows. For player i, (the player for which the strategy s_i^* has been eliminated), let $\tilde{\mu}_i$ be the element in $\Delta(S_i)$ which assigns probability zero to s_i^*, and coincides with μ_i otherwise. Let $\hat{\mu}$ be some strictly positive mixed conjecture profile in Γ. For player i, we define $\mu_i^* = (1-\lambda)\tilde{\mu}_i + \lambda\hat{\mu}_i$ where $\lambda \in (0,1)$. For all other players j, let $\mu_j^* = (1-\lambda)\mu_j + \lambda\hat{\mu}_j$. Since E satisfies property (S) for the original game Γ, the perturbed game Γ^{δ,μ^*} has a Nash equilibrium $e(\delta,\lambda)$ close to E whenever δ is small enough. The parameters δ and λ indicate that the choice of the Nash equilibrium depends on the values of δ and λ chosen. Since s_i^* is weakly dominated, μ^* is strictly positive and $\delta > 0$, it follows that s_i^* is never a best response in Γ^{δ,μ^*}, and hence $e(\delta,\lambda)$ puts probability zero on s_i^*. Let $\tilde{e}(\delta,\lambda)$ be the projection of $e(\delta,\lambda)$ on $\tilde{\Gamma}$. Since $\tilde{e}(\delta,\lambda)$ is member of some compact set for all λ, we may assume without loss of generality that $\lim_{\lambda\downarrow 0}\tilde{e}(\delta,\lambda)$ exists. Say $\tilde{e}(\delta) = \lim_{\lambda\downarrow 0}\tilde{e}(\delta,\lambda)$. Since the perturbed game Γ^{δ,μ^*} approximates the perturbed game $\tilde{\Gamma}^{\delta,\mu}$ as $\lambda \to 0$, it follows that $\tilde{e}(\delta)$ is a Nash equilibrium in $\tilde{\Gamma}^{\delta,\mu}$. By construction, every $e(\delta,\lambda)$ is chosen close to E if δ is small enough. Hence every $\tilde{e}(\delta,\lambda)$ is close to $\tilde{E}$ and thus $\tilde{e}(\delta)$ is close to $\tilde{E}$ if δ is small enough. This construction shows that for every perturbation $\tilde{\Gamma}^{\delta,\mu}$ we can find a Nash equilibrium in $\tilde{\Gamma}^{\delta,\mu}$ close to $\tilde{E}$ if δ is small enough, which implies that $\tilde{E}$ satisfies property (S). Hence, $\tilde{E}$ contains a stable set of $\tilde{\Gamma}$.

The proof for the never a best response property is similar to the one above. If E is a stable set of Γ and s_i^* is an inferior response in every Nash equilibrium of E, let $\tilde{\Gamma}$ be the game in which s_i^* is eliminated. Clearly, s_i^* is assigned probability zero in every Nash equilibrium of E. Let $\tilde{E}$ be the projection of E on $\tilde{\Gamma}$. We show that $\tilde{E}$ satisfies property (S). Let $\tilde{\Gamma}^{\delta,\mu}$ be a perturbation of $\tilde{\Gamma}$. Let Γ^{δ,μ^*} be the corresponding perturbation of the original game Γ in which μ_i^* assigns probability zero to s_i^* and coincides with μ_i otherwise, and μ_{-i}^* coincides with μ_{-i}. Since E satisfies property (S) for the original game Γ, the perturbed game Γ^{δ,μ^*} has a Nash equilibrium $e(\delta)$ close to E whenever δ is small enough. If $e(\delta)$ is close enough to E and δ is close enough to zero, the strategy s_i^* is an inferior response to $e(\delta)$ in the game Γ^{δ,μ^*} and is thus assigned probability zero in $e(\delta)$. Let $\tilde{e}(\delta)$ be the projection of $e(\delta)$ on $\tilde{\Gamma}^{\delta,\mu}$. Then, $\tilde{e}(\delta)$ is a Nash equilibrium in $\tilde{\Gamma}^{\delta,\mu}$ close to $\tilde{E}$. This construction shows that for every perturbation $\tilde{\Gamma}^{\delta,\mu}$ we can find a Nash equilibrium in $\tilde{\Gamma}^{\delta,\mu}$ close to $\tilde{E}$ if δ is small enough, which implies that $\tilde{E}$ satisfies property (S). Hence, $\tilde{E}$ contains a stable set of $\tilde{\Gamma}$. ■

As a consequence, the following result is obtained.

Lemma 5.3.12 *Let Γ be an extensive form game, $\mathbb{P} \in \Delta(Z)$ a stable outcome in Γ induced by some stable set E and s_i^* a weakly dominated strategy or inferior response to E. Then, $\mathbb{P}$ remains a stable outcome after eliminating s_i^* from the normal form $N(\Gamma)$.*

Proof. Let $\mathbb{P}$ be a stable outcome in the game. By definition, there exists a stable set E such that every Nash equilibrium in E induces $\mathbb{P}$. Let s_i^* be a weakly dominated strategy or inferior response to E. Then, s_i^* is assigned probability zero in every Nash equilibrium of E. Let $N'(\Gamma)$ be the normal form game obtained after deleting s_i^* from the normal form $N(\Gamma)$. Let E' be the projection of E on the game $N'(\Gamma)$. By the theorem above, E' contains a stable set E'' of the game $N'(\Gamma)$. Since every Nash equilibrium in E'' induces the outcome $\mathbb{P}$, the result follows. ∎

This result can be used in order to detect non-stable outcomes in a game. Consider again the Battle of the Sexes game in Figure 5.1. The set of Nash equilibria consists of two connected components, $A = \{(a, d)\}$ and $B = \{(c, \lambda d + (1 - \lambda)e|\ \lambda \leq 2/3\}$, inducing outcomes $(3, 1)$ and $(2, 0)$ respectively. Here, we denote outcomes by their induced utilities as to reduce notation. There are thus two candidates for stable outcomes; $(3, 1)$ and $(2, 0)$. If we eliminate the weakly dominated strategy b from the game, strategy e becomes weakly dominated. The reduced game obtained after eliminating both b and e has a unique Nash equilibrium (a, d) inducing the outcome $(3, 1)$. By applying twice the lemma above, $(2, 0)$ can be ruled out as a stable outcome. On the other hand, we know that at least one stable set is contained in a connected component of Nash equilibria. Since $(2, 0)$ is not a stable outcome, component B cannot contain a stable set. Therefore, A must contain a stable set, which implies that $\{(a, d)\}$ is the unique stable set in the game, and $(3, 1)$ is the unique stable outcome.

A similar argument can be used to show that the Beer-Quiche game of Figure 5.5 contains a unique stable outcome. In this game, the set of Nash equilibria consists of two connected components A and B. In component A, player 1 is believed to always choose beer, player 2 is believed to avoid a duel if he observes beer and is believed to challenge with probability greater or equal than 0.5 if he observes quiche. In component B, player 1 is believed to always choose quiche, player 2 is believed to challenge with probability greater or equal than 0.5 if he observes beer and is believed to avoid a duel if he observes quiche. We show that the outcome induced by B cannot be stable. Suppose the opposite. Then, B must contain some stable set E. Both strategies for player 1 in which the weak type chooses beer is an inferior response to E. After eliminating these strategies from the game, the player 2 strategies in which he challenges after observing beer become weakly dominated. If these strategies are eliminated also, player 1's strong type should be believed to choose beer in any Nash equilibrium, contradicting the outcome in B. By applying the lemma above twice, it follows that the outcome induced by B cannot be stable.

Since at least one stable set is contained in a connected component of Nash equilibria and the outcome of B is not stable, it follows that component A should contain a stable set. Therefore, the outcome induced by A is the unique stable outcome in the game.

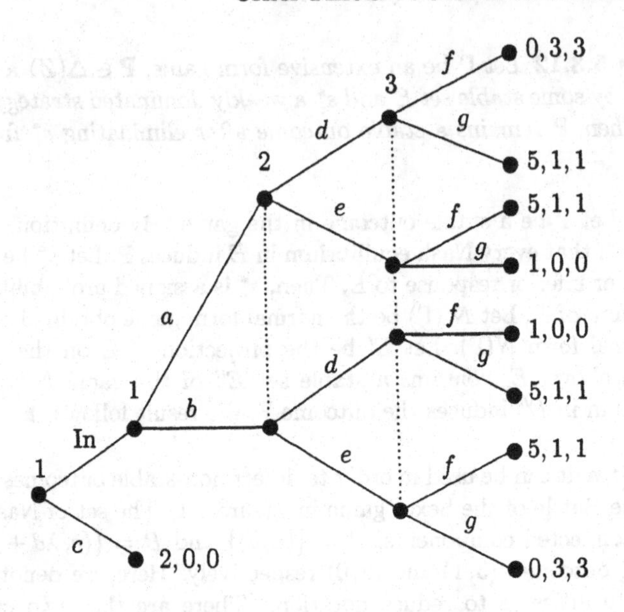

Figure 5.9

A drawback of stable sets is that they may fail to contain a sequential equilibrium. Consider the game in Figure 5.9, which is due to Faruk Gul and appears in Kohlberg and Mertens (1986). It is left to the reader to check that $\{(c, d, f)\} \cup \{(c, e, g)\}$ is a stable set and that $(0.5(In, a) + 0.5(In, b), 0.5d + 0.5e, 0.5f + 0.5g)$ is the unique sequential equilibrium in the game. The stable set does therefore not contain the sequential equilibrium.

We conclude this section with two examples, due to van Damme (1989), which illustrate that stable sets may contain Nash equilibria that do not respect the spirit of backward and/or forward induction. Consider first the game with perfect information in Figure 5.10. The unique stable set in this game contains both the mixed conjecture profile (a, c) satisfying backward induction, and the profile $(a, 0.5c + 0.5d)$. Clearly,

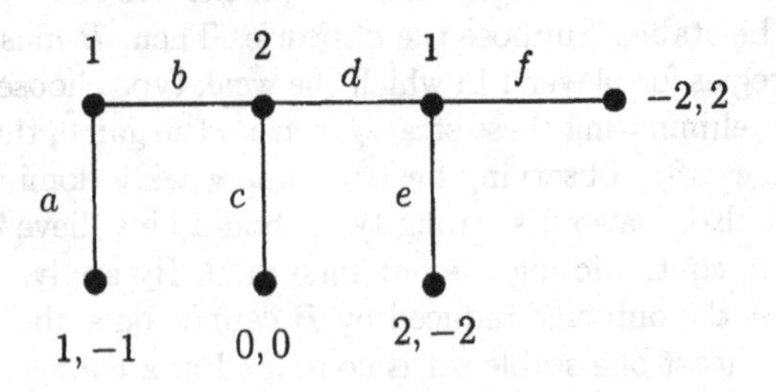

Figure 5.10

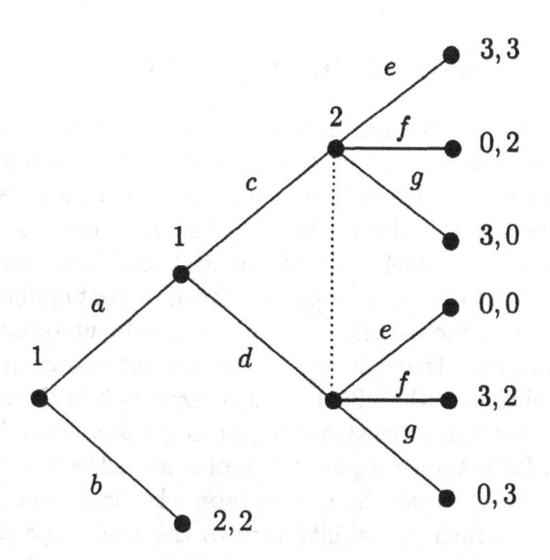

Figure 5.11

the second conjecture profile is not compatible with backward induction but is needed to make the outcome $(1, -1)$ stable. The reason is that by appropriately perturbing the normal form of the game, the perturbed game contains a unique Nash equilibrium in which the conjecture about player 2 is $0.5c + 0.5d$.

Finally, consider the game in Figure 5.11. It can be shown that the outcome $(2, 2)$ is stable. However, none of the Nash equilibria in the corresponding stable set induces a Nash equilibrium in the subgame that starts at player 1's second decision node. (For a precise derivation of this result, the reader is referred to van Damme (1989)). In other terms, the Nash equilibria that are needed to make the outcome $(2, 2)$ stable are no longer viable whenever the above mentioned subgame is reached. In fact, van Damme argues that $(3, 3)$ is the only outcome in this example which is compatible with forward induction. His argument runs as follows. The subgame has three Nash equilibria, (c, e), $(2/3c + 1/3d, 1/2e + 1/2f)$ and $(1/3c + 2/3d, 1/2f + 1/2g)$, yielding player 1 utilities $3, 3/2$ and $3/2$ respectively. Reaching the subgame is a credible signal of player 1 that he is planning to choose c since (c, e) is the only Nash equilibrium in the subgame that would give him more than his outside option b. Given this, player 2 should respond with e and player 1 should choose a. Hauk and Hurkens (1999) provide a detailed discussion of outside option games similar to Figure 5.11, and analyze the performance of different stability concepts, including some not discussed in this section, in such games.

After Kohlberg and Mertens' paper, alternative notions of stable sets of equilibria have been proposed in the literature. See, for instance, Mertens (1989, 1991), Hillas (1990), McLennan (1989c) and Vermeulen, Potters and Jansen (1997). It is outside the scope of this monograph, however, to discuss these alternative stability concepts. For an overview of the relation between the different stability concepts, the reader is referred to Hillas, Jansen, Potters and Vermeulen (2000).

5.4 Forward Induction Equilibrium

In many examples, the never a best response property of stable sets, or some variation of it, may be applied as a criterion to rule out certain "unreasonable" assessments. We have seen that the Beer-Quiche game in Figure 5.5 contains sequential equilibria in which player 1 is believed to always choose quiche and player 2 is believed to respond to quiche with avoiding a duel. Such sequential equilibria can be ruled out on the basis of the never a best response property. The argument applied can be rephrased in terms of restrictions on the beliefs: if player 2 faces the unforeseen event of observing beer, he should conclude that the weak type did not choose it, since choosing beer would never give him more than 2 whereas he expects 3 in the sequential equilibrium at hand. On the other hand, the strong type can get more than his equilibrium utility by choosing beer, for instance if player 2 avoids a duel, which is a best response for him if he believes that it was the strong type who chose beer. This criterion thus requires player 2 to attach probability zero to the weak type at the information set following beer. Given this restriction on the beliefs, player 2 should avoid a duel when observing beer. But then, choosing quiche is no longer optimal for the strong type and hence the sequential equilibria mentioned above should be ruled out.

The key feature in this reasoning is that player 2 compares the possible strategies by player 1 that could have led to his information set, and discards those strategies that, from player 1's viewpoint, are dominated by his equilibrium utility. This corresponds to the so-called *intuitive criterion* (Cho and Kreps, 1987) for signaling games. Cho (1987) generalizes this criterion to the general class of extensive form games. He proposes a refinement of sequential equilibrium, *forward induction equilibrium,* which is based on the principle of discarding player i actions that, from player i's viewpoint, are dominated by his equilibrium utility. In order to formalize the concept, we need the following notation.

For a given information set $h \in H_i$, let $\tilde{B}(h)$ be the set of actions $a \in A(h)$ such that a is a local best response against (σ, β) for some assessment (σ, β). Let $B(h) = \Delta(\tilde{B}(h))$. For a given sequential equilibrium (σ, β), let $u_i(\sigma)$ be the expected utility induced by σ, as defined in Section 3.2. Since σ is a Nash equilibrium, we know that $u_i(\sigma) = \max_{s_i \in S_i} u_i(s_i, \sigma_{-i})$. We refer to $u_i(\sigma)$ as the *equilibrium utility* for player i. Let σ be a behavioral conjecture profile, $h \in H_i$ and a an action at h. By $H(a)$ we denote the information sets that follow the action a. We say that a is a *bad action* at σ if for all behavioral conjecture profiles $\tilde{\sigma}_{H(a)} \in \times_{h' \in H(a)} B(h')$ it holds that

$$u_i(a, \sigma_{-H(a) \cup h}, \tilde{\sigma}_{H(a)}) < u_i(\sigma).$$

Here, $(\sigma_{-H(a) \cup h}, \tilde{\sigma}_{H(a)})$ is the behavioral conjecture profile which coincides with $\tilde{\sigma}_{H(a)}$ at information sets in $H(a)$, and coincides with σ at information sets in $H \backslash (H(a) \cup h)$. Hence, by choosing a at h player i can never get more than his equilibrium utility $u_i(\sigma)$, provided that player i believes that his opponents use local best responses at future information sets. Let $X^{bad}(h, \sigma)$ be the set of nodes x at information set h for which the path to x contains at least one bad action at σ.

Definition 5.4.1 *An assessment (σ, β) is called a forward induction equilibrium if* *it is a sequential equilibrium and, in addition, at every information set h where $X^{bad}(h,\sigma)$ is a strict subset of h, the beliefs β_{ih} put probability zero on every node in $X^{bad}(h,\sigma)$.*

Note that the extra condition imposed by forward induction equilibrium is automatically satisfied at information sets reached with positive probability under σ. Namely, if h is reached with positive probability under σ, then, by local sequential rationality of (σ,β), there is at least one node in h that is not preceded by bad actions at σ. Every bad action leading to h has probability zero under σ and hence every node at h preceded by a bad action has belief zero, by Bayesian consistency of (σ,β). Note also that the concept of forward induction equilibrium does not put restrictions on the beliefs at information sets h where every node is preceded by some bad action at σ.

Let us illustrate the concept by the following example, which is taken from Selten (1978) and Kreps and Wilson (1982b), and is further analyzed in Cho (1987). The game is played by three firms: two potential entrants and an incumbent. At the beginning, a chance move decides whether the incumbent will be weak or strong, with probabilities 1/3 and 2/3 respectively. The incumbent's type is not revealed to any of the entrants. There are two periods in the game. In period 1, the first entrant decides whether or not to enter the market. If it enters, the incumbent chooses whether or not to fight the entrant. In period 2, the same situation is repeated for the second entrant. At the beginning of period 2, the actions chosen by the players in period 1 are common knowledge. The second entrant's uncertainty about the incumbent's type remains, however. The game played in each period is depicted in Figure 5.12. The first utility corresponds to the entrant and the second utility to the incumbent. In this figure, we distinguish the utilities faced by the weak and strong incumbent. The utility for the incumbent in the overall game is simply the sum of the utilities in each period. This game is also known as the *chain store game*. The reader may verify that the following assessment is a sequential equilibrium. In period 1, the entrant is believed to enter the market, and the incumbent is believed not to fight, irrespective of his type. In period 2, the entrant is believed to enter if the incumbent has chosen to fight in period 1, and is believed to stay out if the incumbent has chosen not to fight. The weak incumbent is believed not to fight whereas the strong incumbent is believed to fight at the end of period 2. The entrant's beliefs at the beginning of period 2 are as follows. If the incumbent has chosen not to fight, he attaches belief 1/3 to the weak type. If the incumbent has chosen to fight, he attaches belief 3/4 to the weak type.

The entrant's beliefs at the beginning of period 2 do not seem very intuitive. If he observes that the incumbent has fought in period 1, one would expect him to believe that the incumbent is strong. Indeed, we show that the sequential equilibrium is not a forward induction equilibrium. Let (σ,β) the the sequential equilibrium under consideration, and let h be the entrant's information set at the beginning of period 2, after observing that the incumbent has fought in period 1. This information set contains two nodes, x_w and x_s, where $x_w(x_s)$ is preceded by the entrant's action *In* and the action in which the weak (strong) incumbent chooses fight. We show that fighting for the weak incumbent is a bad action at σ and that fighting for the strong

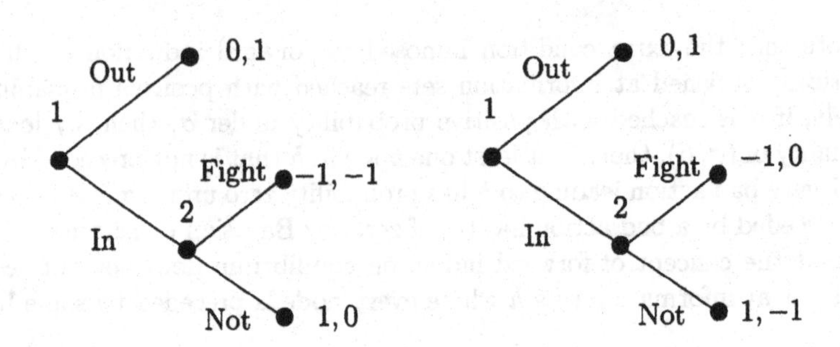

Figure 5.12

incumbent is not a bad action at σ. At σ, the weak incumbent gets $0+1=1$. If he chooses to fight at period 1, he can expect at most $-1+1=0$, so this is a bad action. The strong incumbent gets $-1+1=0$ at σ. If he chooses to fight at period 1, he can possibly obtain $0+1=1$ if the entrant chooses *Out* in period 2. Moreover, *Out* at period 2 is a potential local best response for the entrant. This implies that fighting for the strong incumbent is not a bad action at σ. Hence, if (σ, β) is a forward induction equilibrium, it should attach belief zero to x_w. However, this a contradiction since $\beta(x_w) = 3/4$. Even stronger, the behavioral conjecture profile σ can not be extended to a forward induction equilibrium by any belief system. As we have seen above, in any forward induction equilibrium (σ, β'), the entrant should attach belief zero to the weak type after observing fight in period 1. But then, the entrant should not be believed to enter in period 2, which contradicts σ.

In order to prove existence of forward induction equilibria for every extensive form game, we proceed in two steps. In step 1 we show that for a given extensive form structure, a forward induction equilibrium exists for generic utilities at the terminal nodes. The second step states that the correspondence of forward induction equilibrium outcomes is upper hemi-continuous with respect to the utilities in the game. A precise statement of this property is given in Lemma 5.4.3.

In order to formalize step 1, we need to be explicit about the class of games for which we directly prove existence of forward induction equilibria. For a given extensive form game Γ, let $A(\Gamma)$ be the agent extensive form as defined in the proof of Theorem 3.4.2. Recall that $A(\Gamma)$ is obtained from Γ by assigning each information set to a different player, and giving each player in $A(\Gamma)$ the payoff of the player in Γ to whom he originally belonged. The normal form of $A(\Gamma)$ is called the *agent normal form* (cf.

Selten, 1975) and is denoted by $AN(\Gamma)$. For proving the existence of forward induction equilibria, we concentrate on the agent normal form rather than the normal form of Γ. In the lemma below, we first show the existence in games for which the agent extensive form is a regular game (i.e. has finitely many Nash equilibrium outcomes).

Lemma 5.4.2 *Let Γ be an extensive form game such that the agent extensive form $A(\Gamma)$ is regular. Then, Γ has a forward induction equilibrium.*

Proof. Let Γ be an extensive form game such that $A(\Gamma)$ is regular. Hence, the agent normal form $AN(\Gamma)$ has finitely many Nash equilibrium outcomes. Note that mixed conjecture profiles in $AN(\Gamma)$ coincide with behavioral conjecture profiles in Γ. By the previous section, we know that there exists some stable set E in $AN(\Gamma)$ that is contained within a connected component of Nash equilibria in $AN(\Gamma)$. By regularity of $A(\Gamma)$, all conjecture profiles in E induce the same outcome $\mathbb{P}$. We also know from the previous section that every conjecture profile in E is a normal form perfect equilibrium in $AN(\Gamma)$. By definition, every normal form perfect equilibrium of the agent normal form $AN(\Gamma)$ is a perfect equilibrium in Γ. We may thus conclude that all members in E are perfect equilibria in Γ, and hence, every behavioral conjecture profile $\sigma \in E$ can be extended to a sequential equilibrium (σ, β) in Γ.

Next, we show that all behavioral conjecture profiles in E induce the same set of bad actions. Let $\sigma^1, \sigma^2 \in E$. Then, σ^1 and σ^2 induce the same outcome $\mathbb{P}$. Let h be an information set reached with positive probability in $\mathbb{P}$. Then, by construction, an action a at h is a bad action at σ^1 if and only if it is a bad action at σ^2. If h is reached with probability zero in $\mathbb{P}$, it is reached with probability zero in σ^1 and σ^2, and hence for every action $a \in A(h)$ we have that $u_i(a, \sigma^1_{-h}) = u_i(\sigma^1)$ and $u_i(a, \sigma^2_{-h}) = u_i(\sigma^2)$. This implies that every action at h is not bad at σ^1 and not bad at σ^2. We may thus speak about the set of *bad actions at* E, which we denote by $A^{bad}(E)$.

Let $a \in A^{bad}(E)$. Since all behavioral conjecture profiles in E can be extended to a sequential equilibrium, it follows that a is an inferior response to E. By the never a best response property of stable sets, we have that E contains a stable set of the game which remains after deleting the strategy a from $AN(\Gamma)$. Let $AN^*(\Gamma)$ be the game obtained from $AN(\Gamma)$ after deleting *all* strategies $a \in A^{bad}(E)$. By applying the never a best response iteratively, E contains a stable set E^* for $AN^*(\Gamma)$.

Let σ^* be a behavioral conjecture profile in E^*. Then, σ^* is a normal form perfect equilibrium in $AN^*(\Gamma)$, hence there is a sequence $(\sigma^{*n})_{n\in\mathbb{N}}$ of strictly positive mixed conjecture profiles in $AN^*(\Gamma)$ such that $\sigma^*_{ih}(a) > 0$ only if a is a best response against σ^{*n} for every n. For every n, let $\tilde{\sigma}^n$ be the behavioral conjecture profile in Γ that assigns probability zero to all actions $a \in A^{bad}(E)$, and coincides with σ^{*n} elsewhere. Let $\tilde{\sigma}$ be a behavioral conjecture profile in Γ that assigns strictly positive probability to all actions in $A^{bad}(E)$. For every n, we define the strictly positive behavioral conjecture profile σ^n in Γ by $\sigma^n = (1-\lambda^n)\,\tilde{\sigma}^n + \lambda^n \tilde{\sigma}$, where $(\lambda^n)_{n\in\mathbb{N}}$ is some sequence of strictly positive numbers converging to zero. For every n, let β^n be the beliefs induced by σ^n. Without loss of generality, we assume that $(\beta^n)_{n\in\mathbb{N}}$ converges to some belief system β. For a given information set h, let $X^{bad}(h, E)$ be the set of nodes at h that are preceded by some action $a \in A^{bad}(E)$. By choosing the numbers λ^n small enough, one can achieve that $\beta_{ih}(x) = 0$ whenever $x \in X^{bad}(h, E)$ and $X^{bad}(h, E)$ is a strict subset of h. Let σ be the behavioral conjecture profile that assigns probability zero to all actions in $A^{bad}(E)$, and coincides with σ^* elsewhere. We prove that (σ, β) is a forward induction equilibrium in Γ.

By construction, $(\sigma^n, \beta^n)_{n\in\mathbb{N}}$ converges to (σ, β) and hence (σ, β) is consistent. Moreover, since $X^{bad}(h, \sigma) = X^{bad}(h, E)$ for all h, we have that $\beta_{ih}(x) = 0$ whenever $x \in X^{bad}(h, \sigma)$ and $X^{bad}(h, \sigma)$ is a strict subset of h. It remains to show that (σ, β) is sequentially rational. Since consistent assessments satisfy the one-deviation property, it suffices to show that (σ, β) is locally sequentially rational. Suppose that a is some action at h which is not a local best response against (σ, β). We prove that $\sigma_{ih}(a) = 0$. We distinguish two cases.

Case 1. If $a \in A^{bad}(E)$. Then, by construction, $\sigma_{ih}(a) = 0$.

Case 2. If $a \notin A^{bad}(E)$. Since a is not a local best response against (σ, β), we have that $u_i(a, \sigma^{*n}_{-h}) < u_i(a', \sigma^{*n}_{-h})$ for some action $a' \in A(h)\backslash A^{bad}(E)$ if n large enough. Since σ^* is a normal form perfect equilibrium in $AN^*(\Gamma)$, justified by the sequence $(\sigma^{*n})_{n\in\mathbb{N}}$, it follows that $\sigma^*_{ih}(a) = 0$, and hence $\sigma_{ih}(a) = 0$. This completes the proof. ■

The following lemma states that the correspondence of forward induction equilibrium *outcomes* is upper-hemi continuous with respect to the utilities in the game.

Lemma 5.4.3 *Let S be an extensive form structure, $u = (u_i)_{i\in I}$ a profile of utility functions at the terminal nodes and $(u^n)_{n\in\mathbb{N}}$ a sequence of utility function profiles converging to u. Let $\Gamma = (S, u)$ and $\Gamma^n = (S, u^n)$ be the extensive form games induced by u and u^n, respectively. Suppose that for every n there is an outcome $\mathbb{P}^n \in \Delta(Z)$ induced by some forward induction equilibrium in Γ^n, and $\mathbb{P} = \lim_{n\to\infty} \mathbb{P}^n$. Then, the outcome $\mathbb{P}$ is induced by some forward induction equilibrium in Γ.*

For the proof of this result, the reader is referred to Cho (1986, 1987).

Corollary 5.4.4 *Every extensive form game has a forward induction equilibrium.*

Proof. Let $\Gamma = (S, u)$ be an extensive form game. Since we know that for a given extensive form structure S' the game $\Gamma' = (S', u')$ is regular for generic utilities u' at the terminal nodes, the game $\Gamma = (S, u)$ is the limit of some sequence $\Gamma^n = (S, u^n)$ of games for which the agent extensive form is regular. By Lemma 5.4.2, every game Γ^n contains a forward induction equilibrium (σ^n, β^n) inducing an outcome $\mathbb{P}^n$. Without loss of generality, we may assume that the sequence $(\mathbb{P}^n)_{n\in\mathbb{N}}$ converges to some outcome $\mathbb{P}$. By Lemma 5.4.3, we know that $\mathbb{P}$ is induced by some forward induction equilibrium in Γ, and hence Γ contains at least one forward induction equilibrium. ■

We conclude this section by pointing out a few drawbacks of forward induction equilibrium. First of all, the concept is somewhat asymmetric in the sense that it only puts restrictions on the beliefs over nodes at information sets, and not on conjectures about future play following an information set. Compare, for instance, the Battle of the Sexes games in Figures 5.1 and 5.2. In the game of Figure 5.1, (c, e) is not a forward induction equilibrium. Clearly, b is a bad action at (c, e) and a is not, and therefore player 2 should put probability zero on the lower node, which makes

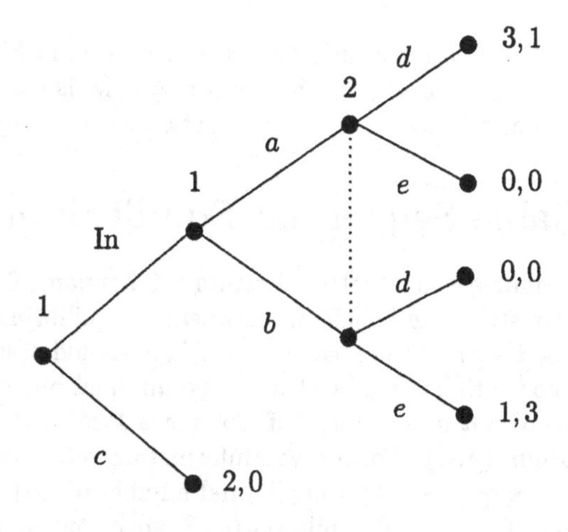

Figure 5.13

e suboptimal. In the game of Figure 5.2, where the order of the players is reversed after player 1 has not chosen the outside option, the corresponding profile $((c, b), e)$ *is* a forward induction equilibrium.

Secondly, the concept of bad actions may be higly unrestrictive if the game is large in the sense that it contains many consecutive information sets. Recall that an action a at information set h is a bad action only if it does worse than the equilibrium outcome for *every* possible combination of potential local best responses at future information sets. The definition thus allows for a broad range of future conjectures that may prevent an action from being bad, especially if there are many information sets to come. Consider for instance the game in Figure 5.7 which is a variation of the game in Figure 5.1 in which the utilities $(0, 0)$ after b and d are replaced by some zero-sum game in which the unique Nash equilibrium leads to expected utilities $(0, 0)$. We have seen that (c, e) is not a forward induction equilibrium in Figure 5.1 since b is a bad action at (c, e). However, b is no longer a bad action at the corresponding sequential equilibrium $((c, 0.5f + 0.5g), (e, 0.5h + 0.5i))$ in Figure 5.7, since it yields more than the equilibrium utility 2 for player 1 if b is followed by the potential local best responses d, f and h. Responsible for this phenomenon is the fact that the action b may be "justified" by future conjectures that do not constitute a Nash equilibrium in the subgame which follows action d.

Another problem is that the concept is not robust against splitting a player's decision into two consecutive decisions. Consider the game in Figure 5.13, which is obtained from Figure 5.1 after splitting player 1's decision (a, b or c) into two consecutive decisions (first, choose In or c, and after choosing In, choose between a and b). As we have seen, (c, e) is not a forward induction equilibrium in Figure 5.1.

However, $\sigma = ((c,b),e)$ *is* a forward induction equilibrium in Figure 5.13. Note that player 1's second information set, at which b is available, is reached with probability zero under $((c,b),e)$ and hence b is not a bad action at $((c,b),e)$.

5.5 Justifiable Sequential Equilibrium

The concept of justifiable sequential equilibrium (McLennan, 1985) provides an alternative criterion to restrict the beliefs in a sequential equilibrium. It is based on the notion of *useless actions* and the idea is that beliefs should assign positive probability only to those nodes that are preceded by the minimal number of useless actions. An action a is called *first order useless* if it is not a local best response against any sequential equilibrium (σ,β). For a given information set h, let $X^1_{\min}(h)$ be the set of nodes in h that are preceded by the minimal number of first order useless actions. Let SE^1 be the set of sequential equilibria (σ,β) such that at every information set h, the beliefs β_{ih} assign positive probability only to nodes in $X^1_{\min}(h)$. Say that an action is *second order useless* if it not a local best response against any sequential equilibrium of SE^1. For every information set h, let $X^2_{\min}(h) \subseteq X^1_{\min}(h)$ be the set of nodes in $X^1_{\min}(h)$ that, among nodes in $X^1_{\min}(h)$, are preceded by the minimal number of second order useless actions. Let SE^2 be the set of sequential equilibria in which the beliefs attach positive probability only to nodes in $X^2_{\min}(h)$. In the same manner, we define $X^k_{\min}(h)$ and SE^k for $k \geq 2$, as long as SE^k is nonempty. We will show, in the theorem below, that indeed SE^k is nonempty for all k, and hence SE^k and $X^k_{\min}(h)$ are well-defined for all k. By construction, $X^{k+1}_{\min}(h) \subseteq X^k_{\min}(h)$ and $SE^{k+1} \subseteq SE^k$ for all k. Since the set of nodes is finite, there exists some K such that $X^k_{\min}(h) = X^K_{\min}(h)$ for all $k \geq K$ and all h. Hence, $SE^k = SE^K$ for all $k \geq K$. Write $SE^\infty = SE^K$.

Definition 5.5.1 *A sequential equilibrium is said to be justifiable if it belongs to* SE^∞.

The final restriction on the beliefs is thus the result of an iterative procedure which ends after finitely many steps. In order to illustrate this procedure, consider the game in Figure 5.14. Clearly, the action b is first order useless, since d is always better. It is the only first order useless action, however. To see this, note that (a,e) together with beliefs $(1,0,0)$ is a sequential equilibrium, and so are (d,f) with beliefs $(0,0,1)$ and $(d, 0.5f + 0.5g)$ with beliefs $(0,0.8,0.2)$. Moreover, c is a local best response against the latter sequential equilibrium, so all actions other than b are local best responses against some sequential equilibrium. By definition, every sequential equilibrium in SE^1 should assign belief zero to the middle node. Given these beliefs, g is an inferior response against every sequential equilibrium of SE^1, which implies that also c is an inferior response against every sequential equilibrium of SE^1. Hence, both c and g are second order useless and the only sequential equilibrium in SE^2 is (a,e) with beliefs $(1,0,0)$. The iterative procedure stops here and therefore (a,e) is the unique justifiable sequential equilibrium in the game. Note that the sequential equilibrium (d,f) with beliefs $(0,0,1)$ survives the first step but is eliminated in the second since c is second order useless but not first order useless.

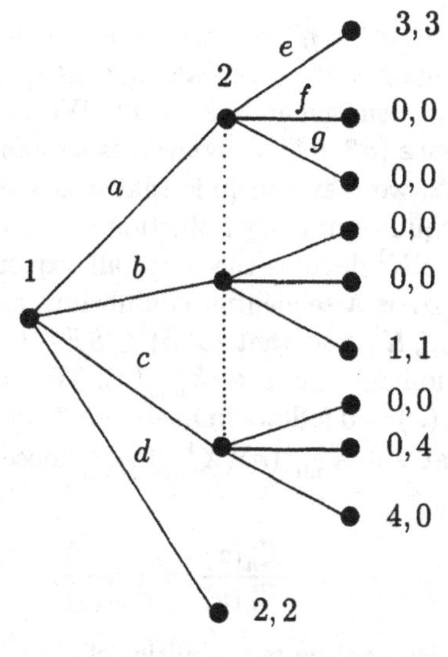

Figure 5.14

We now prove that a justifiable sequential equilibrium always exists.

Theorem 5.5.2 *Every extensive form game has at least one justifiable sequential equilibrium.*

Proof. The proof uses techniques similar to those applied in the proof of Theorem 3.4.2 where we have shown the existence of perfect equilibria. Let Γ be an extensive form game and let $X^k_{\min}(h)$ and SE^k be the sets defined above. As we have seen, there exists some K such that $SE^k = SE^K$ for all $k \geq K$. Let K be minimal with this property. We show that SE^K is nonempty, implying the existence of justifiable sequential equilibria. For the proof of this result, the following notation will be helpful. Let A^1 be the set of first order useless actions and for every $k \in \{2, ..., K\}$, let A^k be the set of k-th order useless actions which are not $(k-1)$-th order useless. Let A^{K+1} be the set of actions which are not K-th order useless (and hence not k-th order useless for any $k < K$). By construction, the set of actions A can be partitioned into the sets $A^1, ..., A^{K+1}$.

Choose some $\epsilon \in (0, 1)$ and define $m = |A| + 1$, where $|A|$ is the total number of actions in the game. We define a sequence $(\eta^n)_{n\in\mathbb{N}}$ of strictly positive vectors where $\eta^n = (\eta^n(a))_{a\in A}$, and A is the set of all actions in the game. For every $k \in \{1, ..., K+1\}$ and every action $a \in A^k$, set

$$\eta^n(a) = (\epsilon^{m^{K+1-k}})^n,$$

where ϵ should be chosen so small that $\sum_{a\in A(h)} \eta^n(a) < 1$ for all h and all n. Let $B(\eta^n)$ be the space of behavioral conjecture profiles σ with $\sigma_{ih}(a) \geq \eta^n(a)$ for all $a \in A$. As in the proof of Theorem 3.4.2, we can show for every n that there is some

$\sigma^n \in B(\eta^n)$ such that $\sigma^n_{ih}(a) > \eta^n(a)$ only if a is a local best response against σ^n. By a local best response against σ^n, we mean that $u_i(a, \sigma^n_{-h}) = \max_{a' \in A(h)} u_i(a', \sigma^n_{-h})$. Let β^n be the belief system generated by σ^n. Without loss of generality, we may assume that the sequence $(\sigma^n, \beta^n)_{n\in\mathbb{N}}$ converges to some assessment (σ, β). As in the proof of Theorem 3.4.2, we may conclude that σ is a perfect equilibrium, and hence (σ, β) is a sequential equilibrium. By induction on k, we prove that $(\sigma, \beta) \in SE^k$ for all $k = 0, ..., K$, where SE^0 denotes the set of all sequential equilibria in Γ.

We know that (σ, β) is a sequential equilibrium and hence $(\sigma, \beta) \in SE^0$. Now, suppose that $k \in \{1, ..., K\}$ and that $(\sigma, \beta) \in SE^{k-1}$. We show that $(\sigma, \beta) \in SE^k$. Let h be an information set and $x \notin X^k_{\min}(h)$. We must show that $\beta_{ih}(x) = 0$. If $x \notin X^{k-1}_{\min}(h)$, then $\beta_{ih}(x) = 0$ follows immediately from the fact that $(\sigma, \beta) \in SE^{k-1}$. Assume, therefore, that $x \in X^{k-1}_{\min}(h) \backslash X^k_{\min}(h)$. Choose some $y \in X^k_{\min}(h)$. For every $n \in \mathbb{N}$,

$$\frac{\beta^n_{ih}(x)}{\beta^n_{ih}(y)} = \frac{\mathbb{P}_{\sigma^n}(x)}{\mathbb{P}_{\sigma^n}(y)},$$

where $\mathbb{P}_{\sigma^n}(x)$ and $\mathbb{P}_{\sigma^n}(y)$ are the probabilities of reaching x and y given σ^n. In order to show that $\beta_{ih}(x) = 0$, it suffices to prove that

$$\lim_{n\to\infty} \frac{\mathbb{P}_{\sigma^n}(x)}{\mathbb{P}_{\sigma^n}(y)} = 0.$$

By definition,

$$\mathbb{P}_{\sigma^n}(x) = \prod_{a \in A_x} \sigma^n(a) \prod_{c \in C_x} \tau(c),$$

where A_x is the set of actions on the path to x and C_x is the set of outcomes of chance moves on the path to x. Hence,

$$\mathbb{P}_{\sigma^n}(x) = \prod_{l=1}^{K+1} \prod_{a \in A_x \cap A^l} \sigma^n(a) \prod_{c \in C(x)} \tau(c).$$

Since $(\sigma, \beta) \in SE^{k-1}$, all k-th order useless actions are not local best responses against (σ, β) and are therefore inferior responses against all (σ^n, β^n) if n is large enough. Without loss of generality, we may assume that all k-th order useless actions are inferior responses against all (σ^n, β^n). Then, each k-th order useless action is an inferior response against every σ^n. Hence, by construction of σ^n we have that $\sigma^n(a) = \eta^n(a)$ for all $a \in A_x \cap A^l$ and all $l = 1, ..., k$. Consequently,

$$\mathbb{P}_{\sigma^n}(x) = \left[\prod_{l=1}^{k} \prod_{a \in A_x \cap A^l} \eta^n(a)\right] \left[\prod_{l=k+1}^{K+1} \prod_{a \in A_x \cap A^l} \sigma^n(a)\right] \prod_{c \in C_x} \tau(c).$$

Similarly,

$$\mathbb{P}_{\sigma^n}(y) = \left[\prod_{l=1}^{k} \prod_{a \in A_y \cap A^l} \eta^n(a)\right] \left[\prod_{l=k+1}^{K+1} \prod_{a \in A_y \cap A^l} \sigma^n(a)\right] \prod_{c \in C_y} \tau(c).$$

Since $x \in X_{\min}^{k-1}(h) \backslash X_{\min}^{k}(h)$ and $y \in X_{\min}^{k}(h)$, x and y are preceded by the same number of l-th order useless actions for $l = 1, ..., k-1$, but y is preceded by less k-th order useless actions than x. Formally, $|A_x \cap A^l| = |A_y \cap A^l|$ for all $l = 1, ..., k-1$ and $|A_x \cap A^k| > |A_y \cap A^k|$. Since $\eta^n(a)$ is constant among actions of the same set A^l and $|A_x \cap A^l| = |A_y \cap A^l|$ for all $l = 1, ..., k-1$ we have that

$$\prod_{l=1}^{k-1} \prod_{a \in A_x \cap A^l} \eta^n(a) = \prod_{l=1}^{k-1} \prod_{a \in A_y \cap A^l} \eta^n(a),$$

and hence

$$\frac{\mathbb{P}_{\sigma^n}(x)}{\mathbb{P}_{\sigma^n}(y)} = \frac{\prod_{a \in A_x \cap A^k} \eta^n(a) \left[\prod_{l=k+1}^{K+1} \prod_{a \in A_x \cap A^l} \sigma^n(a) \right] \prod_{c \in C_x} \tau(c)}{\prod_{a \in A_y \cap A^k} \eta^n(a) \left[\prod_{l=k+1}^{K+1} \prod_{a \in A_y \cap A^l} \sigma^n(a) \right] \prod_{c \in C_y} \tau(c)}.$$

Say $b = \prod_{c \in C_x} \tau(c) / \prod_{c \in C_y} \tau(c)$. It follows

$$\begin{aligned}
\frac{\mathbb{P}_{\sigma^n}(x)}{\mathbb{P}_{\sigma^n}(y)} &\leq b \frac{\prod_{a \in A_x \cap A^k} \eta^n(a)}{\prod_{a \in A_y \cap A^k} \eta^n(a) \left[\prod_{l=k+1}^{K+1} \prod_{a \in A_y \cap A^l} \eta^n(a) \right]} \\
&= b \frac{\prod_{a \in A_x \cap A^k} (\epsilon^{m^{K+1-k}})^n}{\prod_{a \in A_y \cap A^k} (\epsilon^{m^{K+1-k}})^n \left[\prod_{l=k+1}^{K+1} \prod_{a \in A_y \cap A^l} (\epsilon^{m^{K+1-l}})^n \right]} \\
&= b \frac{(\epsilon^{m^{K+1-k}})^{|A_x \cap A^k| n}}{(\epsilon^{m^{K+1-k}})^{|A_y \cap A^k| n} \left[\prod_{l=k+1}^{K+1} (\epsilon^{m^{K+1-l}})^{|A_y \cap A^l| n} \right]}.
\end{aligned}$$

Since $\epsilon^{m^{K+1-l}} \geq \epsilon^{m^{K+1-(k+1)}} = \epsilon^{m^{K-k}}$ for all $l = k+1, ..., K+1$, it follows that

$$\begin{aligned}
\prod_{l=k+1}^{K+1} (\epsilon^{m^{K+1-l}})^{|A_y \cap A^l| n} &\geq \prod_{l=k+1}^{K+1} (\epsilon^{m^{K-k}})^{|A_y \cap A^l| n} \geq \epsilon^{m^{K-k} |A_y| n} \\
&\geq \epsilon^{m^{K-k}(m-1)n} = \epsilon^{(m^{K+1-k} - m^{K-k})n},
\end{aligned}$$

where the third inequality follows from the fact that $|A_y| \leq |A| = m-1$. Consequently,

$$\begin{aligned}
\frac{\mathbb{P}_{\sigma^n}(x)}{\mathbb{P}_{\sigma^n}(y)} &\leq b \frac{(\epsilon^{m^{K+1-k}})^{|A_x \cap A^k| n}}{(\epsilon^{m^{K+1-k}})^{|A_y \cap A^k| n} \epsilon^{(m^{K+1-k} - m^{K-k})n}} \\
&= b (\epsilon^{m^{K+1-k}})^{(|A_x \cap A^k| - |A_y \cap A^k| - 1)n} \epsilon^{m^{K-k} n}.
\end{aligned}$$

Since $|A_x \cap A^k| > |A_y \cap A^k|$, it follows that $|A_x \cap A^k| - |A_y \cap A^k| - 1 \geq 0$ and hence

$$\lim_{n\to\infty} b(\epsilon^{m^{K+1-k}})^{(|A_x \cap A^k| - |A_y \cap A^k| - 1)n} \epsilon^{m^{K-k}n} = 0.$$

This implies that

$$\lim_{n\to\infty} \frac{\mathbb{P}_{\sigma^n}(x)}{\mathbb{P}_{\sigma^n}(y)} = 0,$$

which in turn guarantees that $\beta_{ih}(x) = 0$. Hence $(\sigma, \beta) \in SE^k$. By induction on k we may conclude that $(\sigma, \beta) \in SE^K$, which completes the proof. ■

The concept of justifiable sequential equilibrium basically suffers from the same drawbacks as forward induction equilibrium: it only puts restrictions on beliefs over nodes at information sets, and not on conjectures over future play, and it is sensitive to splitting a player's action into two consecutive actions. In the Battle of the Sexes game of Figure 5.1, for instance, the sequential equilibrium (c, e) is not justifiable since the action b is first order useless and a is not, and hence player 2 should attach belief zero to his lower node. Hence, (a, d) with beliefs $(1, 0)$ is the only justifiable sequential equilibrium in this game. However, the sequential equilibrium $((c, b), e)$ *is* justifiable in the game of Figure 5.2, which is obtained from Figure 5.1 by reversing the order of player 1's and player 2's decision after player 1 has not chosen the outside option. In the game of Figure 5.13, obtained from Figure 5.1 by splitting player 1's decision into two consecutive decisions, the sequential equilibrium $((c, b), e)$ is justifiable since there are no longer first order useless actions in this game.

5.6 Stable Sets of Beliefs

The concepts of forward induction equilibrium and justifiable sequential equilibrium both restrict the beliefs at information sets that have probability zero under the sequential equilibrium (σ, β) at hand. At such information sets, both concepts compare the possible histories that could have led to this information set, and rule out histories that may be viewed "less plausible" than others. An important difference between these two concepts is that the former compares the possible utilities induced by a certain history with the equilibrium utility in (σ, β), and thus takes the sequential equilibrium at hand as unique reference point, whereas the latter tests the "usefulness" of actions by evaluating their performance in other sequential equilibria as well. Hillas (1994) develops a concept, termed *stable sets of beliefs,* that is somewhat related to justifiable sequential equilibrium in the sense that it considers other sequential equilibria in order to test the players' beliefs in a given sequential equilibrium. However, it differs from both concepts mentioned above on two important grounds. First, it uses a different definition of beliefs which at each information set not only specifies the player's conjecture about past play, but also his conjecture about future play. The concept is therefore able to restrict both the conjectures about past and future play at zero probability information sets. In the two apparently equivalent games of Figure 5.1 and 5.2, it rejects in both cases the sequential equilibria in which player 1 chooses

c, whereas forward induction equilibrium and justifiable sequential equilibrium are only able to reject these equilibria in Figure 5.1, but not in Figure 5.2. Secondly, the concept of stable sets of beliefs puts restrictions on *sets of sequential equilibria* rather than on individual sequential equilibria.

In order to formalize the notion of stable sets of beliefs, we introduce some notation and definitions. An *extended belief system* is a function μ which assigns to every information set h and every action a at h a conditional probability distribution $\mu(\cdot| h, a)$ on the set $Z(h,a)$ of terminal nodes following h and a. It thus captures the players' conjectures about past play and future play simultaneously. A pair (σ, μ) of a behavioral conjecture profile σ and an extended belief system is called an *extended assessment*. A "classical" assessment (σ, β) induces a unique extended assessment (σ, μ) as follows:

$$\mu(z| h, a) = \beta_{ih}(x)\, \mathbb{P}_\sigma(z| y)$$

for all $z \in Z(h,a)$, where x is the unique node at h that precedes z and y is the node that directly follows node x and action a.

An extended assessment (σ, μ) is *consistent* if there is a sequence $(\sigma^n)_{n\in\mathbb{N}}$ of strictly positive behavioral conjecture profiles such that $(\sigma^n)_{n\in\mathbb{N}}$ converges to σ and the sequence $(\mu^n)_{n\in\mathbb{N}}$ of induced extended belief systems converges to μ. We say that $a \in A(h)$ is a local best response against μ if $u_i(\mu(\cdot| h, a)) = \max_{a'\in A(h)} u_i(\mu(\cdot| h, a'))$. Here, $u_i(\mu(\cdot| h, a))$ is the expected utility induced by $\mu(\cdot| h, a)$ conditional on h being reached and a being chosen. We say that (σ, μ) is *locally sequentially rational* if for every player i and every $h \in H_i$, we have that $\sigma_{ih}(a) > 0$ only if a is a local best response against μ. An extended assessment is called a *sequential equilibrium* if it is consistent and locally sequentially rational. It can be verified easily that an extended assessment (σ, μ) is a sequential equilibrium if and only if there is a sequential equilibrium (σ, β) in the "classical" sense that induces (σ, μ). Hence, there is a one-to-one relationship between "classical" sequential equilibria and "extended" sequential equilibria. From now on, everything is stated in terms of extended assessments.

Let M be a set of sequential equilibria. For an information set h, we denote by $M(h)$ the set of sequential equilibria (σ, μ) in M that reach h with positive probability, that is, for which $\mathbb{P}_\sigma(h) > 0$. Let $B(h, M) = \{\mu(\cdot| h, \cdot)|\ \exists(\sigma, \mu') \in M(h) \text{ with } \mu'(\cdot| h, \cdot) = \mu(\cdot| h, \cdot)\}$ be the set of belief vectors at h which are possible in sequential equilibria in M that reach h with positive probability. By $S(h, M) = \{\sigma_{ih}|\ \exists(\sigma', \mu) \in M(h) \text{ with } \sigma'_{ih} = \sigma_{ih}\}$ we denote the set of local conjectures at h that are possible in sequential equilibria in M that reach h with positive probability.

Definition 5.6.1 *Let (σ, μ) be a sequential equilibrium and let M be a set of sequential equilibria. Let $h \in H_i$ be an information set with $\mathbb{P}_\sigma(h) = 0$ and $M(h) \neq \emptyset$.*
(1) We say that (σ, μ) is believed in the first sense relative to M at h if: $\sigma_{ih}(a) > 0$ only if a is a local best response against some $\mu'(\cdot| h, \cdot)$ in the convex hull of $B(h, M)$.
(2) We say that (σ, μ) is believed in the second sense relative to M at h if $\mu(\cdot| h, \cdot)$ is in the convex hull of $B(h, M)$.
(3) We say that (σ, μ) is believed in the third sense relative to M at h if σ_{ih} is in the convex hull of $S(h, M)$.

The intuition behind these three notions is that, whenever an information set $h \in H_i$ is reached which contradicts the players' initial conjectures in σ, player i's opponents believe that player i holds some belief at h which is part of some sequential equilibrium $(\tilde{\sigma}, \tilde{\mu}) \in M(h)$ that is compatible with the event of reaching h. Of course, the latter is only possible if $M(h)$ is nonempty. The sequential equilibrium (σ, μ) is said to be *believed in the first sense* relative to M whenever it is believed in the first sense relative to M at all information sets h with $\mathbb{P}_\sigma(h) = 0$ and $M(h) \neq \emptyset$. Similar for the second and third sense.

Definition 5.6.2 *Let M be a set of sequential equilibria. We say that M is a first order internally stable set of beliefs if every sequential equilibrium in M is believed in the first sense relative to M. We say that M is a first order externally stable set of beliefs if every sequential equilibrium not in M is not believed in the first sense relative to M. We say that M is a first order stable set of beliefs if it is both first order internally stable and first order externally stable.*

Similar for the second and the third sense. A stable set of beliefs M can thus be viewed as a consistent theory about the play of the game: if a sequential equilibrium (σ, μ) is selected from M and the course of action contradicts the initial conjectures in σ, the future conjectures and beliefs prescribed by (σ, μ) may be rationalized by the assumption that some other assessment in M is being held by the players. Moreôver, no sequential equilibrium (σ, μ) outside M can be rationalized by the theory that some assessment in M is being held by the players after reaching an information set h wich contradicts the initial conjectures in σ.

In general, there is no logical inclusion between stable sets of beliefs of the first, second and third kind, nor is there a general relationship between stable sets of beliefs and other rationality criteria discussed in this chapter. In order to illustrate some properties of stable sets of beliefs, let us consider a few examples. First, consider the game in Figure 5.15. In this game, there are three behavioral conjecture profiles that can be extended to sequential equilibria, namely $\sigma^1 = (b, c, f)$, $\sigma^2 = (a, d, e)$ and $\sigma^3 = (2/3a + 1/3b, 1/2c + 1/2d, 1/2e + 1/2f)$. By consistency, there are unique extended belief systems μ^1, μ^2, μ^3 which extend $\sigma^1, \sigma^2, \sigma^3$ to sequential equilibria. Let $M = \{(\sigma^1, \mu^1), (\sigma^2, \mu^2), (\sigma^3, \mu^3)\}$. We show that M is first order stable, but not second nor third order stable.

Let h^1, h^2, h^3 be the information sets controlled by players 1, 2 and 3 respectively. Since σ^2 and σ^3 reach every information set with positive probability, (σ^2, μ^2) and (σ^3, μ^3) are believed in any sense relative to M. The sequential equilibrium (σ^1, μ^1) is believed in the first sense relative to M since the choice c is a best response for player 2 against the belief that player 3 will play $1/2e + 1/2f$, and this belief is in $B(h^2, M)$. Hence, M is first order stable.

By consistency of (σ^1, μ^1), player 2's belief $\mu^1(\cdot| h^2, \cdot)$ about player 3's behavior should coincide with the behavioral conjecture about player 3's behavior in σ^1, and hence player 2 in $\mu^1(\cdot| h^2, \cdot)$ should believe that player 3 chooses f with probability 1. However, this belief does not belong to the convex hull of $B(h^2, M)$, and hence (σ^1, μ^1) is not believed in the second sense relative to M.

Moreover, the local conjecture c about player 2 does not belong to the convex hull

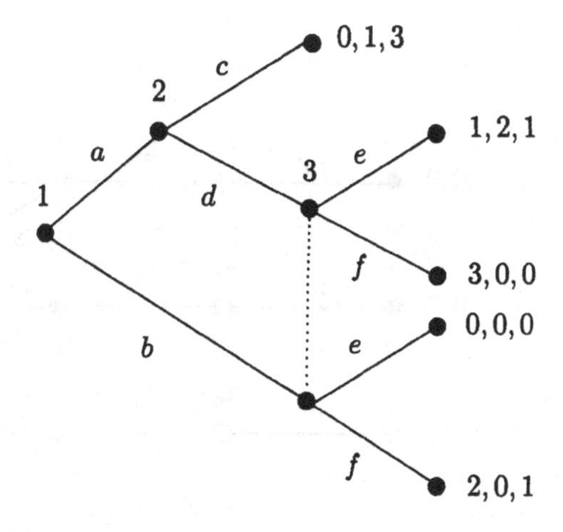

Figure 5.15

of $S(h^2, M)$, and hence (σ^1, μ^1) is not believed in the third sense relative to M. We may thus conclude that M is not second order nor third order stable.

Consider the game of Figure 5.16a. This is a signaling game in which at the beginning a chance move selects the type for player 1. The possible types are t_1, t_2 and t_3 and each has probability 1/3. Player 1, at each of his types, can choose between the messages m and m'. Player 2, upon observing message m, does not know player 1's type, and can choose between the responses r_1, r_2 and r_3. Figure 5.16b depicts the best response for player 2 as a function of his beliefs after observing m, whereas the areas D_{t_i} in Figure 5.16c represent the set of conjectures about player 2 for which sending message m is optimal for type t_i. We show that the second and third order stable sets are different in this game. By analyzing Figures 5.16b and 5.16c, it may be verified that there are three types of sequential equilibria in this game. In the first type, all types of player 1 are believed to choose m', and the conjecture about player 2 is in the area outside D_{t_1}, D_{t_2} and D_{t_3}. More precisely, the sequential equilibria of the first type may be divided into three subtypes. In the first subtype, player 2 has belief β_3 and the conjecture about player 2 is outside D_{t_1}, D_{t_2} and D_{t_3}. In the second subtype, player 2 has a belief on the segment $[\beta_3, \beta_4]$ and the conjecture about player 2 is on the segment $[a, b]$. In the third subtype, player 2 has a belief on the segment $[\beta_1, \beta_3]$ and the conjecture about player 2 is on the segment $[c, d]$. In the second type of sequential equilibria, type t_1 is believed to choose m whereas types t_2 and t_3 are believed to choose m', player 2 has belief $(1, 0, 0)$ and is believed to respond with r_1. In the third type, t_1 is believed to choose m', t_3 is believed to choose m and the conjecture about t_2 is a randomization over m and m' such that the induced beliefs for player 2 coincide with β_2. Player 2 has beliefs β_2 and player 1's conjecture about player 2 is r^{**}. Let $SE(1), SE(2)$ and $SE(3)$ be the sets of sequential equilibria of type 1, 2 and 3 respectively.

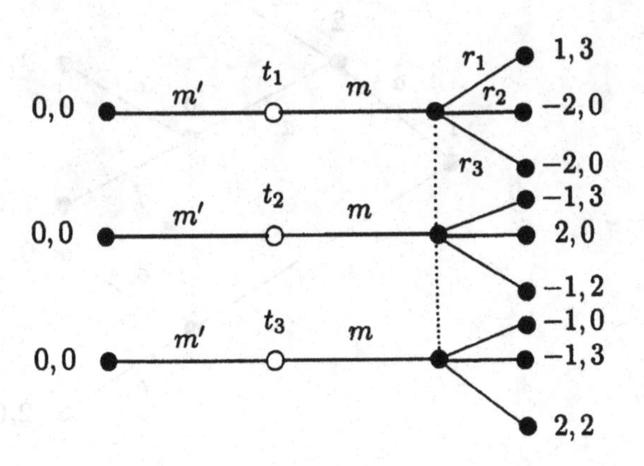

Figure 5.16a

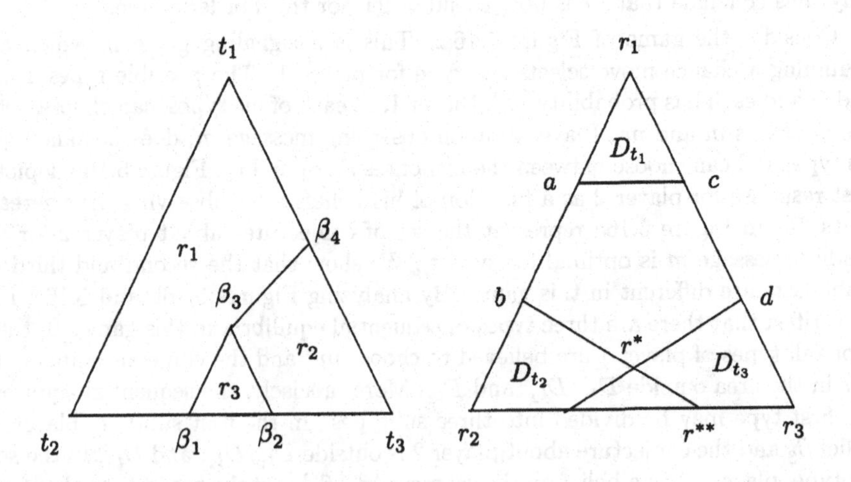

Figures 5.16b and 5.16c

Since $SE(2)$ and $SE(3)$ reach player 2's information set with positive probability, every second and third order stable set should include $SE(2)$ and $SE(3)$. The unique beliefs in $SE(2)$ and $SE(3)$ are $(1,0,0)$ and β_2, respectively, and hence a second order stable set allows for sequential equilibria for which the beliefs are on the segment between $(1,0,0)$ and β_2. From Figure 5.16b, it can be seen that the segment between $(1,0,0)$ and β_2 crosses the segment $[\beta_3,\beta_4]$ in exactly one point, say β^*. Let $SE(1,\beta^*)$ be the sequential equilibria in $SE(1)$ with beliefs β^*. Then, it may be checked that $SE(1,\beta^*) \cup SE(2) \cup SE(3)$ is the unique second order stable set of beliefs.

The unique conjectures about player 2 in $SE(2)$ and $SE(3)$ are r_1 and r^{**}, respectively, and hence a third order stable set allows for sequential equilibria for which the conjecture about player 2 is on the segment $[r_1, r^{**}]$. Let R be the set of all conjectures about player 2 on the segment $[r_1, r^{**}]$ that lie outside D_{t_1}, D_{t_2} and D_{t_3}, and let $SE(1,R)$ be the set of sequential equilibria in $SE(1)$ for which the conjecture about player 2 is in R. Then, it may be checked that $SE(1,R) \cup SE(2) \cup SE(3)$ is the unique third order stable set of beliefs, and this set is different from the unique second order stable set.

Consider again the game in Figure 5.2. In this game, the only sequential equilibrium which reaches player 2's information set is $((In,a),d)$. Since this sequential equilibrium reaches all information sets with positive probability, it should be in every stable set of beliefs (first, second or third order). On the other hand, the sequential equilibrium $((c,b),e)$ cannot be believed (in any sense) relative to a set of sequential equilibria containing $((In,a),d)$, and hence $((c,b),e)$ cannot be in any stable set of beliefs. More precisely, $\{((In,a),d)\}$ is the unique (first, second and third order) stable set of beliefs. Recall that forward induction equilibrium and justifiable sequential equilibrium are not able to rule out the sequential equilibrium $((c,b),e)$.

In the game of Figure 5.11 the concept of stable sets of beliefs uniquely selects the sequential equilibrium $((a,c),e)$ which, according to van Damme's argument, is the only sequential equilibrium compatible with forward induction. In this game, there are three sequential equilibria: $((a,c),e)$, $((b, 2/3c + 1/3d), 1/2e + 1/2f)$ and $((b, 1/3c{+}2/3d), 1/2f{+}1/2g)$. Since $((a,c),e)$ reaches all information sets with positive probability, it belongs to every stable set of beliefs (in any sense). However, the second and third sequential equilibrium cannot be believed in any sense relative to a set of sequential equilibria that contains $((a,c),e)$. Therefore, $\{((a,c),e)\}$ is the unique stable set of beliefs (in any sense). Recall that the concept of stable sets of equilibria is not able to rule out the outcome in which player 1 is believed to choose the outside option b at the beginning of the game.

For other interesting examples, the reader is referred to Hillas (1994). It is shown, for instance, that stable sets of beliefs may fail to exist, and that they may not be unique if they exist. Moreover, stable sets of beliefs and stable sets of equilibria (Kohlberg and Mertens, 1986) may select opposite classes of sequential equilibria in certain games.

Since $SE(2)$ and $SE(3)$ reach player 2's information set with positive probability, every second and third order stable set should include $SE(2)$ and $SE(3)$. The unique beliefs in $SE(2)$ and $SE(3)$ are $(1,0,0)$ and δ_3, respectively, and hence a second order stable set allows for sequential equilibria for which the beliefs are on the segment between $(1,0,0)$ and δ_3. From Figure 5.16b, it can be seen that the segment between $(1,0,0)$ and δ_3 crosses the segment $[\beta_1^*,\delta_2]$ in exactly one point, say β^*. Let $SE(1,\beta^*)$ be the sequential equilibrium in $SE(1)$ with beliefs β^*. Then, it may be checked that $SE(1,\beta^*) \cup SE(2) \cup SE(3)$ is the unique second order stable set of beliefs.

The unique conjectures about player 2 in $SE(2)$ and $SE(3)$ are γ_2 and γ_3^* respectively, and hence a third order stable set allows for sequential equilibria for which the conjectures about player 2 is on the segment $[\gamma_2,\gamma_3^*]$. Let R be the set of all conjectures about player 2 on the segment $[\gamma_2,\gamma_3^*]$ that lie outside D_1, D_2 and D_3, and let $SE(1,R)$ be the set of sequential equilibria in $SE(1)$ for which the conjecture about player 2 is in R. Then, it may be checked that $SE(1,R) \cup SE(2) \cup SE(3)$ is the unique third order stable set of beliefs, and this set is different from the unique second order stable set.

Consider again the game in Figure 5.2. In this game, the only sequential equilibrium which reaches player 2's information set is $((b,c),d)$. Since this sequential equilibrium reaches the information set with positive probability, it should be in every stable set of beliefs (first, second or third order). On the other hand, the sequential equilibrium $((c,b'),e)$ cannot be believed in any sense relative to a set of sequential equilibria containing $((b,c),d)$, and therefore $\{((b,c),d)\}$ is the unique stable set of beliefs. Moreover, $\{((b,c),d)\}$ is the unique (first, second and third order) stable set of beliefs. Recall that forward induction equilibrium and justifiable sequential equilibrium are not able to rule out the sequential equilibrium $((c,b'),e)$.

In the game of Figure 5.[illegible], the concept of stable sets of beliefs uniquely selects the sequential equilibrium (c,e), which, according to van Damme's argument, is the only sequential equilibrium compatible with forward induction. In this game, there are three sequential equilibria: $((a,c),e)$, $((b,2/3a+1/3c),2/3d+1/3e)$ and $((b,1/3c+2/3d),1/2d+1/2e)$. Since $((a,c),e)$ reaches all information sets with positive probability, it belongs to every stable set of beliefs (in any sense). However, the second and third sequential equilibrium cannot be believed in any sense relative to a set of sequential equilibria that contains $((a,c),e)$. Therefore, $\{((a,c),e)\}$ is the unique stable set of beliefs (in any sense). Recall that the concept of stable sets of equilibria is not able to rule out the outcome in which player 1 is believed to choose the outside option b at the beginning of the game.

For other interesting examples, the reader is referred to Hillas (1994). It is shown, for instance, that stable sets of beliefs may fail to exist and that they need not be unique if they exist. Moreover, stable sets of beliefs and stable sets of equilibria (Kohlberg and Mertens, 1986) may select opposite classes of sequential equilibria in certain games.

Chapter 6

Transformations of Games

In this chapter we investigate how rationality criteria for extensive form games react to certain *transformations of games*, that is, functions which map an extensive form game to some new game. In particular, it is analyzed whether the set of conjecture profiles selected by the criterion in the new game "coincides" with the set of conjecture profiles selected in the original game. If this is the case, we say that the criterion is *invariant* to the transformation under consideration. Invariance is a desirable property for a rationality criterion if one believes that the transformation does not affect the "strategic features" of the game.

We concentrate on three classes of transformations: *Thompson-Elmes-Reny transformations, Kohlberg-Mertens transformations* and *player splittings*. The first class of transformations has its origin in a paper by Thompson (1952), and has later been modified in Elmes and Reny (1994). The central result is that two games differ by an iterative application of Thompson-Elmes-Reny transformations if and only if they have the same *pure reduced normal form*. Consequently, a criterion which is invariant with respect to Thompson-Elmes-Reny transformations should only depend on the pure reduced normal form of a game. It turns out that most of the criteria considered up to this point, such as subgame perfect equilibrium, perfect equilibrium, quasi-perfect equilibrium and sequential equilibrium, are not invariant with respect to Thompson-Elmes-Reny transformations. We investigate two particular criteria which *are* invariant with respect to Thompson-Elmes-Reny transformations: proper equilibrium and normal form sequential equilibrium.

The class of Kohlberg-Mertens transformations is an extension of the first class, and it is shown that two games differ by an iterative application of Kohlberg-Mertens transformations if and only if they have the same *mixed reduced normal form*. Invariance with respect to Kohlberg-Mertens transformations turns out to be a very strong requirement, since there is no point-valued solution which is invariant with respect to Kohlberg-Mertens transformations and at the same time assigns to every game a nonempty subset of the set of subgame perfect equilibria. This negative result may be viewed as an additional motivation for considering set-valued solutions.

The class of player splittings, in contrast to the transformations considered above, leaves the structure of actions and information sets unchanged, but instead alters the set of players by assigning different information sets, previously controlled by the same player, to different players in the new game. For different rationality criteria we characterize the class of player splittings to which the criterion is invariant.

6.1 Thompson-Elmes-Reny Transformations

6.1.1 Transformations

The transformations to be considered in this section can be used to transform each extensive form game into some canonical type of game, called a *game in reduced form.* In such a game each player controls exactly one information set, every path from the root to some terminal node crosses each information set and a player's strategy is never equivalent to some other strategy. However, the reduced form of an extensive form game cannot serve as a substitute for the original game unless one believes that these transformations do not change the strategic features of the game. But then, it seems difficult, if not impossible, to objectively define what it means to "leave the strategic features of a game unchanged". Rather than entering this discussion, we simply leave it to the reader to judge whether the transformations, to be defined below, change the strategic features of the game. The main message of this section is then the following: *if* one believes that the Thompson-Elmes-Reny transformations do not change the strategic features of the game, then every extensive form game has an "equivalent" game in reduced form.

We now discuss the four Thompson-Elmes-Reny transformations which are used in this section: *interchange of decision nodes, coalescing of information sets, addition of decision nodes* and *addition of superfluous actions.* The first two have been defined by Thompson (1952), whereas the latter two are taken from Elmes and Reny (1994). Originally, Thompson (1952) proposed, in addition to the first two transformations, two other transformations, called the *inflate-deflate transformation* and *addition of a superfluous move.* A problem with the latter two is that they may transform a game *with* perfect recall into a game *without* perfect recall. In Elmes and Reny (1994) these transformations have been substituted by a single transformation, *addition of decision nodes,* which preserves perfect recall. Their paper shows that Thompson's original result, namely that a game can always be transformed into a game in reduced form by iterative application of Thompson's transformations, still holds if the *inflate-deflate transformation* and *addition of a superfluous move* is substituted by *addition of decision nodes.* It is this result which will be presented in this section. For the sake of convenience, we separate the original definition of *addition of decision nodes* into two different transformations: *addition of decision nodes* and *addition of superfluous actions.* Both transformations are special cases of *addition of decision nodes* as stated by Elmes and Reny.

Instead of providing mathematically rigorous definitions of the four transformations we confine ourselves to verbal descriptions, illustrated by some examples. For the formal definitions, the reader is referred to Elmes and Reny (1994). In this section, we restrict ourselves to extensive form games with perfect recall but without chance moves.

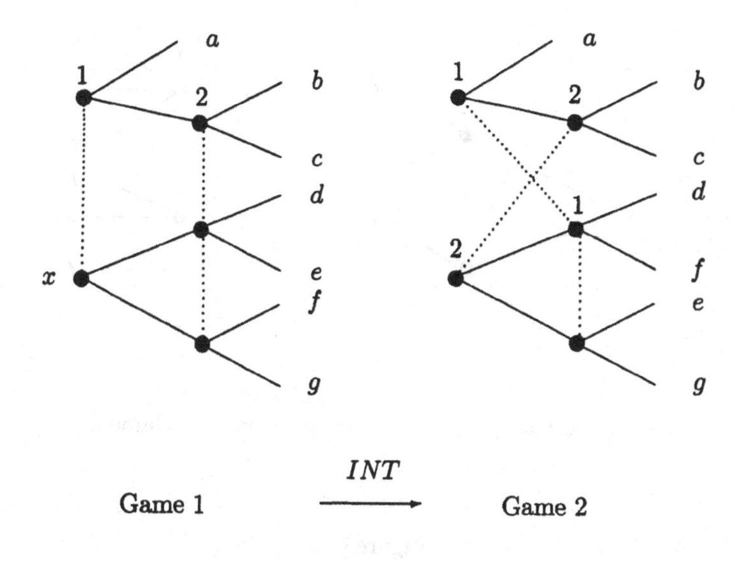

Figure 6.1

Transformation 1: *Interchange of decision nodes* (INT).
Interchange of decision nodes changes the order of play between two players, whenever the player who moves second does not observe the first mover's choice. Consider the example in Figure 6.1. The first picture is a fragment of some game, say game 1, and the second picture represents the corresponding fragment of the transformed game, which we call game 2. The symbols $a, b, ..., g$ represent continuations of the game, rather than terminal nodes. Game 2 is obtained from game 1 by changing the order of player 1 and 2 at node x. This transformation is called an interchange of decision nodes. Note that in the part of the game starting at node x, both player 1 and 2 have two possible choices; *up* and *down*. In the corresponding part of game 2, player 2 first chooses between *up* and *down*, and afterwards player 1 chooses between *up* and *down*. In order to keep the rest of the game unchanged, one should guarantee that in game 2 player 2, when choosing between *up* and *down*, has the same information as when choosing between *up* and *down* in game 1. The same should hold for player 1. The reader can check that the information sets in game 2 are constructed in such a way as to preserve the players' information.

Moreover, if in game 2 players 1 and 2 have both chosen between *up* and *down*, then the continuation of game 2 should coincide with the corresponding continuation in game 1. For instance, if at node x in game 2 player 2 first chooses *up*, and then player 1 chooses *down*, then the continuation is f, which corresponds to player 1 choosing *down* and player 2 choosing *up* in game 1.

In general, interchange of decision nodes may be described as follows. Consider a game, say game 1, an information set h^1 controlled by player i followed by some information set h^2 controlled by player j. Assume that at h^1 and h^2, the corresponding player can choose between two possible actions, say *up* and *down*. Choose a node x

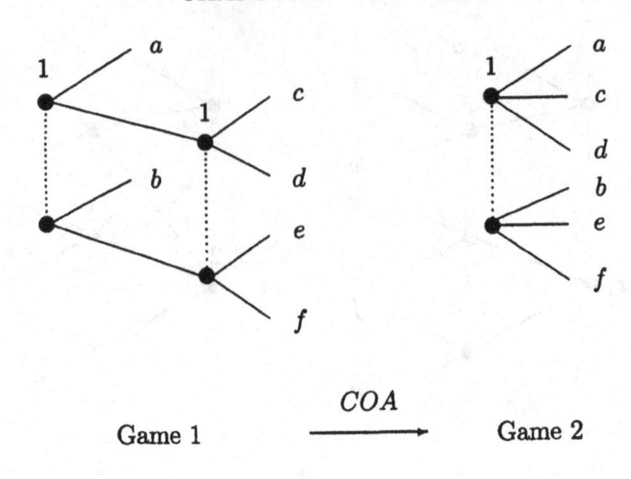

Figure 6.2

in h^1, and two nodes y_1, y_2 in h^2. Assume that y_1 comes directly after x and player i choosing *up*, and y_2 comes directly after x and player i choosing *down*. In the new game, say game 2, player j controls the node x and player i controls y_1 and y_2. The new information set for player j contains x plus the nodes in h^2 minus the nodes $\{y_1, y_2\}$. The new information set for player i contains $\{y_1, y_2\}$ plus the nodes in h^1 minus x. In the new game, the continuation after player j choosing *up* and player i choosing *down* is the same as the continuation in game 1 after player i choosing *down* and player j choosing *up*. The same for the other three combinations. For the parts of game 2 that do not follow node x, the continuations are the same as in game 1.

If game 1 and game 2 have these characteristics, then we say that game 2 is obtained from game 1 by *interchange of decision nodes.*

Transformation 2: *Coalescing of information sets* (*COA*).
Coalescing of information sets reduces two subsequent decisions made by the same player to a single decision. Consider the example in Figure 6.2. In game 1, player 1 controls two subsequent information sets. At the first information set player 1 has two actions: an action which leads to his second information set (*in*) and an action which does not (*out*). At the second information set, he can choose between *up* and *down*. In the new game, game 2, both information sets are reduced to a single information set, and player 1's subsequent decisions at both information sets in game 1 are reduced to a one-shot decision in the new game. More precisely, the subsequent choices *in* and *up* in game 1 are transformed into a one-shot choice (in, up) in game 2, and the continuation after (in, up) in game 2 coincides with the continuation after *in* and *up* in game 1. The same for *in* and *down*. The choice *out* in game 1 remains the same in game 2. Hence, in game 2 there are three possible choices for player 1: *out*, (in, up) and $(in, down)$.

Formally, consider a game, say game 1, and two information sets h^1, h^2 controlled

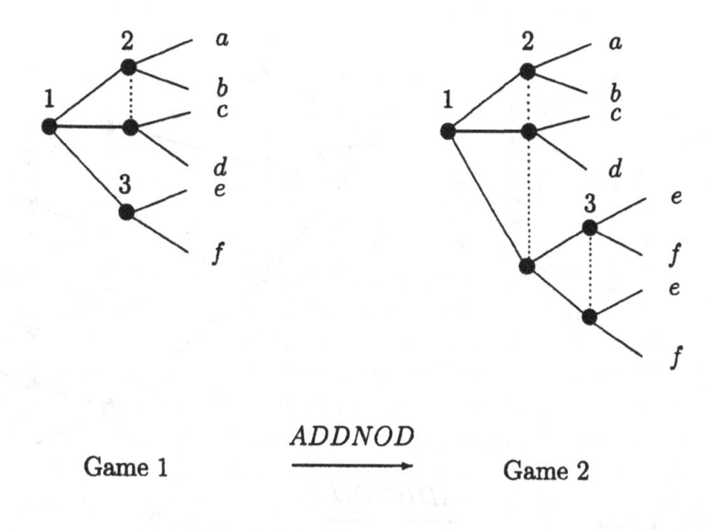

Figure 6.3

by the same player i. Suppose that h^1 and h^2 contain the same number of nodes and that h^2 directly follows h^1. By perfect recall, there is a unique action at h^1 which leads to h^2. Call this action *in*. Let $A(h^1)$ and $A(h^2)$ be the sets of actions available at h^1 and h^2. In the new game, say game 2, h^1 and h^2 are transformed into a single information set h which is a copy of h^1. The set of actions available at h is $A(h) = (A(h^1)\backslash\{in\}) \cup (\{in\} \times A(h^2))$. Every action in $A(h)$ can thus either be written as a^1 with $a^1 \in A(h^1)\backslash\{in\}$ or as (in, a^2) with $a^2 \in A(h^2)$. In game 2, if player i chooses an action a^1 with $a^1 \in A(h^1)\backslash\{in\}$, then the continuation should coincide with the continuation after a^1 in game 1. If in game 2 player i chooses an action (in, a^2) with $a^2 \in A(h^2)$, then the continuation should coincide with the continuation after the subsequent choices *in* and a^2 in game 1.

If game 1 and game 2 have these characteristics, then we say that game 2 is obtained from game 1 by *coalescing of information sets*.

Transformation 3: *Addition of decision nodes* (*ADDNOD*).
In order to explain this transformation, we need the following definition. Say that a player i information set h is *maximal* if for every strategy $s_i \in S_i(h)$ and every strategy profile s_{-i} of the other players, it holds that $s = (s_i, s_{-i})$ reaches h with probability one. Recall that $S_i(h)$ are those player i strategies that choose all player i actions leading to h. Hence, if the player i information set h is maximal, it can only be avoided by player i's own behavior at previous information sets.

Addition of decision nodes can be used in order to transform a non-maximal information set into a maximal information set. Consider the example in Figure 6.3. In game 1, player 2's information set is clearly non-maximal, since player 1 alone can avoid this information set by choosing his lower action. In game 2, player 2's information set is maximal, since it can no longer be avoided by any play of the game. In order to achieve maximality, we have added a decision node to player 2's

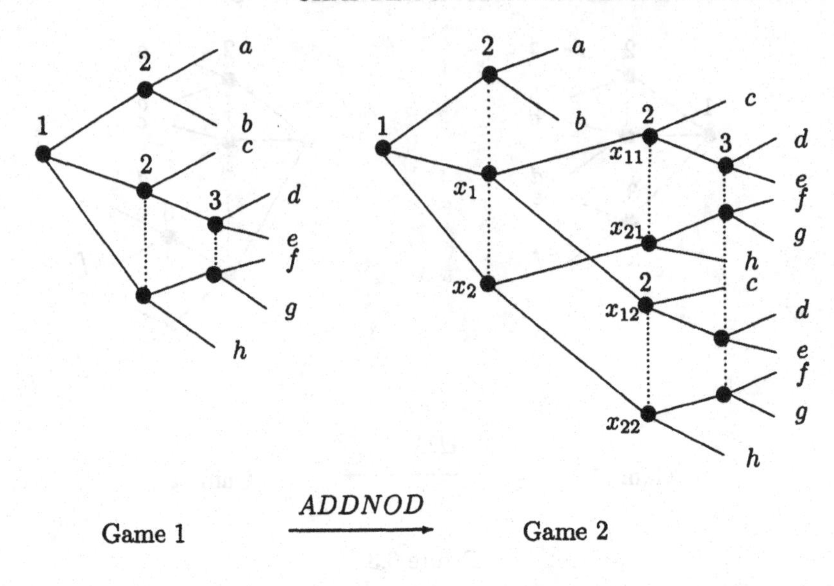

Figure 6.4

information set (the lower node in player 2's information set in game 2). Let us call this extra node x. Player 2's move at node x may be seen as a dummy move, since the continuation after x and player 2 choosing *up* is the same as the continuation after node x and player 2 choosing *down*, and hence player 2's move at x does not influence the rest of the game. Note that player 3 does not observe the outcome of this dummy move.

Things may become more complicated if the player, whose non-maximal information set in game 1 is transformed into a maximal one in game 2, controls some other information set in game 1. In this case, we have to make sure that this player *observes* his own dummy moves, in order to preserve perfect recall. In order to illustrate this complication, consider the example in Figure 6.4. Player 2's upper information set in game 1 is non-maximal. In game 2, it has been transformed into a maximal information set by adding the decision nodes x_1 and x_2 to this information set. Player 2's moves at the extra nodes x_1 and x_2 may be seen as dummy moves, since for the continuation of the game it does not matter whether player 2 chooses *up* or *down* at the nodes x_1 and x_2. Indeed, consider the part of the game which follows the nodes x_1 and x_2 and player 2 choosing *up* at these nodes. Then, one enters the part of the game which starts at nodes x_{11} and x_{21}. On the other hand, the part of the game that follows nodes x_1 and x_2 and player 2 choosing *down* there is the part which starts at nodes x_{12} and x_{22}. Clearly, the part of the game starting at $\{x_{11}, x_{21}\}$ is an exact copy of the part starting at $\{x_{12}, x_{22}\}$. In order to preserve perfect recall, player 2 has to observe his own dummy move at nodes x_1 and x_2. As such, player 2 should know whether he is in the part of the game starting at $\{x_{11}, x_{21}\}$ or the part of the game starting at $\{x_{12}, x_{22}\}$, and hence $\{x_{11}, x_{21}\}$ and $\{x_{12}, x_{22}\}$ are two different information sets for player 2. Player 3, on the other hand, does not observe player 2's dummy move at x_1 and x_2, and hence his information set has been duplicated when moving from game 1 to game 2. We say that game 2 is obtained from game 1 by addition of decision nodes.

In general, consider a game, say game 1, and a player i information set h with nodes $y_1, ..., y_K$ at which this player has two possible actions, *up* and *down*. By perfect recall, there is a unique sequence $a_1, ..., a_L$ of player i actions leading to h. Choose some nodes $\tilde{x}_1, \tilde{x}_2, ..., \tilde{x}_M$ not in h, which may be terminal or non-terminal nodes, with the restriction that every path from the root to some $\tilde{x}_m$ contains the player i actions $a_1, ..., a_L$. In the new game, say game 2, we add nodes $x_1, x_2, ..., x_M$ to the information set h, resulting in a new, larger information set $h^* = \{y_1, ..., y_K, x_1, ..., x_M\}$. At this new information set, player i still has available the choices *up* and *down*. The continuation of game 2 at some node $y \in \{y_1, ..., y_K\}$ coincides with the corresponding continuation in game 1. If the game is at some added node $x_m \in \{x_1, ..., x_M\}$ and player i chooses *up* here, the game moves to a newly added node x_{m1} which is controlled by the player who moves at $\tilde{x}_m$ in game 1. The continuation of game 2 at x_{m1} coincides with the continuation of game 1 at $\tilde{x}_m$. If game 2 is at the node $x_m \in \{x_1, ..., x_M\}$ and player i chooses *down* here, the game moves to a newly added node x_{m2} which is controlled by the same player who moves at $\tilde{x}_m$ in game 1. Again, the continuation of game 2 at node x_{m2} coincides with the continuation of game 1 at node $\tilde{x}_m$. Hence, x_{m1} and x_{m2} are always controlled by the same player, and the continuations of game 2 at nodes x_{m1} and x_{m2} are identical since they both are copies of the continuation of game 1 at $\tilde{x}_m$. Player i's moves at the added nodes $x_1, ..., x_M$ may thus be seen as dummy moves, since they have no influence on the continuation of the game.

Player i, in the continuation of game 2 after the nodes $\{x_1, ..., x_M\}$, observes the dummy move he made at these nodes, and can thus distinguish between nodes that are preceded by some node x_{m1} (that is, preceded by his dummy move *up*) and nodes that are preceded by some node x_{m2} (that is, preceded by his dummy move *down*). Consequently, in the continuation of game 2 after the added nodes $\{x_1, ..., x_M\}$ player 2 has twice as much information sets as in the continuation of game 1 after the nodes $\{\tilde{x}_1, ..., \tilde{x}_M\}$. More precisely, each of these information sets $\tilde{h}$ in game 1 has been duplicated in game 2 into two identical information sets $\tilde{h}^1$ and $\tilde{h}^2$: one following his dummy move *up* and one following his dummy move *down*.

A player $j \neq i$ does not observe the dummy move chosen by player i at the nodes $\{x_1, ..., x_M\}$. Hence, in the continuation of game 2 after the added nodes $\{x_1, ..., x_M\}$, player j has the same number of information sets as in the continuation of game 1 after $\{\tilde{x}_1, ..., \tilde{x}_M\}$. More precisely, every such player j information set $\tilde{h}$ in game 1 is transformed into a larger information set h' in game 2 in which the number of nodes has been duplicated. Every node y in $\tilde{h}$ is mapped to two nodes y_1, y_2 in h', where y_1 is preceded by player i's dummy move *up*, and y_2 is preceded by player i's dummy move *down*.

If game 1 and game 2 have these characteristics, then we say that game 2 is obtained from game 1 by *addition of decision nodes*.

It can easily be seen that a non-maximal information set can always be transformed into a maximal information set by applying *ADDNOD*. Namely, let h be a

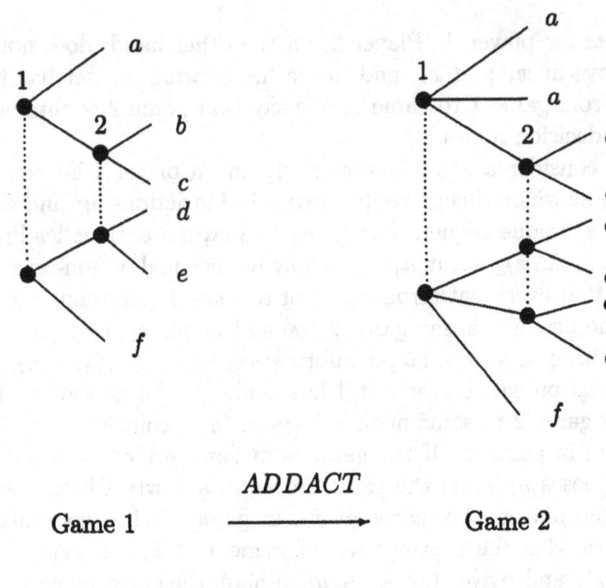

Figure 6.5

non-maximal information set controlled by player i. Then, we can find some nodes $\tilde{x}_1, ..., \tilde{x}_M$ not in h, such that (1) each of these nodes lies on a different path, (2) each path leading to $\tilde{x}_m$ contains the player i actions $a_1, ..., a_L$ leading to h and (3) for every strategy $s_i \in S_i(h)$ and every strategy profile s_{-i} we have that (s_i, s_{-i}) either reaches h or one of the nodes in $\{\tilde{x}_1, ..., \tilde{x}_M\}$. By proceding as above, the information set h is then extended to a larger information set h^*, by adding nodes $x_1, ..., x_M$, such that for every $s_i \in S_i(h)$ and every s_{-i} it holds that (s_i, s_{-i}) reaches h^*, and hence h^* is maximal.

Transformation 4: *Addition of superfluous actions* (*ADDACT*).
Addition of superfluous actions inserts an extra action at an information set which is an exact copy of some other action at this information set. Consider the games in Figure 6.5. Game 2 is obtained from game 1 by inserting an extra action for player 1 which is a duplicate of player 1's upper action in game 1.

Formally, consider a game, say game 1, and an information set h controlled by player i in this game. Let $A^1(h)$ be the actions available at h. In the new game, say game 2, the set of actions available at h is given by $A^2(h) = A^1(h) \cup \{a^{2*}\}$. For every node $x \in h$ and action $a^1 \in A^1(h)$, let $y^1(x, a^1)$ be the node in game 1 which immediately follows x and a^1. For every $x \in h$ and $a^2 \in A^2(h)$, let $y^2(x, a^2)$ be the node which in game 2 immediately follows x and a^2. Suppose that there is some action $a^{1*} \in A^1(h)$ such that the following holds: (1) for every $x \in h$, the continuation in game 2 after node $y^2(x, a^{2*})$ coincides with the continuation in game 1 after node $y^1(x, a^{1*})$, (2) if a player $j \neq i$ could not distinguish between some node $y^1(x, a^{1*})$ and a node y in game 1, then player j cannot distinguish between $y^2(x, a^{2*})$ and y in game 2.

If games 1 and 2 have these characteristics, we say that game 2 is obtained from game 1 by *addition of a superfluous action* (in this case the action a^{2*}, which is a copy of a^{1*}).

For each of the four transformations above we define the corresponding *inverse transformation* in the obvious way. For instance, if game 2 is obtained from game 1 by the transformation $ADDACT$, then we say that game 1 is obtained from game 2 by the inverse transformation $ADDACT^{-1}$. In the same way, we define the inverse transformations INT^{-1}, COA^{-1} and $ADDNOD^{-1}$.

Definition 6.1.1 *The transformations $INT, COA, ADDNOD$ and $ADDACT$ and their inverses are called Thompson-Elmes-Reny transformations.*

In the following subsection it is shown that every extensive form game without chance moves can be transformed into a game in reduced form by iterative application of Thompson-Elmes-Reny transformations.

6.1.2 Games in Reduced Form

Before stating the result, we first provide a formal definition of games in reduced form. We say that a strategy s_i is *equivalent* to some other strategy s_i' if $u_j(s_i, s_{-i}) = u_j(s_i', s_{-i})$ for every player j and every strategy profile s_{-i}. We may now define games in reduced form as follows, using the notion of maximal information sets.

Definition 6.1.2 *An extensive form game Γ without chance moves is said to be in reduced form if: (1) every player has exactly one information set, (2) every information set is maximal and (3) no two different strategies controlled by the same player are equivalent.*

In particular, a game in reduced form has the property that every path from the root to some terminal node crosses each information set. This follows from the fact that every player has one information set, and every information set is maximal. In their paper, Elmes and Reny (1994) have proved the following result.

Theorem 6.1.3 *Let Γ be an extensive form game without chance moves. Then, there is a game in reduced form which can be obtained from Γ by iterative application of Thompson-Elmes-Reny transformations.*

Before sketching the proof for this theorem, we illustrate the proof by an example which demonstrates how and in which order the Thompson-Elmes-Reny transformation can be applied to transform a game into a game in reduced form. When discussing the proof, the reader may use this example as a reference point. Consider the example in Figure 6.6. Denote the five games by game 1, game 2, ..., game 5. The symbols $a, b, ..., g$ now represent utility pairs at the terminal nodes. We assume that the utility pairs $a, ..., g$ are all different. By applying COA^{-1} to game 1, we make sure that every information set contains exactly two actions. Note that player 1's information set at the beginning of game 1 has been separated into two consecutive

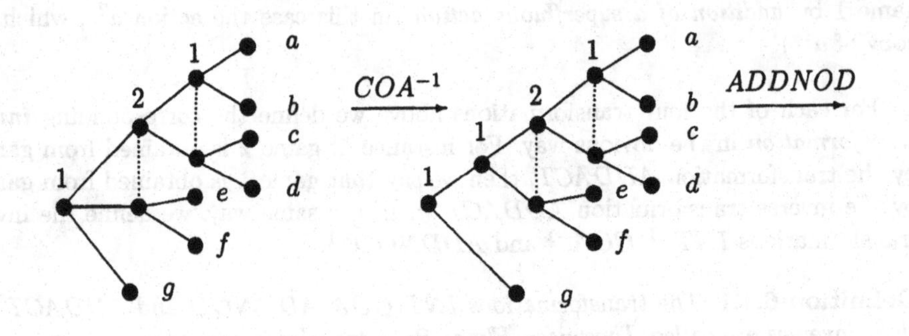
1
2
1
a
b
c
d
e
f
g
COA^{-1}
1
2
1
1
a
b
c
d
e
f
g
ADDNOD

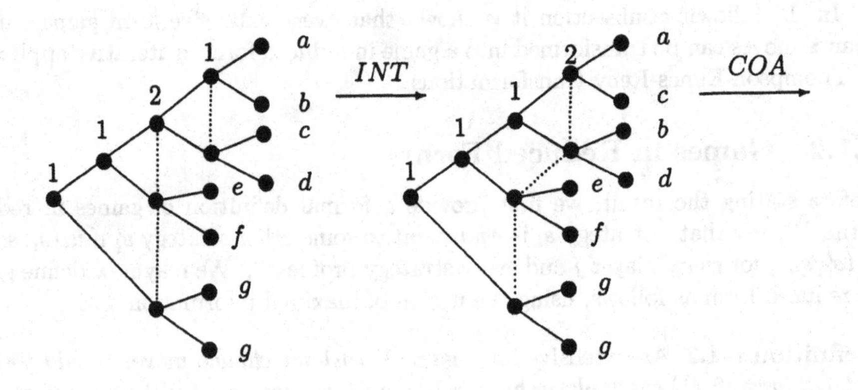
1
2
1
1
a
b
c
d
e
f
g
g
INT
2
1
1
1
a
c
b
d
e
f
g
g
COA

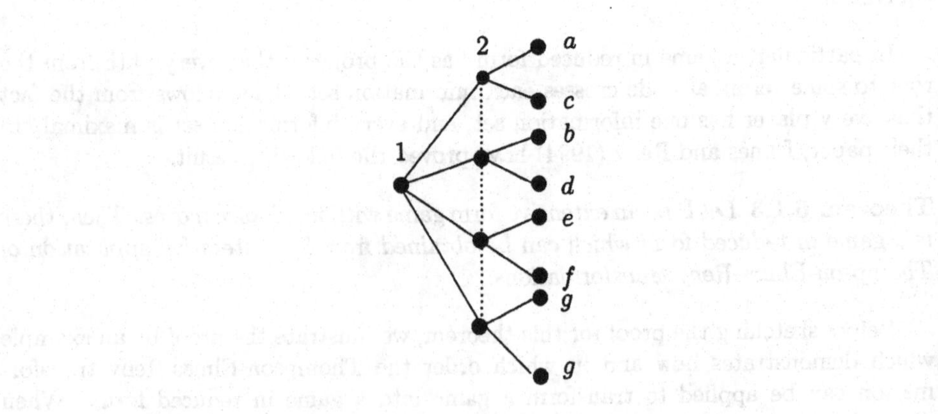
2
1
a
c
b
d
e
f
g
g

Figure 6.6

information sets in game 2. By applying $ADDNOD$ to game 2 we transform player 2's non-maximal information set in game 2 into a maximal information set in game 3. Next, we use INT to change the order of player 1 and player 2 in that part of the game that starts at player 2's upper node in game 3. We finally apply COA to reduce player 1's subsequent information sets into a single information set. Game 5 is a game in reduced form.

We now provide a sketch of the proof. For a formal proof, the reader is referred to Elmes and Reny (1994).

Sketch of proof of Theorem 6.1.3. First of all, one can show that each of the Thompson-Elmes-Reny transformations preserves perfect recall, that is, a game with perfect recall is transformed into a game with perfect recall. Now, consider an arbitrary extensive form game Γ_1 without chance moves.
Step 1. By applying COA^{-1} finitely often, Γ_1 can be transformed into a game Γ_2 in which the number of actions at each information set is two.
Step 2. By applying $ADDNOD$ finitely often, Γ_2 can be transformed into a game Γ_3 in which every information set is maximal. The idea is the following. For a given player i, choose, if possible, a non-maximal information set h which is not preceded by any other of his non-maximal information sets. By applying $ADDNOD$, transform h into a maximal information set. One can show that maximal information sets in the original game remain maximal in the new game. In the new game, choose again, if possible, a non-maximal information set for player i which is not preceded by any other non-maximal player i information set, and so on. It can be shown that this procedure stops after finitely many applications of $ADDNOD$, and the resulting game is a game in which every player i information set is maximal. By applying this procedure for each player i, we end up with a game Γ_3 in which all information sets are maximal.
Step 3. By applying COA and INT finitely many times, we can transform Γ_3 into a game Γ_4 in which every player has exactly one information set, which is maximal. In order to see this, note that the transformations COA and INT preserve the maximality of information sets. The maximal information sets in Γ_3 can thus be rearranged by INT such that all information sets controlled by the same player are never separated by an information set of some other player. Finally, merge all consecutive information sets controlled by the same player to a single information set by COA.
Step 4. By applying $ADDACT^{-1}$ finitely often, Γ_4 can be transformed into a game Γ_5 in which no two strategies controlled by the same player are equivalent. The game Γ_5 is then, by construction, a game in reduced form. ■

6.1.3 Pure Reduced Normal Form

We now characterize, for a given extensive form game, those games that can be obtained from it by an iterative application of Thompson-Elmes-Reny transformations. More precisely, it is shown that game 2 can be obtained from game 1 by an iterative application of Thompson-Elmes-Reny transformations if and only if game 1 and game 2 have the same *pure reduced normal form.* If one believes that Thompson-Elmes-Reny transformation do not change the strategic features of a game, then those classes of games with the same pure reduced normal form may thus be viewed as "equivalence classes".

The pure reduced normal form is formally defined as follows. Let Γ be an extensive form game without chance moves. For every player i and every strategy $s_i \in S_i$ let $[s_i]$ be the class of player i strategies that are equivalent to s_i, and let $S_i^* = \{[s_i] | s_i \in S_i\}$ be the set of equivalence classes in S_i. For every player i, define the utility function $u_i^* : \times_{j\in I} S_j^* \to \mathbb{R}$ by $u_i^*(([s_j])_{j\in I}) = u_i((s_j)_{j\in I})$ for every $([s_j])_{j\in I} \in \times_{j\in I} S_j^*$. The utility function u_i^* is thus the projection of u_i on the equivalence classes of strategies.

Definition 6.1.4 *Let Γ be an extensive form game without chance moves, and for every player i let S_i^* and u_i^* be defined as above. Then, the pair $((S_i^*)_{i\in I}, (u_i^*)_{i\in I})$ is called the pure reduced normal form of Γ.*

The following definition clarifies what we mean by saying that "two games have the same pure reduced normal form".

Definition 6.1.5 *Let Γ_1 and Γ_2 be two extensive form games with the same set of players and without chance moves. Let $((S_i^{1*})_{i\in I}, (u_i^{1*})_{i\in I})$ and $((S_i^{2*})_{i\in I}, (u_i^{2*})_{i\in I})$ be the pure reduced normal forms of Γ_1 and Γ_2, respectively. We say that Γ_1 and Γ_2 have the same pure reduced normal form if for every player i there exists a bijective mapping $f_i : S_i^{1*} \to S_i^{2*}$ such that for $f = \times_{i\in I} f_i$ we have that*

$$u_j^{2*}(f([s])) = u_j^{1*}([s])$$

for every player j and every $[s] \in \times_{i\in I} S_i^{1*}$.

Here, the mapping $f : \times_{i\in I} S_i^{1*} \to \times_{i\in I} S_i^{2*}$ is formally defined as follows: for every $[s] = ([s_i])_{i\in I} \in \times_{i\in I} S_i^{1*}$ we have $f([s]) = (f_i([s_i]))_{i\in I}$. In words, two games are said to have the same pure reduced normal form if the pure reduced normal forms coincide after some appropriate relabeling of the equivalence classes of strategies. In Dalkey (1953), two games having the same pure reduced normal form are called *equivalent*.

Theorem 6.1.6 *Let Γ_1 and Γ_2 be two extensive form games without chance moves. Then, Γ_1 and Γ_2 differ by an iterative application of Thompson-Elmes-Reny transformations if and only if Γ_1 and Γ_2 have the same pure reduced normal form.*

Proof. Suppose first that Γ_2 can be obtained from Γ_1 by an iterative application of Thompson-Elmes-Reny transformations. The reader may verify that each of the Thompson-Elmes-Reny transformations preserves the pure reduced normal form. Formally, it may be checked that if Γ' is obtained from Γ by applying some Thompson-Elmes-Reny transformation, then Γ and Γ' have the same pure reduced normal form. But then, it is clear that Γ_1 and Γ_2 have the same pure reduced normal form.

Now, suppose that Γ_1 and Γ_2 have the same pure reduced normal form. By Theorem 6.1.3, we know that there exist extensive form games Γ_1^* and Γ_2^* in reduced form such that Γ_1^* is obtained from Γ_1 and Γ_2^* is obtained from Γ_2 by Thompson-Elmes-Reny transformations. Since Thompson-Elmes-Reny transformations preserve the pure reduced normal form, it follows that Γ_1^* and Γ_2^* have the same pure reduced normal form. But then, since Γ_1^* and Γ_2^* are both in reduced form, and have the same pure reduced normal form, it may be verified that Γ_2^* can be obtained from Γ_1^* by iteratively applying INT and appropriately relabeling the decision nodes in the game. We now use the convention that two extensive form games are considered *the same* if one can be mapped into the other by a relabeling of the decision nodes. Using this convention, we may thus conclude that Γ_2^* can be obtained from Γ_1^* by Thompson-Elmes-Reny transformations. Hence, Γ_1^* can be obtained from Γ_1, Γ_2^* can be obtained from Γ_1^* and Γ_2 can be obtained from Γ_2^*, and thus Γ_2 can be obtained from Γ_1 by iterative application of Thompson-Elmes-Reny transformations. ■

6.2 Thompson-Elmes-Reny Invariance

We next turn to the question how solutions react to Thompson-Elmes-Reny transformations of a game. Formally, a *(point-valued) solution* is a correspondence φ which assigns to every extensive form game Γ a set $\varphi(\Gamma)$ of behavioral conjecture profiles. For the sake of convenience we do not explictly incorporate beliefs in the solution. However, if we say, for instance, that σ is a sequential equilibrium, then we mean that there is some belief system β such that (σ, β) is a sequential equilibrium. We use the word point-valued in order to distinguish such solutions from *set-valued* solutions assigning to every game a collection of sets of behavioral conjecture profiles. An example of a set-valued solution is the correspondence assigning to every game the collection of stable sets of equilibria (see Section 5.3). If one believes that Thompson-Elmes-Reny transformations do not change the strategic features of a game, then a necessary requirement for an extensive form solution would be the following: if the extensive form games Γ_1 and Γ_2 differ only by an iterative application of Thompson-Elmes-Reny transformations, then the solution applied to Γ_1 should be "equivalent" to the solution in Γ_2. It is not obvious, however, how to formally define that the solutions of both games should be "equivalent", since both games in general have different spaces of behavioral conjectures.

As a possible solution to this problem, we adopt the following approach. We contruct for every player i a correspondence τ_i which maps behavioral conjectures σ_i in Γ_1 to collections $\tau_i(\sigma_i)$ of "equivalent" behavioral conjectures in Γ_2. Below, we formally explain what we mean by "equivalent". These transformations τ_i allow us to compare behavioral conjecture profiles in Γ_1 with those in Γ_2.

We construct the transformations τ_i as follows. Let Γ_1 and Γ_2 be two extensive form games without chance moves, and suppose that Γ_2 can be obtained from Γ_1 by Thompson-Elmes-Reny transformations. From Theorem 6.1.6 we know that Γ_1 and Γ_2 have the same pure reduced normal form. Let $\Gamma_1^* = ((S_i^{1*})_{i\in I}, (u_i^{1*})_{i\in I})$ and $\Gamma_2^* = ((S_i^{2*})_{i\in I}, (u_i^{2*})_{i\in I})$ be the pure reduced normal forms of Γ_1 and Γ_2. Then, by definition, there exist bijective mappings $f_i : S_i^{1*} \to S_i^{2*}$ for every i such that for $f = \times_{i\in I} f_i$ we have that $u_j^{1*} = u_j^{2*} \circ f$ for every j.

For every player i, the mapping f_i can be extended to a mapping $F_i : \Delta(S_i^{1*}) \to \Delta(S_i^{2*})$ by letting, for every $\mu_i^{1*} \in \Delta(S_i^{1*})$,

$$F_i(\mu_i^{1*}) = \mu_i^{2*} \in \Delta(S_i^{2*}) \text{ with } \mu_i^{2*}([s_i^{2*}]) = \mu_i^{1*}(f_i^{-1}([s_i^{2*}]))$$

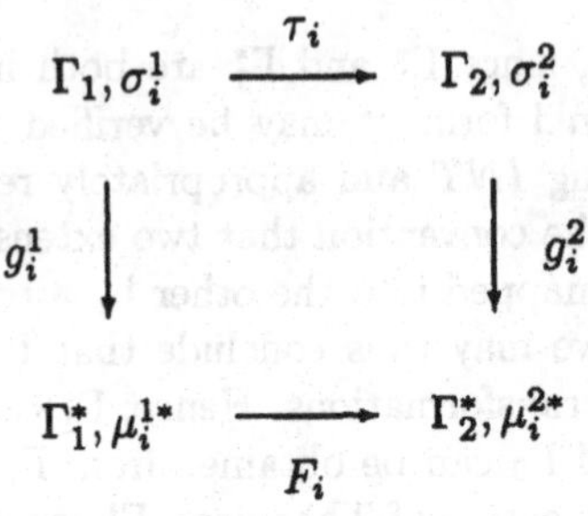

$$
\begin{array}{ccc}
\Gamma_1, \sigma_i^1 & \xrightarrow{\tau_i} & \Gamma_2, \sigma_i^2 \\
\Big\downarrow g_i^1 & & \Big\downarrow g_i^2 \\
\Gamma_1^*, \mu_i^{1*} & \xrightarrow[F_i]{} & \Gamma_2^*, \mu_i^{2*}
\end{array}
$$

Figure 6.7

for every $[s_i^{2*}] \in S_i^{2*}$. Hence, F_i transforms mixed conjectures in Γ_1^* into mixed conjectures in Γ_2^*.

Now, we construct a mapping g_i^1 transforming behavioral conjectures in Γ_1 into behavioral conjectures in the pure reduced normal form Γ_1^*. For every behavioral conjecture σ_i^1 about player i in Γ_1, let μ_i^1 be the mixed conjecture in Γ_1 induced by σ_i, that is,

$$\mu_i^1(s_i^1) = \prod_{h \in H_i(s_i)} \sigma_{ih}^1(s_i^1(h))$$

for every strategy s_i^1 in Γ_1. Now, let $g_i^1(\sigma_i^1)$ be the mixed conjecture μ_i^{1*} in Γ_1^* given by

$$\mu_i^{1*}([s_i^{1*}]) = \sum_{s_i^1 \in [s_i^{1*}]} \mu_i^1(s_i^1)$$

for every equivalence class $[s_i^{1*}] \in S_i^{1*}$. Similarly, we construct a mapping g_i^2 transforming behavioral conjectures in Γ_2 into mixed conjectures in the pure reduced normal form Γ_2^*.

Now, we define the correspondence τ_i mapping behavioral conjectures in Γ_1 to sets of behavioral conjectures in Γ_2. For every behavioral conjecture σ_i^1 in Γ_1, let

$$\tau_i(\sigma_i^1) = \{\sigma_i^2 | \; g_i^2(\sigma_i^2) = F_i(g_i^1(\sigma_i^1))\}.$$

The construction of τ_i is summarized in the table of Figure 6.7. Intuitively, $\tau_i(\sigma_i^1)$ may be interpreted as those behavioral conjectures in Γ_2 that induce the same mixed conjectures on the pure reduced normal form as σ_i^1. Finally, let $\tau = (\tau_i)_{i \in I}$ be the correspondence mapping every behavioral conjecture profile $\sigma^1 = (\sigma_i^1)_{i \in I}$ in Γ_1 to the set $\tau(\sigma^1)$ of behavioral conjecture profiles in Γ_2 given by $\tau(\sigma^1) = \times_{i \in I} \tau_i(\sigma_i^1)$. By construction, the correspondence τ preserves expected utilities, that is, $u_i^1(\sigma^1) = u_i^2(\sigma^2)$ for every $\sigma^2 \in \tau(\sigma^1)$ and every player i, where u_i^1 and u_i^2 denote the expected utilities in Γ_1 and Γ_2 respectively.

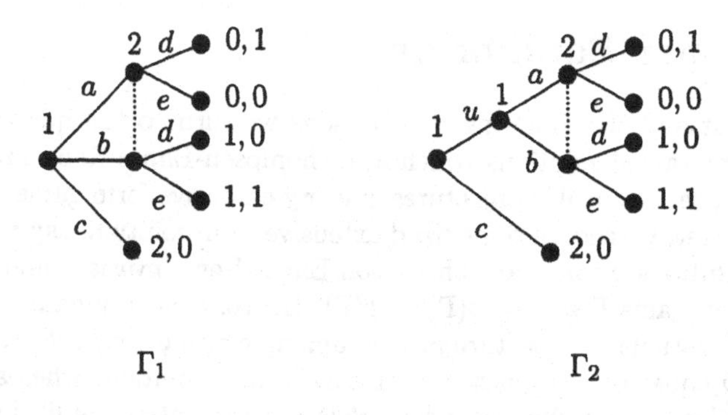

Figure 6.8

Definition 6.2.1 *A (point-valued) solution φ is said to be Thompson-Elmes-Reny invariant if for every pair Γ_1, Γ_2 of extensive form games without chance moves that differ solely by Thompson-Elmes-Reny transformations, the following holds: if $\sigma^1 \in \varphi(\Gamma_1)$ then there is some $\sigma^2 \in \tau(\sigma^1)$ with $\sigma^2 \in \varphi(\Gamma_2)$. Here, τ is the transformation defined above.*

If Γ_1 and Γ_2 differ by Thompson-Elmes-Reny transformations, we may also find a correspondence $\tilde{\tau}$ mapping behavioral conjecture profiles in Γ_2 to sets of "equivalent" behavioral conjecture profiles in Γ_1. Hence, Thompson-Elmes-Reny invariance implies, in addition to the property above, the following: if $\sigma^2 \in \varphi(\Gamma_2)$ then there is some $\sigma^1 \in \tilde{\tau}(\sigma^2)$ with $\sigma^1 \in \varphi(\Gamma_1)$. The property in the definition should thus hold for both directions.

We conclude this section by showing that the subgame perfect equilibrium solution, perfect equilibrium solution, quasi-perfect equilibrium solution and sequential equilibrium solution are not Thompson-Elmes-Reny invariant. By the subgame perfect equilibrium solution we mean the correspondence which assigns to every game the set of subgame perfect equilibria. Similarly for the other solutions. Consider the games Γ_1 and Γ_2 in Figure 6.8. The game Γ_2 can be obtained from Γ_1 by Thompson-Elmes-Reny transformations since both games have the same pure reduced normal form. In fact, Γ_2 can be obtained from Γ_1 by a single application of COA^{-1}. The conjecture profile (c, d) in Γ_1 is a subgame perfect equilibrium, perfect equilibrium, quasi-perfect equilibrium, and can be extended to a sequential equilibrium. However, there is no subgame perfect equilibrium in Γ_2 in which player 2 is believed to choose d, since the unique Nash equilibrium in the subgame following player 1's move u is (b, e). Consequently, there is no perfect equilibrium, quasi-perfect equilibrium nor sequential equilibrium in Γ_2 in which player 2 is believed to play d. It thus follows that these four solutions are not Thompson-Elmes-Reny invariant. In the following two sections we investigate solutions which *are* Thompson-Elmes-Reny invariant.

6.3 Proper Equilibrium

In the present and the following two sections we turn to the question whether we can find point-valued solutions which are Thompson-Elmes-Reny invariant, and, in addition, induce "rational" conjectures in every extensive form game. In order to formalize the latter, we choose some fixed extensive form solution, say ψ, and search for nonempty solutions φ which are Thompson-Elmes-Reny invariant and which for every extensive form game Γ satisfy $\varphi(\Gamma) \subseteq \psi(\Gamma)$. Hence, for every game Γ we view $\psi(\Gamma)$ as the set of "rational" conjecture profiles among which φ may select. In this section, we choose ψ equal to the quasi-perfect equilibrium solution, whereas the following two sections investigate the case where ψ is the sequential equilibrium solution. Of course, one could continue by choosing other solutions ψ, but we restrict ourselves to the two cases mentioned above.

The first question we wish to answer in this section is whether we can find a solution which is Thompson-Elmes-Reny invariant and consists of solely quasi-perfect equilibria in every extensive form game. In the example in Figure 6.8 we have seen that the quasi-perfect equilibrium solution itself is not Thompson-Elmes-Reny invariant. Hence, every refinement of the quasi-perfect equilibrium solution which is Thompson-Elmes-Reny invariant must be a *strict* refinement of quasi-perfect equilibrium. The following theorem, which is based on Theorem 3.6.4, states that the proper equilibrium solution, introduced in Section 3.6, *is* a Thompson-Elmes-Reny invariant refinement of quasi-perfect equilibrium.

Theorem 6.3.1 *The proper equilibrium solution is Thompson-Elmes-Reny invariant and assigns to every game a nonempty subset of the set of quasi-perfect equilibria.*

Proof. Let φ be the proper equilibrium solution. From Theorem 3.6.4 we know that every proper equilibrium is a quasi-perfect equilibrium, and hence it only remains to show that φ is Thompson-Elmes-Reny invariant. Let Γ_1, Γ_2 be two extensive form games without chance moves such that Γ_2 can be obtained from Γ_1 by Thompson-Elmes-Reny transformations. Let $\Gamma_1^* = ((S_i^{1*})_{i\in I}, (u_i^{1*})_{i\in I})$ and $\Gamma_2^* = ((S_i^{2*})_{i\in I}, (u_i^{2*})_{i\in I})$ be the corresponding pure reduced normal forms. Conform Section 6.2, there is some bijective function $F = \times_{i\in I} F_i$ mapping a mixed conjecture profile μ^{1*} in Γ_1^* to some "equivalent" mixed conjecture profile $\mu^{2*} = F(\mu^{1*})$ in Γ_2^*. Let $g^1 = (g_i^1)_{i\in I}$ be the function mapping a behavioral conjecture profile σ^1 in Γ_1 to an "equivalent" mixed conjecture profile $\mu^{1*} = g^1(\sigma^1)$ in Γ^{1*}. Similar for g^2. Let τ be the correspondence mapping behavioral conjecture profiles σ^1 in Γ_1 to a set of "equivalent" behavioral conjecture profiles $\tau(\sigma^1)$ in Γ_2.

Let σ^1 be a proper equilibrium in Γ_1. We show that there is some $\sigma^2 \in \tau(\sigma^1)$ which is a proper equilibrium in Γ_2. Since σ^1 is a proper equilibrium, there is a proper sequence $(\sigma^{1n})_{n\in\mathbf{N}}$ of strictly positive behavioral conjecture profiles converges to σ^1. Let $(\mu^{1n})_{n\in\mathbf{N}}$ be the sequence of induced mixed conjecture profiles. Then, the sequence $(\mu^{1n})_{n\in\mathbf{N}}$ is proper, by which we mean that μ^{1n} is strictly positive for all n, and $u_i(s_i, \mu_{-i}^{1n}) < u_i(t_i, \mu_{-i}^{1n})$ implies $\lim_{n\to\infty} \mu_i^{1n}(s_i)/\mu_i^{1n}(t_i) = 0$. Consider the sequence $(\mu^{1*n})_{n\in\mathbf{N}} = (g^1(\sigma^{1n}))_{n\in\mathbf{N}}$ of mixed conjecture profiles in the pure reduced normal form Γ_1^*. Clearly, every μ^{1*n} is strictly positive. Now, suppose that $u_i^*([s_i^*], \mu_{-i}^{1*n}) <$

$u_i^*([t_i^*], \mu_{-i}^{1*n})$ for some n, some player i and some $[s_i^*], [t_i^*] \in S_i^*$. Then, for every $s_i \in [s_i^*]$ and every $t_i \in [t_i^*]$ we have that $u_i(s_i, \mu_{-i}^{1n}) < u_i(t_i, \mu_{-i}^{1n})$. Since the sequence $(\mu^{1n})_{n\in\mathbf{N}}$ is proper, we have that $\lim_{n\to\infty} \mu_i^{1n}(s_i)/\mu_i^{1n}(t_i) = 0$. This holds for all $s_i \in [s_i^*], t_i \in [t_i^*]$, and hence $\lim_{n\to\infty} \mu_i^{1n*}([s_i^*])/\mu_i^{1n*}([t_i^*]) = 0$. We may thus conclude that the sequence $(\mu^{1*n})_{n\in\mathbf{N}}$ is proper in Γ_1^*. It directly follows that the corresponding sequence $(\mu^{2*n})_{n\in\mathbf{N}} = (F(\mu^{1*n}))_{n\in\mathbf{N}}$ is proper in Γ_2^*. Let $(\mu^{2n})_{n\in\mathbf{N}}$ be the sequence of strictly positive mixed conjecture profiles in Γ_2 such that for every n, every player i, every $[s_i^{2*}] \in S_i^{2*}$ and every $s_i^2 \in [s_i^{2*}]$ we have that

$$\mu_i^{2n}(s_i^2) = \frac{\mu_i^{2*n}([s_i^{2*}])}{||[s_i^{2*}]||},$$

where $||[s_i^{2*}]||$ is the cardinality of the equivalence class $[s_i^{2*}]$. Then, it can easily be verified that the sequence $(\mu^{2n})_{n\in\mathbf{N}}$ is proper. For every n, let σ^{2n} be the behavioral conjecture profile in Γ_2 induced by μ^{2n}. Without loss of generality, we may assume that $(\sigma^{2n})_{n\in\mathbf{N}}$ converges to some behavioral conjecture profile σ^2 in Γ_2. Then, by construction, σ^2 is a proper equilibrium in Γ_2. The behavioral conjecture profiles σ^{2n} have been designed in such a way that $F(g^1(\sigma^{1n})) = F(\mu^{1*n}) = \mu^{2*n} = g^2(\sigma^{2n})$ for every n, and hence $F(g^1(\sigma^1)) = \lim_{n\to\infty} F(g^1(\sigma^{1n})) = \lim_{n\to\infty} g^2(\sigma^{2n}) = g^2(\sigma^2)$. But then, by definition, $\sigma^2 \in \tau(\sigma^1)$. Hence, σ^2 is a proper equilibrium in Γ_2 with $\sigma^2 \in \tau(\sigma^1)$. This completes the proof. ■

The result above raises the following question: Is it true that a Thompson-Elmes-Reny invariant refinement of quasi-perfect equilibrium always solely consists of proper equilibria? The answer is *no*: for a given game Γ there may exist behavioral conjecture profiles σ that are not proper equilibria, but which induce a quasi-perfect equilibrium in every game Γ_2 that can be obtained from Γ by Thompson-Elmes-Reny transformations. An example may be found in Hillas (1996). However, the answer to the question above would be *yes* if, in addition to the conjecture profile σ, we would also fix the sequence $(\sigma^n)_{n\in\mathbf{N}}$ of strictly positive behavioral conjecture profiles "justifying" the properness or quasi-perfectness of σ. Let us be more explicit about this.

Let $(\sigma, (\sigma^n)_{n\in\mathbf{N}})$ be a pair consisting of a behavioral conjecture profile and a sequence of strictly positive behavioral conjecture profiles. We say that the pair $(\sigma, (\sigma^n)_{n\in\mathbf{N}})$ is a *proper equilibrium* if $(\sigma^n)_{n\in\mathbf{N}}$ converges to σ and $(\sigma^n)_{n\in\mathbf{N}}$ is a proper sequence. Here, $(\sigma^n)_{n\in\mathbf{N}}$ is a sequence which "justifies" the properness of σ. Note that the same proper equilibrium σ may have two different sequences $(\sigma^n)_{n\in\mathbf{N}}$ and $(\tilde{\sigma}^n)_{n\in\mathbf{N}}$ which justify the properness of σ. Within the new approach, the pairs $(\sigma, (\sigma^n)_{n\in\mathbf{N}})$ and $(\sigma, (\tilde{\sigma}^n)_{n\in\mathbf{N}})$ are thus viewed as two different proper equilibria.

The same can be done for quasi-perfect equilibrium. Say that a pair $(\sigma, (\sigma^n)_{n\in\mathbf{N}})$ is a *quasi-perfect equilibrium* if $(\sigma^n)_{n\in\mathbf{N}}$ is a sequence of strictly positive behavioral conjecture profiles converging to σ and if for every player i, every information set $h \in H_i$ and every $s_i \in S_i(h)$ it holds that $\sigma_{i|h}(s_i) > 0$ only if s_i is a sequential best response against σ_{-i}^n at h for all n.

A correspondence φ assigning to every extensive form game Γ a set $\varphi(\Gamma)$ of pairs $(\sigma, (\sigma^n)_{n\in\mathbf{N}})$ of behavioral conjecture profiles and sequences of strictly positive behavioral conjecture profiles is called a *sequence solution*. We say that a sequence solution φ is *Thompson-Elmes-Reny invariant* if for every pair Γ_1, Γ_2 of extensive form games without chance moves the following holds: if Γ_2 can be obtained from Γ_1 by Thompson-Elmes-Reny transformations, then for every $(\sigma^1, (\sigma^{1n})_{n\in\mathbb{N}}) \in \varphi(\Gamma)$ there is some $(\sigma^2, (\sigma^{2n})_{n\in\mathbb{N}}) \in \varphi(\Gamma_2)$ with $\sigma^2 \in \tau(\sigma^1)$ and $\sigma^{2n} \in \tau(\sigma^{1n})$ for all n. We can now show the following result, which is based on Mailath, Samuelson and Swinkels (1997) and Hillas (1996).

Theorem 6.3.2 *Let φ be a sequence solution which is Thompson-Elmes-Reny invariant and which to each extensive form game without chance moves assigns a subset of the set of quasi-perfect equilibria. Then, φ assigns to every extensive form game without chance moves a subset of the set of proper equilibria.*

Proof. Let φ be a sequence solution with the properties above. Let Γ be an extensive form game without chance moves, and let $(\sigma, (\sigma^n)_{n\in\mathbb{N}}) \in \varphi(\Gamma)$. We show that $(\sigma, (\sigma^n)_{n\in\mathbb{N}})$ is a proper equilibrium. It clearly suffices to show that the sequence $(\sigma^n)_{n\in\mathbb{N}}$ is proper. To this purpose, consider a player i and strategies $s_i, t_i \in S_i$ such that $u_i(s_i, \sigma^n_{-i}) < u_i(t_i, \sigma^n_{-i})$ for some $n \in \mathbb{N}$. We know that, by applying Thompson-Elmes-Reny transformations iteratively, we can transform Γ into a game Γ_1 in reduced form. Hence, in Γ_1 player i has a unique information set h. Since the strategies s_i and t_i in Γ are not equivalent, player i has two actions at h in Γ_1 which can be identified with s_i and t_i. By applying COA^{-1}, we split player i's unique information set h in Γ_1 into two consecutive information sets h^1 and h^2. At h^1, player i has available the actions $A(h^1) = (A(h)\backslash\{s_i, t_i\}) \cup \{out\}$. If he chooses *out* here, information set h^2 is reached, at which player i can choose between s_i and t_i. After choosing s_i or t_i, the game continues as in Γ_1. If player i does not choose *out* at h^1, the game continues as in Γ_1. Denote this new game by Γ_2. Hence, Γ_2 can be obtained from Γ by Thompson-Elmes-Reny transformations.

Since φ is Thompson-Elmes-Reny invariant, there is some $(\sigma^2, (\sigma^{2k})_{k\in\mathbb{N}}) \in \varphi(\Gamma_2)$ with $\sigma^2 \in \tau(\sigma)$ and $\sigma^{2k} \in \tau(\sigma^k)$ for all k. By assumption, φ consists solely of quasi-perfect equilibria and hence $(\sigma^2, (\sigma^{2k})_{k\in\mathbb{N}})$ is a quasi-perfect equilibrium. For the particular n chosen above we have that $u_i(s_i, \sigma^n_{-i}) < u_i(t_i, \sigma^n_{-i})$ in Γ. For every k we have that σ^{2k} in Γ_2 is "equivalent" to σ^k in Γ. It thus follows that

$$u_i((out, s_i), \sigma^{2n}_{-i}) < u_i((out, t_i), \sigma^{2n}_{-i})$$

in Γ_2 for this particular n, and hence

$$u_i(((out, s_i), \sigma^{2n}_{-i})|\, h^2) < u_i(((out, t_i), \sigma^{2n}_{-i})|\, h^2) \tag{6.1}$$

where (out, s_i) is the player i strategy in Γ_2 in which he chooses *out* at h^1 and s_i at h^2. Similar for (out, t_i). Since $(\sigma^2, (\sigma^{2k})_{k\in\mathbb{N}})$ is a quasi-perfect equilibrium in Γ_2, (6.1) implies that $\sigma^2_{i|h^2}(out, s_i) = 0$, and hence $\sigma^2_{ih^2}(s_i) = 0$. For every k, σ^{2k}_i is equivalent to σ^k_i in Γ and thus

$$\sigma^{2k}_{ih^2}(s_i) = \frac{\sigma^k_i(s_i)}{\sigma^k_i(s_i) + \sigma^k_i(t_i)}.$$

Since $\sigma^2_{ih^2}(s_i) = 0 = \lim_{k\to\infty} \sigma^{2k}_{ih^2}(s_i)$, it follows that $\lim_{k\to\infty} \sigma^k_i(s_i)/\sigma^k_i(t_i) = 0$. Since this holds for every s_i, t_i with $u_i(s_i, \sigma^n_{-i}) < u_i(t_i, \sigma^n_{-i})$ for some n, we may conclude that the sequence $(\sigma^n)_{n\in\mathbb{N}}$ is proper. This implies that $(\sigma, (\sigma^n)_{n\in\mathbb{N}})$ is a proper equilibrium, which completes the proof. ■

By combining the insights delivered by the proofs of Theorems 6.3.1 and 6.3.2, we can be even more explicit about the relationship between proper equilibrium (viewed as a conjecture-sequence pair) and quasi-perfect equilibrium (viewed as a conjecture-sequence pair). In the proof of Theorem 6.3.1, we have in fact shown that a proper equilibrium $(\sigma, (\sigma^n)_{n\in\mathbb{N}})$ in a game Γ induces, for every game Γ_2 that can be obtained from Γ by Thompson-Elmes-Reny transformations, a proper equilibrium $(\sigma^2, (\sigma^{2n})_{n\in\mathbb{N}})$ in Γ_2. Since every proper equilibrium $(\sigma^2, (\sigma^{2n})_{n\in\mathbb{N}})$ is a quasi-perfect equilibrium, it follows that $(\sigma, (\sigma^n)_{n\in\mathbb{N}})$ induces a quasi-perfect equilibrium in every game that can be obtained from Γ by Thompson-Elmes-Reny transformations. The proof of Theorem 6.3.2 demonstrates that every conjecture-sequence pair $(\sigma, (\sigma^n)_{n\in\mathbb{N}})$ which induces a quasi-perfect equilibrium in every game that can be obtained from Γ by Thompson-Elmes-Reny transformations, is necessarily a proper equilibrium. We thus obtain the following corollary.

Corollary 6.3.3 *Let Γ be an extensive form game without chance moves, and let $(\sigma, (\sigma^n)_{n\in\mathbb{N}})$ be a conjecture-sequence pair. Then, the following two statements are equivalent:*
(1) $(\sigma, (\sigma^n)_{n\in\mathbb{N}})$ induces, for every game Γ_2 that can be obtained from Γ by Thompson-Elmes-Reny transformations, a quasi-perfect equilibrium $(\sigma^2, (\sigma^{2n})_{n\in\mathbb{N}})$ in Γ_2,
(2) $(\sigma, (\sigma^n)_{n\in\mathbb{N}})$ is a proper equilibrium.

Proper equilibrium, when considered as a conjecture-sequence pair, may thus be viewed as the Thompson-Elmes-Reny invariant analogue to quasi-perfect equilibrium. The example in Hillas (1996) shows that this equivalence relation would not hold if, instead of considering conjecture-sequence pairs, we would consider behavioral conjecture profiles only. Namely, a behavioral conjecture profile σ which induces a quasi-perfect equilibrium σ^2 in every game Γ_2 that can be obtained from Γ by Thompson-Elmes-Reny transformations, is not necessarily a proper equilibrium. In view of the corollary above, such a situation can only occur if for a given σ there are at least two different quasi-perfect equilibria $\sigma^2, \tilde{\sigma}^2$ induced by σ, for which two different sequences $(\sigma^n)_{n\in\mathbb{N}}$ and $(\tilde{\sigma}^n)_{n\in\mathbb{N}}$ in Γ are needed to justify the quasi-perfectness of σ^2 and $\tilde{\sigma}^2$, respectively.

6.4 Normal Form Information Sets

In the present and the following section, which are based on Mailath, Samuelson and Swinkels (1993), we present a solution which is Thompson-Elmes-Reny invariant and selects a nonempty subset of sequential equilibria in every game. In view of the previous section the proper equilibrium solution satisfies these properties, since it selects in every game a subset of the set of quasi-perfect equilibria, which in turn is a subset of the set of sequential equilibria. However, the solution to be presented in the

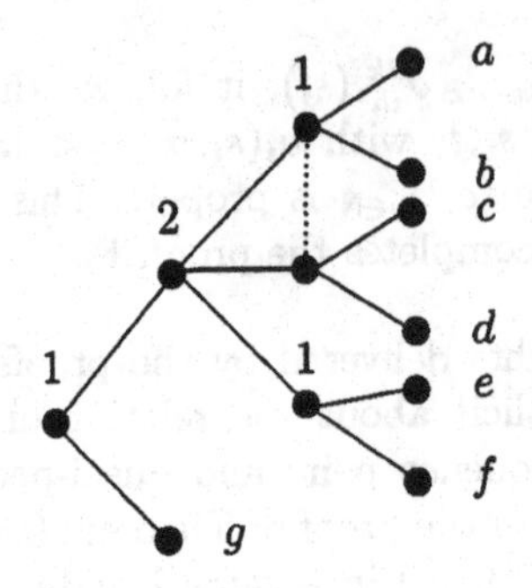

Figure 6.9

following section, termed *normal form sequential equilibrium,* is a strict coarsening of proper equilibrium, and can be seen as the Thompson-Elmes-Reny invariant analogue to sequential equilibrium. The latter is formalized by a result similar to Corollary 6.3.3 in the previous section: a conjecture-sequence pair induces a sequential equilibrium in every game which can be obtained from the original one by Thompson-Elmes-Reny transformations *if and only if* this conjecture-sequence pair is a normal form sequential equilibrium.

In the concept of normal form sequential equilibrium, a central role is played by the notion of *normal form information sets.* Consider an extensive form game Γ without chance moves, in which the set of strategies for each player i is given by S_i. Intuitively, a normal form information set for player i is a set X of strategy profiles with the following property: we can find some other game Γ_2 which can be obtained from Γ by Thompson-Elmes-Reny transformations, and a player i information set h in Γ_2, such that X is the set of strategy profiles which lead to h in Γ_2. By perfect recall, there is a unique sequence of player i actions that lead to h in Γ_2. Let X_i be the set of player i strategies which choose these actions. Then, it is clear that X can be written in the form $X = X_i \times X_{-i}$, where X_{-i} is a set of strategy profiles for the opponents.

Consider now the game in Figure 6.9. Here, $a, ..., g$ denote utility pairs at the terminal nodes. We assume that all utility pairs are different. The pure reduced normal form is given by the table in Figure 6.10, in which $s_1, ..., s_5$ are the strategies controlled by player 1, and t_1, t_2, t_3 are controlled by player 2. Let h be the player 1 information set in Figure 6.9 which contains two nodes. Then, the set of strategy profiles in the pure reduced normal form that reaches h is given by $X = \{s_1, s_2, s_3, s_4\} \times \{t_1, t_2\} = X_1 \times X_2$. Now, consider the player 1 strategies $s_1, s_4 \in X_1$. Then, we can find some player 1 strategy in X_1 which agrees with s_1 on X_2 and agrees with s_4 on $S_2 \backslash X_2$; namely s_2. Formally we mean the following: if player 2 chooses a strategy in X_2, that is, chooses among $\{t_1, t_2\}$, then playing s_1

	t_1	t_2	t_3
s_1	a	c	e
s_2	a	c	f
s_3	b	d	e
s_4	b	d	f
s_5	g	g	g

Figure 6.10

	L	R
T	a	b
B	d	d

M_1

	L	R
T	a	c
B	e	e

M_2

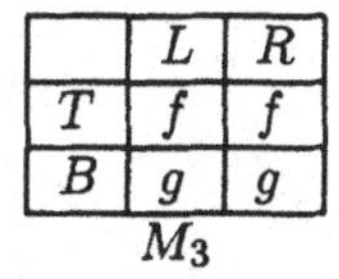

	L	R
T	f	f
B	g	g

M_3

Figure 6.11

induces the same utilities as playing s_2. If player 2 chooses in $S_2 \backslash X_2$, that is, chooses t_3, then playing s_4 induces the same utilities as playing s_2. Hence, s_2 is a player 1 strategy in X_1 which mimicks s_1's behavior if the opponent plays in X_2 and mimicks s_4's behavior if the opponent plays outside X_2. The same can be shown for every other pair of strategies in X_1. The set X thus has the following property: for every pair of strategies $x_1, y_1 \in X_1$ there is some strategy $z_1 \in X_1$ which agrees with x_1 on X_2 and agrees with y_1 on $S_2 \backslash X_2$. A set of strategy profiles with this property is called a *normal form information set* for player 1.

In general, let Γ be an extensive form game without chance moves, i some player and $X_{-i} \subseteq \times_{j \neq i} S_j$ a set of strategy profiles controlled by the other players. Let $s_i, t_i \in S_i$. We say that s_i *agrees* with t_i on X_{-i} if for every $s_{-i} \in X_{-i}$ we have that $u_j(s_i, s_{-i}) = u_j(t_i, s_{-i})$ for every player j. Similarly, we define the event that s_i agrees with t_i on $S_{-i} \backslash X_{-i}$.

Definition 6.4.1 *A set of strategy profiles X is called a normal form information set for player i if X can be written in the form $X = X_i \times X_{-i}$, and if for every $s_i, r_i \in X_i$ there exists some $t_i \in X_i$ such that t_i agrees with s_i on X_{-i} and t_i agrees with r_i on $S_{-i} \backslash X_{-i}$.*

In order to illustrate the definition, consider the pure reduced normal form of some 3-player game, represented by the tables in Figure 6.11. Here, player 1 can choose between M_1, M_2 and M_3 (matrix 1, matrix 2 and matrix 3), player 2 can choose between T and B (top and bottom) and player 3 can choose between L and R (left and right). The reader may verify that the set $X = \{(M_1, M_2)\} \times \{(T, R), (B, L), (B, R)\}$ is a normal form information set for player 1. Note that X_{-1} can not be written as a cartesian product $X_2 \times X_3$. We can construct an extensive form game with the pure reduced normal form as indicated above, and a player 1 information set h in

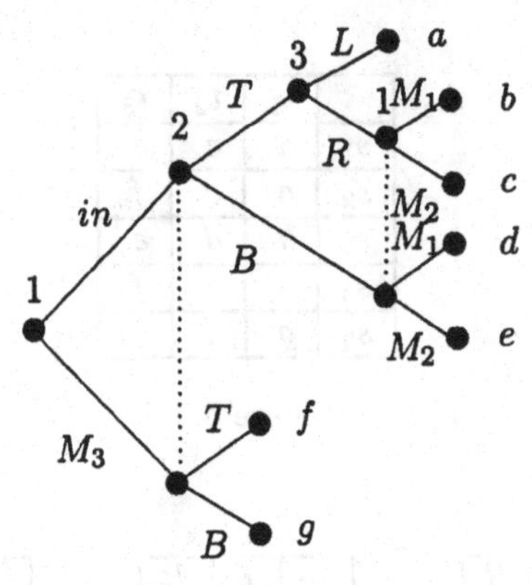

Figure 6.12

this game, such that X is exactly the set of strategy profiles which reach h. Consider namely the extensive form game in Figure 6.12. Let h be the second information controlled by player 1 in this game. Then, the set of strategy profiles leading to h is $\{(in, M_1), (in, M_2)\} \times \{(T, R), (B, L), (B, R)\}$, and hence coincides with X.

In the following lemma, it is shown that this property holds generally for normal form information sets: for a given normal form information set for player i, we can always find a game, obtained from the original game by Thompson-Elmes-Reny transformations, and a player i information set h in this new game, such that the set of strategy profiles leading to h coincides exactly with the normal form information set. We first formally define what we mean by strategy profiles reaching an information set. Consider an extensive form game without chance moves and an information set h in this game. We say that a strategy profile $s = (s_i)_{i \in I}$ *reaches* the information set h if $\mathbb{P}_s(h) = 1$. By $S(h)$ we denote the set of strategy profiles that reach h.

Lemma 6.4.2 *Let Γ be an extensive form game without chance moves and X a set of strategy profiles in Γ. Then, X is a normal form information set for player i if and only if there is some game without chance moves Γ_2 obtained from Γ by Thompson-Elmes-Reny transformations, and some player i information set h in Γ_2, such that $S^2(h) = \tau(X)$.*

Here, $S^2(h)$ is the set of strategy profiles in Γ_2 that reach h, τ is the correspondence mapping behavioral conjecture profiles in Γ to sets of "equivalent" behavioral conjecture profiles in Γ_2, and $\tau(X) = \cup_{s \in X} \tau(s)$. As an abuse of notation, we identify each strategy s_i with an arbitrary behavioral conjecture σ_i for which $\sigma_i(s_i) = 1$. The expressions $\tau(s)$ and $\tau(X)$ should be read in this sense.

Proof. (a) Assume first that X is a normal form information set for player i in Γ. Without loss of generality, we assume that $I = \{1, ..., n\}$ and $i = 1$. We then know that X can be written in the form $X = X_1 \times X_{-1}$. Choose some fixed $s_1^* \in S_1$. Let K be the set of those $s_1 \in X_1$ which agree with s_1^* on $S_{-1} \backslash X_{-1}$, and let L be the set of those $s_1 \in X_1$ which agree with s_1^* on X_{-1}. Now, define the game Γ_2 as follows.

At the beginning, player 1 chooses between *in* and *out*. After this move, player 2 chooses some $s_2 \in S_2$ without having observed player 1's move. Then, if there are more than two players, player 3 chooses some $s_3 \in S_3$ without having observed player 1's nor player 2's move, and so on, until the last player n is reached. Hence, all players $j \in \{2, ..., n\}$ move in ignorance of the previous moves. After player n's move, it is player 1's turn again.

If player 1 previously played *in*, then there are two information sets for player 1; h_1 and h_2, where h_2 may possibly be empty. The nodes in h_1 coincide with those previous strategy choices where player 1 chose *in* and players $j \neq 1$ chose some $s_{-1} \in X_{-1}$. At h_1, player 1 can choose an action $k \in K$ (see above), after which the game ends. The terminal node in Γ_2 after $(in, s_{-1}, k) \in \{in\} \times X_{-1} \times K$ has the same utilities as the terminal node following (k, s_{-1}) in Γ. The nodes in h_2 coincide with those previous strategy choices where player 1 chose *in* and players $j \neq i$ chose some $s_{-1} \in S_{-1} \backslash X_{-1}$. Hence, h_2 is empty if $X_{-1} = S_{-1}$. At h_2, player 1 can choose a strategy $l \in L$ (see above), after which the game ends. The terminal node in Γ_2 after $(in, s_{-1}, l) \in \{in\} \times (S_{-1} \backslash X_{-1}) \times L$ has the same utilities as the terminal node following (l, s_{-1}) in Γ.

If player 1 previously played *out*, then there is only one information set for player 1; h_3. The nodes in h_3 coincide with those previous strategy choices where player 1 chose *out* and players $j \neq 1$ chose some arbitrary $s_{-1} \in S_{-1}$. At h_3, player 1 can choose some action $s_1 \in S_1 \backslash X_1$, after which the game ends. The terminal node in Γ_2 after $(out, s_{-1}, s_1) \in \{out\} \times S_{-1} \times (S_1 \backslash X_1)$ has the same utilities as the terminal node following (s_1, s_{-1}) in Γ. We show that Γ_2 can be obtained from Γ by Thompson-Elmes-Reny transformations, and that $\tau(X) = S^2(h_1)$.

In order to show the first claim, we prove that Γ_2 and Γ have the same pure reduced normal form. By comparing Γ_2 and Γ, it is sufficient to investigate player 1's strategies only. We first show that every strategy in Γ_2 has an equivalent strategy in Γ. Every strategy for player 1 in Γ_2 can either be written in the form (in, k, l) with $k \in K$ and $l \in L$, or in the form (out, s_1) with $s_1 \in S_1 \backslash X_1$. Let s_1^2 be a strategy in Γ_2 of the form (in, k, l). Since X is a normal form information set for player 1 and $k, l \in X_1$, we can find some $t_1 \in X_1$ such that t_1 agrees with k on X_{-1} and agrees with l on $S_{-1} \backslash X_{-1}$. But then, by construction of the utilities in Γ_2, we have that t_1 always induces the same utilities in Γ as (in, k, l) in Γ_2. Hence, every strategy $s_1^2 = (in, k, l)$ in Γ_2 has an equivalent strategy $t_1 \in X_1$ in Γ. Now, let s_1^2 be a strategy in Γ_2 of the form (out, s_1) with $s_1 \in S_1 \backslash X_1$. Then, the strategy s_1 in Γ is equivalent. Hence, every strategy $s_1^2 = (out, s_1)$ has an equivalent strategy in Γ. We may thus conclude that every strategy for player 1 in Γ_2 has an equivalent strategy in Γ.

We next prove that every strategy for player 1 in Γ has an equivalent strategy in Γ_2. Let $s_1 \in S_1$ be a strategy for player 1 in Γ. If $s_1 \notin X_1$, then the strategy $s_1^2 = (out, s_1)$ in Γ_2 is equivalent to s_1. If $s_1 \in X_1$, then, since X is a normal form information set for player 1, there is some $k \in K$ such that k agrees with s_1 on X_{-1}, and some $l \in L$ such that l agrees with s_1 on $S_{-1} \backslash X_{-1}$. Then, the strategy (in, k, l) in Γ_2 is equivalent to s_1. Hence, every strategy for player 1 in Γ has an equivalent strategy in Γ_2. It thus follows that Γ and Γ_2 have the same pure reduced normal form, and hence Γ_2 can be obtained from Γ by Thompson-Elmes-Reny transformations.

In the game Γ_2, we have that $S^2(h_1) = Y_1 \times Y_{-1}$ where Y_1 contains those player 1 strategies where player 1 chooses *in*, and Y_{-1} contains those strategy profiles where players $j \neq 1$ choose a strategy profile $s_{-1} \in X_{-1}$. Since $\tau(X) = Y_1 \times Y_{-1}$, we have that $S^2(h_1) = \tau(X)$, which was to prove.

(b) Assume, next, that there is some game Γ_2 without chance moves, which can be obtained from Γ by Thompson-Elmes-Reny transformations, and some player i information set h in Γ_2, such that $\tau(X) = S^2(h)$. We show that X is a normal form information set for player i.

By perfect recall, there is a unique sequence $a_1, ..., a_K$ of player i actions which leads to h. Let Y_i be the set of player i strategies in Γ_2 which choose the actions $a_1, ..., a_K$. For every node $x \in h$, let $A_{-i}(x)$ be the set of actions on the path to x, not controlled by player i. For every node $x \in h$, let $S^2_{-i}(x)$ be the set of strategy profiles s^2_{-i} in Γ_2 which choose all the actions in $A_{-i}(x)$. Define $Y_{-i} = \cup_{x \in h} S^2_{-i}(x)$. Since every path to some node in h contains the player i actions $a_1, ..., a_K$, it follows that $S^2(h) = Y_i \times Y_{-i}$. Since $\tau(X) = S^2(h)$, it follows that X can be written in the form $X = X_i \times X_{-i}$.

Now, suppose that $s_i, r_i \in X_i$. We show that there is some $t_i \in X_i$ which agrees with s_i on X_{-i} and agrees with r_i on $S_{-i} \backslash X_{-i}$. Let $s^2_i \in \tau_i(s_i)$ and $r^2_i \in \tau_i(r_i)$. Then, since $\tau_i(X_i) = Y_i$, we have that $s^2_i, r^2_i \in Y_i$. Hence, both s^2_i and r^2_i choose all the player i actions $a_1, ..., a_K$ leading to h. Let $\tilde{H}_i$ be the set of player i information sets that either precede h or follow h, and let $H^*_i = \tilde{H}_i \cup \{h\}$. Define the strategy t^2_i as follows: $t^2_i(h') = s^2_i(h')$ for all $h' \in H^*_i$, and $t^2_i(h') = r_i(h')$ for all h' not in H^*_i. Then, since t^2_i chooses the actions $a_1, ..., a_K$, we have that $t^2_i \in Y_i$. Now, let $s^2_{-i} \in Y_{-i}$. Then, (t^2_i, s^2_{-i}) reaches h, and hence (t^2_i, s^2_{-i}) only reaches player i information sets in H^*_i. But then, by construction, (t^2_i, s^2_{-i}) reaches the same terminal node as (s^2_i, s^2_{-i}). We may thus conclude that t^2_i agrees with s^2_i on Y_{-i}. Suppose now that $s^2_{-i} \in S^2_{-i} \backslash Y_{-i}$. Then, (t^2_i, s^2_{-i}) does not reach h and hence (t^2_i, s^2_{-i}) only reaches player i information sets that either precede h or belong to $H_i \backslash H^*_i$. But then, by construction, (t^2_i, s^2_{-i}) reaches the same terminal node as (r^2_i, s^2_{-i}). Hence, t^2_i agrees with r^2_i on $S^2_{-i} \backslash Y_{-i}$. Now, let $t_i \in X_i$ be such that $t^2_i \in \tau_i(t_i)$. Since the transformation τ preserves utilities, it follows that t_i agrees with s_i on X_{-i} and agrees with r_i on $S_{-i} \backslash X_{-i}$. We may thus conclude that X is a normal form information set for player i. ■

In order to illustrate the construction of the game Γ_2 in part (a) of the proof of Lemma 6.4.2, consider the game Γ in Figure 6.13. The reader may check that the set $X = X_1 \times X_2 = \{CH, CI, DH, DI\} \times \{AF, AG\}$ is a normal form information set for player 1. Note that there is no player 1 information set h in Γ such that $X = S(h)$. The proof of Lemma 6.4.2 demonstrates that we can construct some game Γ_2, obtained from Γ by Thompson-Elmes-Reny transformations, in which there is some player 1 information set h_1 such that X corresponds to $S^2(h_1)$. The game Γ_2 constructed according to this proof is depicted in Figure 6.14. Let the upper player 1 information set following *In* be denoted by h_1, the lower player 1 information set following *In* be denoted by h_2, and the player 1 information set following *Out* be denoted by h_3. Conform the proof of Lemma 6.4.2 we have chosen some fixed $s^*_1 \in X_1$, in this case $s^*_1 = CH$. Hence, we have that $K = \{CH, DH\}$ and $L = \{CH, CI\}$. Moreover, $S_1 \backslash X_1 = \{EH, EI\}$. Player 1 can thus choose between CH and DH at h_1, between CH and CI at h_2 and between EH and EI at h_3. In Γ_2 it holds that $S^2(h_1) = Y_1 \times Y_2$

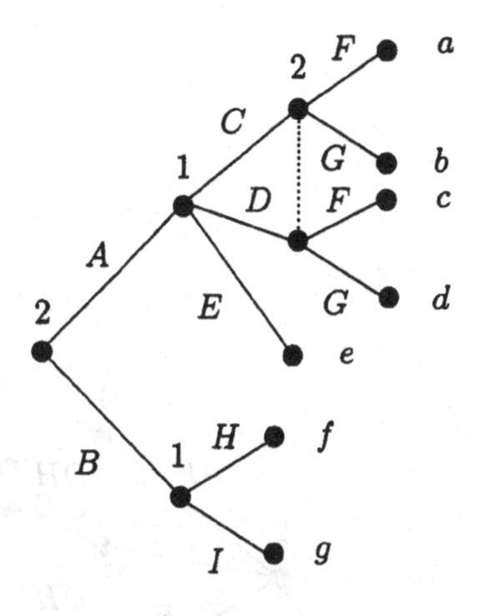

Figure 6.13

where

$$Y_1 = \{(In, CH, CH), (In, CH, CI), (In, DH, CH), (In, DH, CI)\}$$

and $Y_2 = \{AF, AG\}$. Since the correspondence τ_1 is such that

$$\begin{aligned} \tau_1(CH) &= \{(In, CH, CH)\}, \tau_1(CI) = \{(In, CH, CI)\}, \\ \tau_1(DH) &= \{(In, DH, CH)\}, \tau_1(DI) = \{(In, DH, CI)\} \end{aligned}$$

it follows that $\tau_1(X_1) = Y_1$ and thus $S^2(h_1) = \tau(X)$.

6.5 Normal Form Sequential Equilibrium

We are now able to introduce the concept of normal form sequential equilibrium. Consider an extensive form game Γ without chance moves and some strictly positive mixed conjecture profile μ. Let $X = X_i \times X_{-i}$ be some normal form information set for player i. We define the conditional probability distributions $\mu_i(\cdot|\, X) \in \Delta(X_i)$ and $\mu_{-i}(\cdot|\, X) \in \Delta(X_{-i})$ by

$$\mu_i(s_i|\, X) = \frac{\mu_i(s_i)}{\sum_{r_i \in X_i} \mu_i(r_i)}$$

for all $s_i \in X_i$, and

$$\mu_{-i}(s_{-i}|\, X) = \frac{\mu_{-i}(s_{-i})}{\sum_{r_{-i} \in X_{-i}} \mu_{-i}(r_{-i})}$$

for all $s_{-i} \in X_{-i}$. Here, $\mu_{-i}(r_{-i}) = \prod_{j \neq i} \mu_j(r_j)$ for every $r_{-i} = (r_j)_{j \neq i} \in X_{-i}$.

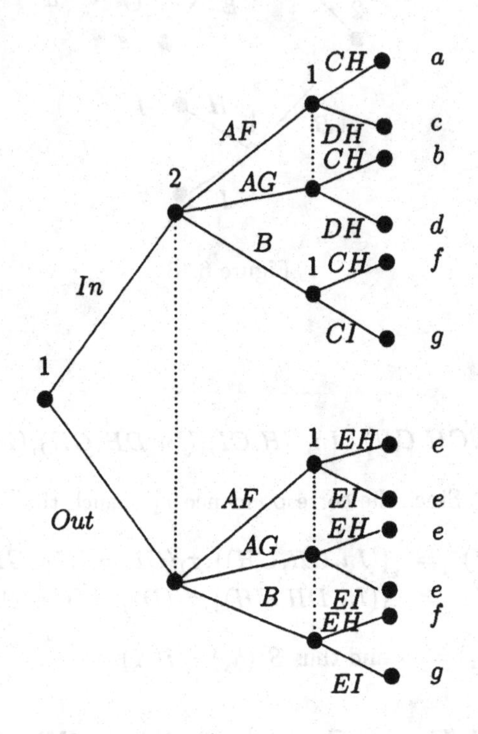

1
In
Out
2
AF
AG
B
1
CH
DH
CH
DH
1
CH
CI
a
c
b
d
f
g
1
EH
EI
EH
EI
EH
EI
e
e
e
e
f
g

Figure 6.14

Definition 6.5.1 *Let $(\mu^n)_{n\in\mathbb{N}}$ be a sequence of strictly positive mixed conjecture profiles. The sequence $(\mu^n)_{n\in\mathbb{N}}$ is called* sequential *if for every player i and every normal form information set X for player i the following holds:*
(1) the conditional probability distributions $\mu_i^n(\cdot|\, X)$ and $\mu_{-i}^n(\cdot|\, X)$ converge to some limit distributions $\mu_i(\cdot|\, X)$ and $\mu_{-i}(\cdot|\, X)$, and
(2) for every $s_i \in X_i$, $\mu_i(s_i|\, X) > 0$ only if

$$u_i(s_i, \mu_{-i}(\cdot|\, X)) = \max_{s_i' \in X_i} u_i(s_i', \mu_{-i}(\cdot|X)).$$

The condition (2) can be seen as a sequential rationality requirement for the limit distributions at the normal form information set X. The limit distribution $\mu_{-i}(\cdot|\, X)$ may be viewed as player i's conjecture about the opponents' behavior, conditional on the event that some strategy profile in X is being played. The limit distribution $\mu_i(\cdot|\, X)$ reflects the common conjecture about player i's behavior, conditional on X. The condition (1) thus puts a restriction on the players' conjectures, comparable to consistency of assessments as defined by Kreps and Wilson (1982a). A behavioral conjecture profile induced by some sequential sequence of strictly positive mixed conjecture profiles is called a normal form sequential equilibrium.

Definition 6.5.2 *A behavioral conjecture profile σ is called a normal form sequential equilibrium if there is some sequential sequence $(\mu^n)_{n\in\mathbb{N}}$ of strictly positive mixed conjecture profiles, such that the induced sequence $(\sigma^n)_{n\in\mathbb{N}}$ of strictly positive behavioral conjecture profiles converges to σ.*

We will now demonstrate that the relationship between normal form sequential equilibrium and sequential equilibrium is analogous to the relationship between proper equilibrium and quasi-perfect equilibrium. First of all, it can be shown that every normal form sequential equilibium is extendable to a sequential equilibrium.

Lemma 6.5.3 *Let Γ be an extensive form game without chance moves and σ a normal form sequential equilibrium in Γ. Then, σ can be extended to a sequential equilibrium (σ, β).*

Proof. Let Γ be an extensive form game without chance moves and σ a normal form sequential equilibrium. Then, by construction, there is a sequential sequence $(\mu^n)_{n\in\mathbb{N}}$ of strictly positive mixed conjecture profiles which induces a sequence $(\sigma^n)_{n\in\mathbb{N}}$ of strictly positive behavioral conjecture profiles converging to σ. For every n, let β^n be the belief system induced by σ^n. Without loss of generality, we assume that $(\beta^n)_{n\in\mathbb{N}}$ converges to some belief system β. Hence, (σ, β) is by construction consistent. It remains to show that (σ, β) is sequentially rational. Let i be a player and h an information set controlled by i. We know that the set $X = S(h)$ of strategy profiles reaching h is a normal form information set for player i. Since the sequence $(\mu^n)_{n\in\mathbb{N}}$ is sequential, the conditional distributions $\mu_i^n(\cdot|\, X)$ and $\mu_{-i}^n(\cdot|\, X)$ converge to some distributions $\mu_i(\cdot|\, X)$ and $\mu_{-i}(\cdot|\, X)$, and

$$\mu_i(s_i|\, X) > 0 \text{ only if } u_i(s_i, \mu_{-i}(\cdot|\, X)) = \max_{s_i' \in X_i} u_i(s_i', \mu_{-i}(\cdot|\, X)). \tag{6.2}$$

Since σ^n and β^n are induced by μ^n for every n, it may be verified that

$$u_i(s_i, \mu_{-i}(\cdot| X)) = u_i((s_i, \sigma_{-i})| h, \beta_{ih})$$

for every $s_i \in X_i$. Moreover, it may be verified that $\mu_i(s_i| X) = \sigma_{i|h}(s_i)$ for all $s_i \in X_i = S_i(h)$. Hence (6.2) implies that $\sigma_{i|h}(s_i) > 0$ only if

$$u_i((s_i, \sigma_{-i})| h, \beta_{ih}) = \max_{s_i' \in S_i(h)} u_i((s_i', \sigma_{-i})| h, \beta_{ih}),$$

and thus (σ, β) is sequentially rational at h. Since this holds for every player i and every $h \in H_i$, the assessment (σ, β) is sequentially rational. ■

Not every sequential equilibrium is a normal form sequential equilibrium. Consider again the game Γ_1 in Figure 6.8. The behavioral conjecture profile $\sigma = (c, d)$ can be extended to a sequential equilibrium, for instance by assigning beliefs $(1, 0)$ to player 2's information set. However, (c, d) is not a normal form sequential equilibrium in Γ_1. In fact, we prove that (c, e) is the unique normal form sequential equilibrium in the game. Let $(\mu^n)_{n\in\mathbb{N}}$ be a sequential sequence of strictly positive mixed conjecture profiles, and $(\sigma^n)_{n\in\mathbb{N}}$ the sequence of induced behavioral conjecture profiles which we assume to converge to some $\tilde{\sigma}$. Note that the set of strategy profiles $X = \{a, b\} \times \{d, e\}$ is a normal form information set for player 1. Since $(\mu^n)_{n\in\mathbb{N}}$ is sequential, the conditional distributions $\mu_1^n(\cdot| X)$ and $\mu_2^n(\cdot| X)$ converge to some distributions $\mu_1(\cdot| X)$ and $\mu_2(\cdot| X)$, and $\mu_1(s_1| X) > 0$ only if $u_1(s_1, \mu_2(\cdot| X)) = \max_{s_1' \in X_1} u_1(s_1', \mu_2(\cdot| X))$. The latter implies that $\mu_1(a| X) = 0$.

On the other hand, X is also a normal form information set for player 2, and hence $\mu_2(s_2| X) > 0$ only if $u_2(s_2, \mu_1(\cdot| X)) = \max_{s_2' \in X_2} u_2(s_2', \mu_1(\cdot| X))$. Since $\mu_1(a| X) = 0$, this implies that $\mu_2(d| X) = 0$, and hence $\tilde{\sigma}_2 = e$. Since in every sequential equilibrium player 1 is believed to choose c with probability one, it follows that player 1 is believed to choose c with probability one in every normal form sequential equilibrium. Hence, $\tilde{\sigma} = (c, e)$ is the only candidate for a normal form sequential equilibrium. Since it can be checked that $\tilde{\sigma}$ is indeed a normal form sequential equilibrium, it must be the only one.

In the following theorem, we prove that every proper equilibrium σ is a normal form sequential equilibrium. Since we know that a proper equilibrium always exists, it follows that every game has at least one normal form sequential equilibrium.

Theorem 6.5.4 *Let Γ be an extensive form game without chance moves and σ a proper equilibrium in Γ. Then, σ is a normal form sequential equilibrium. In particular, every extensive form game without chance moves has a normal form sequential equilibrium.*

Proof. Let σ be a proper equilibrium in Γ. Then, there is a proper sequence $(\sigma^n)_{n\in\mathbb{N}}$ of strictly positive behavioral conjecture profiles converging to σ. Let $(\mu^n)_{n\in\mathbb{N}}$ be the proper sequence of induced mixed conjecture profiles. We can find a subsequence $(\tilde{\mu}^n)_{n\in\mathbb{N}}$ of $(\mu^n)_{n\in\mathbb{N}}$ such that for every player i and every normal form information set X for player i, the conditional distributions $\tilde{\mu}_i^n(\cdot| X)$ and $\tilde{\mu}_{-i}^n(\cdot| X)$ converge. Without loss of generality, we assume that the sequence $(\mu^n)_{n\in\mathbb{N}}$ itself already satisfies this property. We prove that the sequence $(\mu^n)_{n\in\mathbb{N}}$ is sequential, which would imply that σ is a normal form sequential equilibrium.

Consider a player i and a normal form information set $X = X_i \times X_{-i}$ for player i. Let $\mu_i(\cdot|\, X)$ and $\mu_{-i}(\cdot|\, X)$ be the limit conditional distributions induced by $(\mu^n)_{n\in\mathbb{N}}$. We show that

$$\mu_i(s_i|\, X) > 0 \text{ only if } u_i(s_i, \mu_{-i}(\cdot|\, X)) = \max_{s_i' \in X_i} u_i(s_i', \mu_{-i}(\cdot|\, X)).$$

Let $s_i \in X_i$ be such that $u_i(s_i, \mu_{-i}(\cdot|\, X)) < \max_{s_i' \in X_i} u_i(s_i', \mu_{-i}(\cdot|\, X))$. Then, there is some $r_i \in X_i$ such that $u_i(s_i, \mu_{-i}(\cdot|\, X)) < u_i(r_i, \mu_{-i}(\cdot|\, X))$. Hence,

$$u_i(s_i, \mu_{-i}^n(\cdot|X)) < u_i(r_i, \mu_{-i}^n(\cdot|X))$$

if n is large enough. By definition,

$$u_i(s_i, \mu_{-i}^n) = \mathbb{P}_{\mu_{-i}^n}(X_{-i})\; u_i(s_i, \mu_{-i}^n(\cdot|\, X)) + \mathbb{P}_{\mu_{-i}^n}(S_{-i}\backslash X_{-i})\; u_i(s_i, \mu_{-i}^n(\cdot|\, S_{-i}\backslash X_{-i})),$$

where $\mathbb{P}_{\mu_{-i}^n}(X_{-i})$ is the probability that some strategy profile in X_{-i} is chosen, given μ_{-i}^n, and $\mu_{-i}^n(\cdot|\, S_{-i}\backslash X_{-i})$ is the conditional probability distribution on $S_{-i}\backslash X_{-i}$, induced by μ_{-i}^n. Since $\mathbb{P}_{\mu_{-i}^n}(X_{-i}) > 0$ and $u_i(s_i, \mu_{-i}^n(\cdot|X)) < u_i(r_i, \mu_{-i}^n(\cdot|X))$ if n is large enough, we have that

$$u_i(s_i, \mu_{-i}^n) < \mathbb{P}_{\mu_{-i}^n}(X_{-i})\; u_i(r_i, \mu_{-i}^n(\cdot|\, X)) + \mathbb{P}_{\mu_{-i}^n}(S_{-i}\backslash X_{-i})\; u_i(s_i, \mu_{-i}^n(\cdot|\, S_{-i}\backslash X_{-i})) \tag{6.3}$$

if n large enough. Since X is a normal form information set for player i, we can find some $t_i \in X_i$ such that t_i agrees with r_i on X_{-i}, and t_i agrees with s_i on $S_{-i}\backslash X_{-i}$. Hence, $u_i(r_i, \mu_{-i}^n(\cdot|\, X)) = u_i(t_i, \mu_{-i}^n(\cdot|\, X))$ and $u_i(s_i, \mu_{-i}^n(\cdot|\, S_{-i}\backslash X_{-i})) = u_i(t_i, \mu_{-i}^n(\cdot|\, S_{-i}\backslash X_{-i}))$. By (6.3), it follows that

$$\begin{aligned} u_i(s_i, \mu_{-i}^n) &< \mathbb{P}_{\mu_{-i}^n}(X_{-i})\; u_i(t_i, \mu_{-i}^n(\cdot|\, X)) + \mathbb{P}_{\mu_{-i}^n}(S_{-i}\backslash X_{-i})\; u_i(t_i, \mu_{-i}^n(\cdot|\, S_{-i}\backslash X_{-i})) \\ &= u_i(t_i, \mu_{-i}^n) \end{aligned}$$

if n is large enough. But then, since the sequence $(\mu^n)_{n\in\mathbb{N}}$ is proper, we have that $\lim_{n\to\infty} \mu_i^n(s_i)/\mu_i^n(t_i) = 0$. Since $t_i \in X_i$, this implies that

$$\mu_i(s_i|\, X) = \lim_{n\to\infty} \mu_i^n(s_i|\, X) = 0,$$

which completes the proof. ■

The correspondence which assigns to every game without chance moves the set of normal form sequential equilibria is called the *normal form sequential equilibrium solution.* We prove that the normal form sequential equilibrium solution is Thompson-Elmes-Reny invariant.

Theorem 6.5.5 *The normal form sequential equilibrium solution is Thompson-Elmes-Reny invariant, and assigns to every game without chance moves a nonempty subset of the set of sequential equilibria.*

Here, we say that a behavioral conjecture profile σ is a *sequential equilibrium* if it can be extended to a sequential equilibrium (σ, β).

Proof. In view of Lemma 6.5.3, we only have to show that the normal form sequential equilibrium solution is Thompson-Elmes-Reny invariant. Let Γ_1, Γ_2 be two extensive form games without chance moves, such that Γ_2 can be obtained from Γ_1 by Thompson-Elmes-Reny transformations. Let τ be the correspondence mapping behavioral conjecture profiles σ^1 in Γ_1 to sets $\tau(\sigma^1)$ of "equivalent" behavioral conjecture profiles in Γ_2. From Lemma 6.4.2 it immediately follows that Γ_1 and Γ_2 have "equivalent" collections of normal form information sets, that is, X is a normal form information set for player i in Γ_1 if and only if $\tau(X)$ is a normal form information set for player i in Γ_2. Using this insight, and adapting the techniques used in the proof of Theorem 6.3.1, one can show that the sets of normal form sequential equilibria in Γ_1 and Γ_2 are "equivalent", and hence the normal form sequential equilibrium solution is Thompson-Elmes-Reny invariant. ■

In Section 6.3 we have seen that every sequence solution which is Thompson-Elmes-Reny invariant and selects solely quasi-perfect equilibria (viewed as conjecture-sequence pairs) in every game, necessarily is a refinement of proper equilibrium. A similar result holds for normal form sequential equilibrium and sequential equilibrium, as the following theorem will demonstrate. We first need to define normal form sequential equilibrium and sequential equilibrium in terms of conjecture-sequence pairs. Let $(\sigma, (\sigma^n)_{n\in\mathbf{N}})$ be a strategy-sequence pair, consisting of a behavioral conjecture profile σ and a sequence $(\sigma^n)_{n\in\mathbf{N}}$ of strictly positive behavioral conjecture profiles converging to σ. We say that $(\sigma, (\sigma^n)_{n\in\mathbf{N}})$ is a *normal form sequential equilibrium* if there is a sequential sequence $(\mu^n)_{n\in\mathbf{N}}$ of strictly positive mixed conjecture profiles which induces $(\sigma^n)_{n\in\mathbf{N}}$. We say that $(\sigma, (\sigma^n)_{n\in\mathbf{N}})$ is a *sequential equilibrium* if (1) $(\sigma^n)_{n\in\mathbf{N}}$ is induced by a sequence $(\mu^n)_{n\in\mathbf{N}}$ of strictly positive mixed conjecture profiles such that for every player i and every normal form information set X for player i the conditional distributions $\mu_i^n(\cdot|\, X)$ and $\mu_{-i}^n(\cdot|\, X)$ converge, and (2) the sequence $(\beta^n)_{n\in\mathbf{N}}$ of belief systems induced by $(\sigma^n)_{n\in\mathbf{N}}$ converges to some belief system β, such that (σ, β) is a sequential equilibrium.

Theorem 6.5.6 *Let φ be a sequence solution which is Thompson-Elmes-Reny invariant and which to every game without chance moves assigns a subset of the set of sequential equilibria. Then, φ assigns to every game without chance moves a subset of the set of normal form sequential equilibria.*

Proof. Let φ be a sequence solution with the properties above. Let Γ be an extensive form game without chance moves, and let $(\sigma, (\sigma^n)_{n\in\mathbf{N}}) \in \varphi(\Gamma)$. We show that $(\sigma, (\sigma^n)_{n\in\mathbf{N}})$ is a normal form sequential equilibrium. By definition, $(\sigma^n)_{n\in\mathbf{N}}$ is induced by some sequence $(\mu^n)_{n\in\mathbf{N}}$ of strictly positive mixed conjecture profiles, such that for every player i and every normal form information set X for player i we have that the conditional distributions $\mu_i^n(\cdot|\, X)$ converge to some $\mu_i(\cdot|\, X)$ and the conditional distributions $\mu_{-i}^n(\cdot|\, X)$ converge to some $\mu_{-i}(\cdot|\, X)$. We prove that for every such X,

$$\mu_i(s_i|\ X) > 0 \text{ only if } u_i(s_i, \mu_{-i}(\cdot|\ X)) = \max_{s_i' \in X_i} u_i(s_i', \mu_{-i}(\cdot|\ X)).$$

Let such a normal form information set $X = X_i \times X_{-i}$ for player i be given. By Lemma 6.4.2 we know that we can find some game Γ_2 without chance moves, obtained from Γ by Thompson-Elmes-Reny transformations, and some player i information set h in Γ_2, such that $\tau(X) = S^2(h)$. Since, by assumption, φ is Thompson-Elmes-Reny invariant, and consists solely of sequential equilibria, we can find some conjecture-sequence pair $(\sigma^2, (\sigma^{2n})_{n\in\mathbf{N}})$ with $\sigma^2 \in \tau(\sigma)$ and $\sigma^{2n} \in \tau(\sigma^n)$ for all n such that $(\sigma^2, (\sigma^{2n})_{n\in\mathbf{N}})$ is a sequential equilibrium in Γ_2. Then, by definition, the sequence of belief systems $(\beta^{2n})_{n\in\mathbf{N}}$ induced by $(\sigma^{2n})_{n\in\mathbf{N}}$ converges to some belief system β^2 and (σ^2, β^2) is a sequential equilibrium. In particular, we have at information set h that $\sigma^2_{i|h}(s_i) > 0$ only if

$$u_i((s_i, \sigma^2_{-i})|\ h, \beta_{ih}) = \max_{s_i' \in S_i(h)} u_i((s_i', \sigma^2_{-i})|\ h, \beta_{ih}). \tag{6.4}$$

Moreover, the sequence $(\sigma^{2n})_{n\in\mathbf{N}}$ is induced by some sequence $(\mu^{2n})_{n\in\mathbf{N}}$ of strictly positive mixed conjecture profiles in Γ_2 such that for every player i and every normal form information set Y for player i, we have that the conditional distributions $\mu_i^{2n}(\cdot|\ Y)$ and $\mu_{-i}^{2n}(\cdot|\ Y)$ converge to some $\mu_i^2(\cdot|\ Y)$ and $\mu_{-i}^2(\cdot|\ Y)$. Now, let $Y = S^2(h)$. Then, $Y = Y_i \times Y_{-i}$ is a normal form information set for player i in Γ_2. From (6.4) it follows that

$$\mu_i^2(s_i|\ Y) > 0 \text{ only if } u_i(s_i, \mu_{-i}^2(\cdot|\ Y)) = \max_{s_i' \in Y_i} u_i(s_i', \mu_{-i}^2(\cdot|\ Y)),$$

which in turn implies that

$$\mu_i(s_i|\ X) > 0 \text{ only if } u_i(s_i, \mu_{-i}(\cdot|\ X)) = \max_{s_i' \in X_i} u_i(s_i', \mu_{-i}(\cdot|\ X)).$$

This completes the proof. ■

Theorem 6.5.5 can be sharpened as to show that a normal form sequential equilibrium in a game Γ, viewed as a conjecture-sequence pair, induces a normal form sequential equilibrium (viewed as conjecture-sequence pair) in every game Γ_2 that can be obtained from the original one by Thompson-Elmes-Reny transformations. In particular, every (conjecture-sequence) normal form sequential equilibrium in Γ induces a (conjecture-sequence) sequential equilibrium in Γ_2. On the other hand, the proof of Theorem 6.5.6 shows that every conjecture-sequence pair in Γ which induces a (conjecture-sequence) sequential equilibrium in every game Γ_2 obtained from Γ by Thompson-Elmes-Reny transformations, is necessarily a normal form sequential equilibrium. We thus have the following equivalence.

Corollary 6.5.7 *Let Γ be an extensive form game without chance moves, and let $(\sigma, (\sigma^n)_{n\in\mathbf{N}})$ be a conjecture-sequence pair. Then, the following two statements are equivalent:*

(1) $(\sigma, (\sigma^n)_{n\in\mathbb{N}})$ *induces, for every game* Γ_2 *that can be obtained from* Γ *by Thompson-Elmes-Reny transformations, a sequential equilibrium* $(\sigma^2, (\sigma^{2n})_{n\in\mathbb{N}})$ *in* Γ_2,
(2) $(\sigma, (\sigma^n)_{n\in\mathbb{N}})$ *is a normal form sequential equilibrium.*

In view of this result, the concept of normal form sequential equilibrium, when stated in terms of conjecture-sequence pairs, may be viewed as the Thompson-Elmes-Reny invariant analogue to sequential equilibrium. As in Section 6.3, the equivalence relation stated above no longer applies if normal form sequential equilibrium and sequential equilibrium are stated solely in terms of conjecture profiles, without specifying the supporting sequences. An example in Mailath, Samuelson and Swinkels (1993) shows that a behavioral conjecture profile which induces a sequential equilibrium in every game obtained from the original one by Thompson-Elmes-Reny transformations, is not necessarily a normal form sequential equilibrium.

6.6 Kohlberg-Mertens Transformations

6.6.1 Superfluous Chance Moves

So far in this chapter, we concentrated solely on extensive form games without chance moves. In this section we extend the concept of invariance to games with chance moves, and introduce, in addition to the Thompson-Elmes-Reny transformations, two new transformations which involve the addition or deletion of chance moves: *addition of superfluous chance moves* and *truncation*. These transformations were first introduced in Kohlberg and Mertens (1986), and in their paper it is argued that applying such transformations does not alter the strategic features of the game.

Transformation 5: *Addition of superfluous chance moves* $(ADDCHA)$.
Addition of superfluous chance moves is a transformation which inserts a chance move immediately after some information set $h \in H_i$. Every possible outcome c of the chance move corresponds to an action $a(c)$ at h, and every player other than player i does not observe whether the outcome of the chance move is the result of nature choosing c or player i choosing the action $a(c)$. As an illustration, consider Figure 6.15. In game 2, the upper outcome of the chance move corresponds to player 1 choosing his upper action in game 1, whereas the lower outcome of the chance move corresponds to player 1 choosing his lower action in game 1.

Formally, consider an extensive form game, say game 1, which may contain chance moves. Consider a player i and an information set h controlled by this player, with set of actions $A^1(h)$. In the new game, say game 2, the set of actions at h is extended to $A(h^2) = A(h^1) \cup \{chance\}$. Let x be some node in h. If player i chooses *chance* at x, a newly introduced chance move τ_x occurs with set of possible outcomes C, which is the same for every $x \in h$. For every $c \in C$, let $\tau_x(c)$ be the probability that c occurs. These probabilities do not depend upon the node $x \in h$ that precedes the chance move, hence $\tau_x(c) = \tau_{x'}(c)$ for every $c \in C$ and every $x, x' \in h$. Every outcome $c \in C$ corresponds to some action $a(c) \in A^1(h)$ in the following way. For every node $x \in h$ and action $a^1 \in A^1(h)$, let $y^1(x, a^1)$ be the node in game 1 which immediately follows x and a^1. If nature chooses c at the chance move following node $x \in h$, then

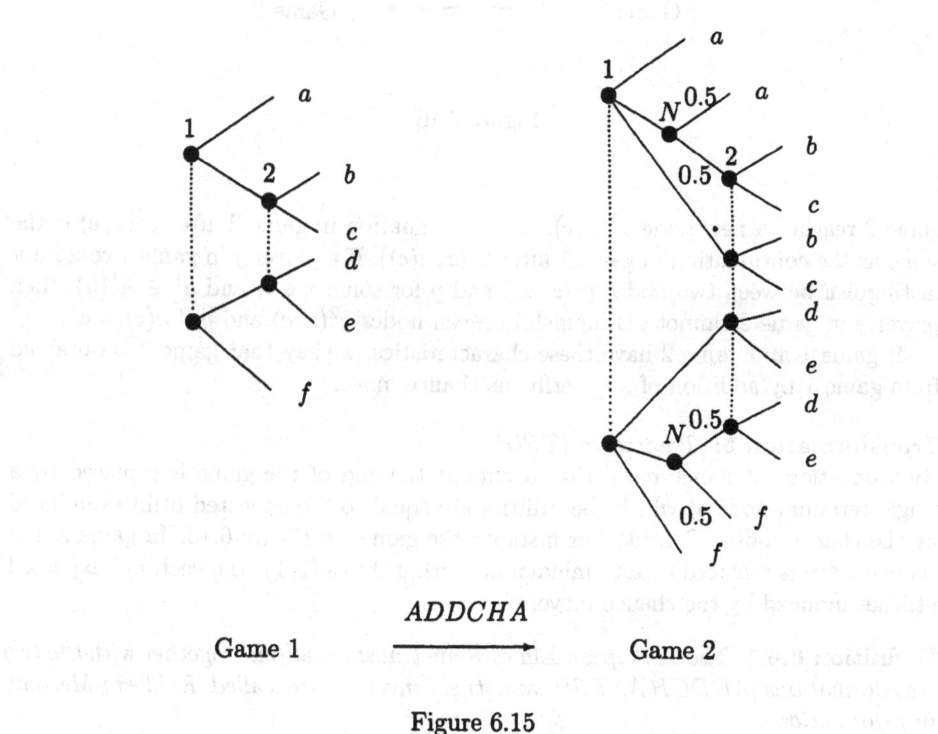

Figure 6.15

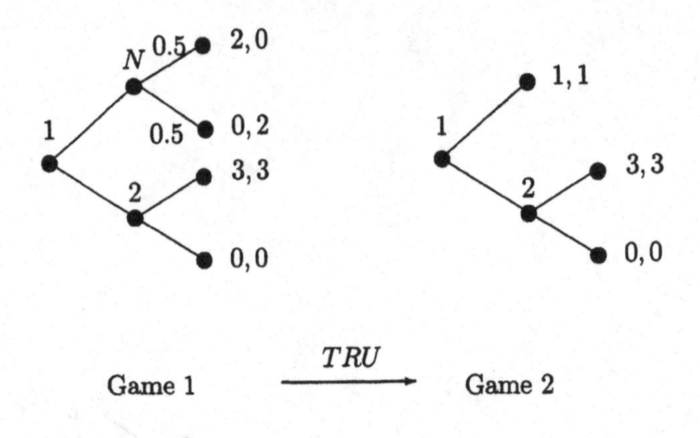

Figure 6.16

game 2 reaches a new node $y^2(x,c)$. The continuation in game 2 after $y^2(x,c)$ is the same as the continuation in game 1 after $y^1(x, a(c))$. If a player j in game 1 could not distinguish between two nodes $y^1(x, a^1)$ and y for some $x \in h$ and $a^1 \in A^1(h)$, then player j in game 2 cannot distinguish between nodes $y^2(x,c)$ and y if $a(c) = a^1$.

If game 1 and game 2 have these characteristics, we say that game 2 is obtained from game 1 by addition of a superfluous chance move.

Transformation 6: *Truncation* (TRU).
By truncation, a chance move that occurs at the end of the game is replaced by a single terminal node at which the utilities are equal to the expected utilities induced by the chance move. Consider for instance the games in Figure 6.16. In game 2, the chance move is replaced by a terminal node with utilities $(1,1)$: the vector of expected utilities induced by the chance move.

Definition 6.6.1 *The Thompson-Elmes-Reny transformations, together with the two transformations $ADDCHA$, TRU and their inverses are called Kohlberg-Mertens transformations.*

In Section 6.1 we have seen that a game Γ_1 without chance moves can be transformed into a game Γ_2 by Thompson-Elmes-Reny transformations if and only if these games have the same pure reduced normal form. In the following subsection we prove a similar result for Kohlberg-Mertens transformations by using the notion of *mixed* reduced normal form instead of *pure* reduced normal form.

6.6.2 Mixed Reduced Normal Form

We first provide the definition of a game in *mixed reduced form,* as an analogue to a game in reduced form, and then prove that every game can be transformed into a game in mixed reduced form by an iterative application of Kohlberg-Mertens transformations.

Definition 6.6.2 *An extensive form game is said to be in mixed reduced form if (1) the game contains no chance moves, (2) the game is in reduced form and (3) for every player i and every strategy s_i we can find no randomization $\mu_i \in \Delta(S_i)$ with $\mu_i(s_i) = 0$ such that $u_j(s_i, s_{-i}) = u_j(\mu_i, s_{-i})$ for all players j and all strategy profiles s_{-i}.*

Condition (3) states that no strategy s_i should be equivalent to some randomization over strategies which does not use s_i. Suppose now that the Thompson-Elmes-Reny transformations are extended, in an appropriate way, to extensive form games with chance moves. We then obtain the following theorem, which may be seen as an analogue to Theorem 6.1.3, and is due to Kohlberg and Mertens (1986).

Theorem 6.6.3 *Let Γ be an extensive form game, with or without chance moves. Then, there is a game in mixed reduced form which can be obtained from Γ by iterative application of Kohlberg-Mertens transformations.*

Sketch of proof. Let Γ be an extensive form game, with or without chance moves.
Step 1. By an iterative application of Thompson-Elmes-Reny transformations, Γ can be transformed into a game Γ_1 with the following properties: (1) every player controls exactly one information set and (2) every path from the root to a terminal node crosses each information set. The proof for this fact is similar to the proof of Theorem 6.1.3. Note that the game Γ_1 may still contain chance moves.
Step 2. By an iterative application of INT, we may transform Γ_1 into a game Γ_2 in which all chance moves occur at the end. Then, by TRU, we transform Γ_2 into a game Γ_3 without chance moves.
Step 3. Suppose that in Γ_3 some player i contains a strategy s_i^* and a randomization $\mu_i^* \in \Delta(S_i)$ such that $\mu_i^*(s_i^*) = 0$ and μ_i^* is equivalent to s_i^*. Let $\tilde{S}_i = \{s_i \in S_i | \mu_i^*(s_i) > 0\}$. By applying INT, we can make sure that player i's information set h is the last information set in the game. Call this game Γ_4. By TRU^{-1}, player i's action s_i^* may be replaced by a collection of chance moves $(\tau_x)_{x \in h}$ with set of outcomes $\tilde{S}_i$ such that $\tau_x(s_i) = \mu_i^*(s_i)$ for all $x \in h$ and all $s_i \in \tilde{S}_i$. For every x and every outcome of the chance move $\tilde{s}_i$ let the utility vector at the corresponding terminal node coincide with the utility vector in Γ_4 after x and action $\tilde{s}_i$. Now, by $ADDCHA^{-1}$, the chance moves $(\tau_x)_{x \in h}$ may all be deleted since they coincide with probability distributions over actions at h. Hence, the strategy s_i^* has been removed from the game. By proceding in this way, Γ_4 can be transformed into a game Γ^* in which (1) every player controls exactly one information set, (2) every path from the root to a terminal node crosses each information set, (3) no chance moves are present and (4) no strategy s_i is equivalent to some randomization $\mu_i \in \Delta(S_i)$ with $\mu_i(s_i) = 0$. But then, by definition, Γ^* is in mixed reduced form. ■

In order to illustrate how a game in reduced form can be transformed into a game in mixed reduced form by Kohlberg-Mertens transformations, consider Figure 6.17. Let the four games be denoted by game 1, ..., game 4. Game 1 is in reduced form, but not in mixed reduced form, since player 1's lower action is equivalent to the randomization which chooses his upper and middle action both with probability one

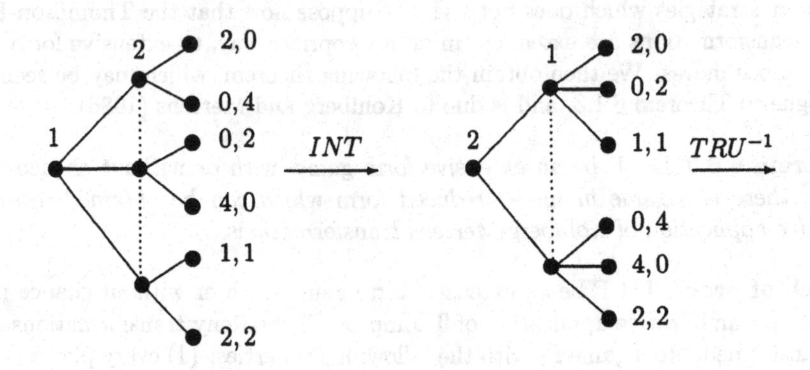
1
2
2,0
0,4
0,2
4,0
1,1
2,2
INT
2
1
2,0
0,2
1,1
0,4
4,0
2,2
TRU⁻¹

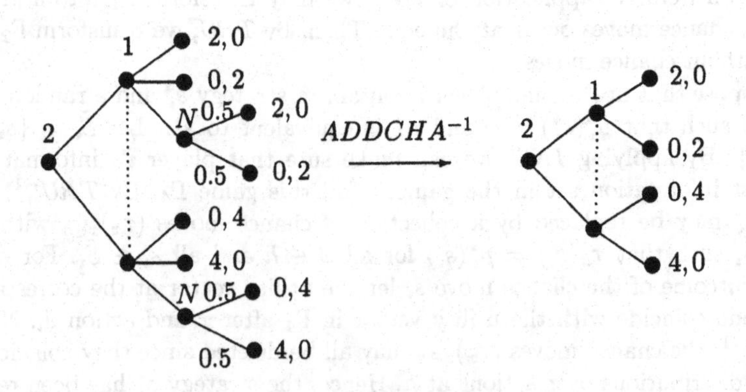
2
1
2,0
0,2
N
0.5
2,0
0.5
0,2
0,4
4,0
N
0.5
0,4
0.5
4,0
ADDCHA⁻¹
2
1
2,0
0,2
0,4
4,0

Figure 6.17

half. By INT the order of player 1 and 2 has been exchanged such that player 1 comes last. By TRU^{-1} player 1's lower action has been replaced by a chance move which mimicks the above mentioned randomization. By $ADDCHA^{-1}$ the chance move is deleted, which results in a game in mixed reduced form.

We now introduce the *mixed reduced normal form* of a game, as an analogue to the pure reduced normal form, and show that a game Γ_1 can be obtained from a game Γ_2 by Kohlberg-Mertens transformations if and only if they have the same mixed reduced normal form. Let Γ be an extensive form game, with or without chance moves, and let $((S_i^*)_{i\in I}, (u_i^*)_{i\in I})$ be the pure reduced normal form of Γ. Formally, we have defined the pure reduced normal form only for games without chance moves, but the definition can also be applied to games with chance moves. For every player i delete from S_i^* those equivalence classes $[s_i^*]$ that are equivalent to some randomization μ_i with $\mu_i(s_i) = 0$ for all $s_i \in [s_i^*]$. By saying that $[s_i^*]$ is equivalent to μ_i, we mean that s_i is equivalent to μ_i for all $s_i \in [s_i^*]$. Let $(S_i^{**})_{i\in I}$ be the sets of remaining equivalence classes, and let $(u_i^{**})_{i\in I}$ be the restrictions of $(u_i^*)_{i\in I}$ on $\times_{j\in I} S_j^{**}$.

Definition 6.6.4 *Let Γ be an extensive form game and for every player i let S_i^{**} and u_i^{**} be defined as above. Then, the pair $((S_i^{**})_{i\in I}, (u_i^{**})_{i\in I})$ is called the mixed reduced normal form of Γ.*

In the same way as for the pure reduced normal form, we may formally define the event that two games have the *same* mixed reduced normal form. The following theorem, which is due to Kohlberg and Mertens (1986), characterizes the "equivalence classes" of games that are closed under the application of Kohlberg-Mertens transformations.

Theorem 6.6.5 *Let Γ_1 and Γ_2 be two extensive form games. Then, Γ_2 can be obtained from Γ_1 by an iterative application of Kohlberg-Mertens transformation if and only if Γ_1 and Γ_2 have the same mixed reduced normal form.*

Proof. The result follows easily from Theorem 6.6.3 and the observation that each of the Kohlberg-Mertens transformations preserves the mixed reduced normal form of a game. ■

6.6.3 Kohlberg-Mertens Invariance

In Section 6.2 we have formally defined Thompson-Elmes-Reny invariance of solutions. The intuition of Thompson-Elmes-Reny invariance is that a solution should be insensitive to each of the Thompson-Elmes-Reny transformations. Formally, we constructed for each pair of games Γ_1, Γ_2 that differ solely by Thompson-Elmes-Reny transformations, a correspondence τ which maps conjecture profiles in Γ_1 to sets of "equivalent" conjecture profiles in Γ_2, making use of the pure reduced normal forms of Γ_1 and Γ_2. The key step in the construction of τ is to define functions g_i^1 and g_i^2 which transform behavioral conjectures in Γ_1 and Γ_2 into "equivalent" mixed conjectures in the corresponding pure reduced normal form. Since we know that Γ_1 and Γ_2 have the same pure reduced normal form, we are able to "compare" mixed conjectures in the two pure reduced normal forms, and therefore, by using the functions g_i^1 and g_i^2, are able to "compare" behavioral conjectures in Γ_1 and Γ_2.

In view of Theorem 6.6.5 such a construction may be extended to games Γ_1, Γ_2 which differ by Kohlberg-Mertens transformations, by making use of the mixed reduced normal forms of Γ_1 and Γ_2. Formally, we define functions $\tilde{g}_i^1$ and $\tilde{g}_i^2$ which transform behavioral conjectures in Γ_1 and Γ_2 into "equivalent" mixed conjectures in the corresponding *mixed* reduced normal forms. Such a function $\tilde{g}_i^1$ can be designed in two steps. Take a behavioral conjecture σ_i^1 in Γ_1. By applying the function g_i^1 of Section 6.2, σ_i^1 is transformed into an equivalent mixed conjecture μ_i^{1*} of the pure reduced normal form. Now, suppose that some equivalence class $[s_i^*]$ in the pure reduced normal form is deleted when turning to the mixed reduced normal form, hence there is some randomization μ_i with $\mu_i(s_i) = 0$ for all $s_i \in [s_i^*]$ such that $[s_i^*]$ is equivalent to μ_i. Let μ_i^{1**} be the mixed conjecture in the pure reduced normal form obtained from μ_i^{1*} by putting all the weight that μ_i^{1*} previously assigned to $[s_i^*]$ on the randomization μ_i (or, to be more precise, on the probability distribution on the equivalence classes induced by μ_i). In this way, μ_i^{1*} is transformed into an equivalent mixed conjecture μ_i^{1**} which puts no weight on the deleted equivalence class $[s_i^*]$. By repeating this transformation until no further equivalence class $[s_i^*]$ can be deleted, we obtain a mixed conjecture μ_i^{1**} in the mixed reduced normal form which is equivalent to μ_i^{1*}, and thus is equivalent to σ_i^1. Let $\tilde{g}_i^1$ be the function which transforms every behavioral conjecture σ_i^1 in Γ_1 into this mixed conjecture μ_i^{1**} of the mixed reduced normal form. In the same way, we construct $\tilde{g}_i^2$.

Since by Theorem 6.6.5 we know that Γ_1 and Γ_2 have the same mixed reduced normal form, we can find a function $\tilde{F}_i$ which maps every mixed conjecture μ_i^{1**} in the mixed reduced normal form of Γ_1 to the corresponding mixed conjecture μ_i^{2**} of the mixed reduced normal form of Γ_2.

Finally, let $\tilde{\tau}_i$ be the correspondence mapping behavioral conjectures in Γ_1 to sets of "equivalent" behavioral conjectures in Γ_2. Formally,

$$\tilde{\tau}_i(\sigma_i^1) = \{\sigma_i^2 | \; \tilde{g}_i^2(\sigma_i^2) = \tilde{F}_i(\tilde{g}_i^1(\sigma_i^1))\}$$

for every behavioral conjecture σ_i^1 in Γ_1. Let $\tilde{\tau} = (\tilde{\tau}_i)_{i \in I}$ be the correspondence mapping behavioral conjecture profiles in Γ_1 to sets of "equivalent" behavioral conjecture profiles in Γ_2. Kohlberg-Mertens invariance is now defined in the obvious way.

Definition 6.6.6 *An extensive form solution φ is said to be Kohlberg-Mertens invariant if for every pair Γ_1, Γ_2 of extensive form games that differ solely by Kohlberg-Mertens transformations the following holds: if $\sigma^1 \in \varphi(\Gamma_1)$ then there is some $\sigma^2 \in \tilde{\tau}(\sigma^1)$ with $\sigma^2 \in \varphi(\Gamma_2)$. Here, $\tilde{\tau}$ is the correspondence defined above.*

Similar to Sections 6.3 and 6.5, we would wish to find solutions which are both Kohlberg-Mertens invariant and assign to every extensive form game some nonempty set of "reasonable" conjecture profiles. The following lemma, however, shows that this is impossible if one restricts attention to point-valued solutions, and substitutes "reasonable" by "subgame perfect".

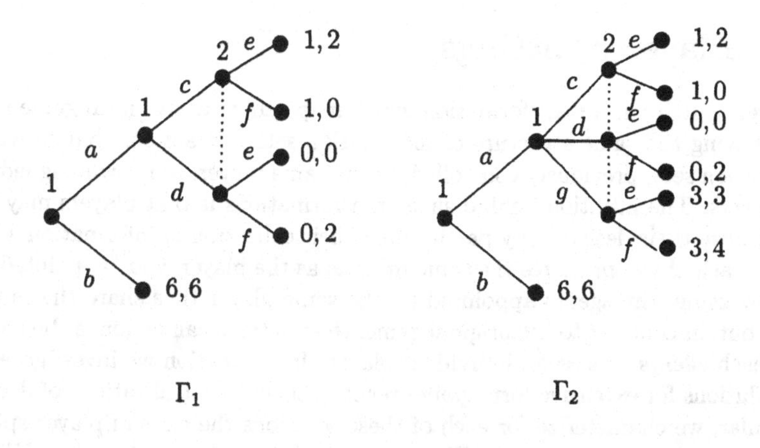

Figure 6.18

Lemma 6.6.7 *There is no point-valued solution for extensive form games which is Kohlberg-Mertens invariant and assigns to every game a nonempty set of subgame perfect equilibria.*

Proof. Suppose that φ is a point-valued solution which to every extensive form game assigns a nonempty set of subgame perfect equilibria. Consider the games Γ_1 and Γ_2 in Figure 6.18, which are taken from Mailath, Samuelson and Swinkels (1993). As may be verified easily, Γ_1 and Γ_2 have the same mixed reduced normal form, and thus they differ by Kohlberg-Mertens transformations. In fact, Γ_2 is obtained from Γ_1 by adding the action g for player 1, which is equivalent to the randomization $0.5(a,d) + 0.5b$ in Γ_1. The unique subgame perfect equilibrium in Γ_1 is $((b,c),e)$ whereas the unique subgame perfect equilibrium in Γ_2 is $((b,g),f)$. Hence, $\varphi(\Gamma_1) = \{((b,c),e)\}$ and $\varphi(\Gamma_2) = \{((b,g),f)\}$. Since the conjectures about player 2 in $\varphi(\Gamma_1)$ and $\varphi(\Gamma_2)$ are obviously not equivalent, it follows that φ is not Kohlberg-Mertens invariant. ■

The above negative result may be seen as another motivation to introduce set-valued solutions, in addition to the two motivations already lined out in Section 5.3.1. Indeed, the notion of Kohlberg-Mertens invariance has been extended in the literature to the class of set-valued solutions, and particular set-valued solutions φ have been developed with the following property: φ is Kohlberg-Mertens invariant, and φ assigns to every game Γ a collection $\varphi(\Gamma)$ of sets of conjecture profiles such that every set $E \in \varphi(\Gamma)$ contains a sequential equilibrium. Solutions with this property are, for instance, the stability concepts developed in Mertens (1989, 1991) and Hillas (1990). A detailed discussion of these stability concepts is outside the scope of this monograph, however.

6.7 Player Splittings

A player splitting is a transformation which maps an extensive form game into a new game having the same structure of information sets and actions, but in which some information sets, previously controlled by the same player, are now assigned to different players. The intuition behind such transformations is that players may decide to decentralize their decisions by partitioning their collection of information sets among *agents*, each of whom shares the same utilities as the player who "appointed" him. In the new game, the agents appointed by the same player thus share the same preferences, but, in contrast to the original game, their actions can no longer be coordinated since each agents acts as an individual player. In his section we investigate how several solutions for extensive form games react to such decentralizations of decisions. In particular, we characterize for each of these solutions the class of player splittings to which this solution is invariant. The results in this section are based on Kline (1997) and Perea, Jansen and Vermeulen (2000). In this section, we restrict ourselves to games with perfect recall. In the two papers above, also games without perfect recall are considered.

We first formally introduce *player splittings* and *invariance* with respect to player splittings. Let $\mathcal{S}$ be an extensive form structure satisfying perfect recall with player set I, in which every player i controls the collection H_i of information sets.

Definition 6.7.1 *A player splitting on* $\mathcal{S}$ *is a pair* $\pi = ((L_i)_{i\in I}, (l_i)_{i\in I})$ *where* $(L_i)_{i\in I}$ *is a disjoint collection of finite sets and* l_i *is a surjective function from* H_i *to* L_i *for every* $i \in I$.

For every information set $h \in H_i$, the image $l_i(h) \in L_i$ denotes the *agent* to which h is assigned. Hence, the new set of players (or agents) is given by $L = \cup_{i\in I} L_i$, and each agent $l \in L_i$ controls the collection $H_l = \{h \in H_i | \; l_i(h) = l\}$ of information sets previously controlled by player i. For every agent $l \in L$, let $i(l) \in I$ be the player of the original game to which l belongs, that is, $l \in L_{i(l)}$. The extensive form structure $\mathcal{S}$ is thus transformed into a new extensive form structure, denoted by $\pi(\mathcal{S})$, which differs from $\mathcal{S}$ only by the player labels at the information sets. Let Z be the set of terminal nodes in $\mathcal{S}$ and $u : Z \to \mathbb{R}^I$ a utility function to $\mathcal{S}$. By $\pi(u) : Z \to \mathbb{R}^L$ we denote the utility function to $\pi(\mathcal{S})$ given by $(\pi(u))(z) = (v_l(z))_{l\in L}$ with $v_l(z) = u_{i(l)}(z)$ for all $z \in Z$. Hence, agent l in the new game receives the same utility as player $i(l)$ in the original game at every terminal node. The pair $\pi(\Gamma) = (\pi(\mathcal{S}), \pi(u))$ is thus a new extensive form game, to which we refer as the game induced by the player splitting π. An extreme case of a player splitting is to assign each information set to a different agent. In this case, the resulting game is what we have called the *agent extensive form* of a game.

Consider a game Γ and a player splitting π on the extensive form structure in Γ. Let $\pi(\Gamma)$ be the new game induced by π. In order to define invariance of a solution φ under the player splitting π, we need to compare the sets of behavioral conjecture profiles selected by φ in Γ and $\pi(\Gamma)$. Note, however, that the space of behavioral conjecture *profiles* in Γ is the same as in $\pi(\Gamma)$, and therefore, unlike with Thompson-Elmes-Reny and Kohlberg-Mertens invariance, no additional transformation is needed to compare $\varphi(\Gamma)$ with $\varphi(\pi(\Gamma))$. We simply say that the solutions in Γ and $\pi(\Gamma)$ coincide if $\varphi(\Gamma) = \varphi(\pi(\Gamma))$. In this section we restrict our attention to point-valued solutions. An extension of player splitting invariance to set-valued solutions can be found in Perea, Jansen and Vermeulen (2000).

Definition 6.7.2 *Let φ be a (point-valued) solution for extensive form games. Let S be an extensive form structure with perfect recall and π a player splitting on S. We say that φ is invariant with respect to π if for every game Γ with extensive form structure S the following holds: $\sigma \in \varphi(\Gamma)$ if and only if $\sigma \in \varphi(\pi(\Gamma))$.*

Note that we require the solutions of the original and the new game to coincide for *all possible utilities* at the terminal node, and not just for one particular utility function. The reason is that a player splitting π may not make a difference for a particular utility function u, while changing the solution for another utility function u'. Invariance thus makes a statement about the impact of a player splitting *independent* of the utilities at the terminal nodes.

In Perea, Jansen and Vermeulen (2000), a stronger version of invariance is considered. In their notion, not only the solution of $\pi(\Gamma)$ should coincide with the solution at Γ, but also the solution of every game Γ' which is *equivalent* to $\pi(\Gamma)$. By *equivalent* we mean that $\pi(\Gamma)$ and Γ' have the same extensive form structure, and that the utility for agent l at a terminal node z in Γ' coincides with the utility for agent l at z in $\pi(\Gamma)$ whenever agent l appears on the path to z. The games $\pi(\Gamma)$ and Γ' may thus yield different utilities for agent l at those terminal nodes that are not preceded by agent l. This stronger version of invariance is then a generalization of the *player splitting property* as introduced in Mertens (1989). The results in this section would not change, however, if this stronger version of invariance were used.

The solutions investigated in this section are the Nash equilibrium solution, the subgame perfect equilibrium solution, the perfect equilibrium solution, the normal form perfect equilibrium solution, the quasi-perfect equilibrium solution, the proper equilibrium solution and the sequential equilibrium solution. The sequential equilibrium solution denotes the correspondence which always selects the set of behavioral conjecture profiles which can be extended to a sequential equilibrium. Among these solutions, it can easily be shown that the perfect equilibrium solution and sequential equilibrium solution are invariant against *every* player splitting.

Lemma 6.7.3 *The perfect equilibrium solution and the sequential equilibrium solution are invariant with respect to every player splitting.*

Proof. Let Γ be an arbitrary extensive form game, and π an arbirary player splitting on the extensive form structure of Γ. Let $\pi(\Gamma)$ be the extensive form game obtained after the player splitting. From the definition of perfect equilibrium, it follows immediately that the set of perfect equilibria in Γ and $\pi(\Gamma)$ coincide. By Lemma 4.3.6 we know that an assessment (σ, β) is a sequential equilibrium if and only if it is consistent and locally sequentially rational. Since the set of consistent and locally sequentially rational assessments in Γ and $\pi(\Gamma)$ coincide, it follows that the set of sequential equilibria in Γ and $\pi(\Gamma)$ are the same. ■

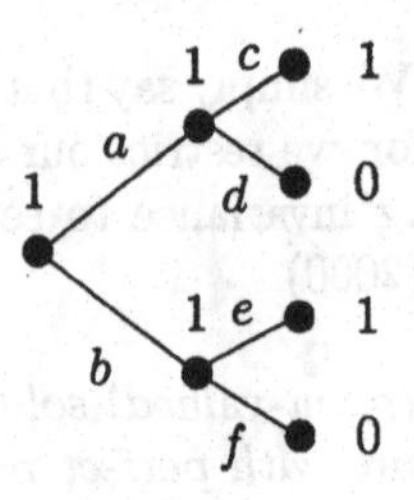

Figure 6.19

In the remainder, it is shown that each of the other solutions is not invariant with respect to all player splittings. The following lemma provides a reason why some of the above mentioned solutions cannot be invariant with respect to all player splittings. It turns out, namely, that invariance with respect to all player splittings does not allow for solutions which solely select behavioral conjectures that exclude weakly dominated strategies. A solution φ is called *undominated* if for every game Γ, every $(\sigma_i)_{i\in I} \in \varphi(\Gamma)$ and every player i the conjecture σ_i assigns probability zero to all weakly dominated strategies. A solution is called *nonempty* if it assigns to every game a nonempty set of behavioral conjecture profiles.

Lemma 6.7.4 *There is no nonempty solution φ which is undominated and invariant with respect to all player splittings.*

Proof. Suppose that the solution φ is nonempty and undominated. Consider the game Γ in Figure 6.19. Let h_1, h_2, h_3 denote the left, upper-right and lower-right information set, respectively. Let π_1 be the player splitting which assigns h_1 and h_2 to the same agent, and assigns h_3 to a different agent. Since (a, c) and e are the unique strategies in $\pi_1(\Gamma)$ which are not weakly dominated, it follows that $\varphi(\pi_1(\Gamma)) = \{((a, c), e)\}$. Let π_2 be an alternative player splitting which assigns h_1 and h_3 to the same agent, and h_2 to some other agent. In $\pi_2(\Gamma)$, the unique strategies which are not weakly dominated are (b, e) and c, and hence $\varphi(\pi_2(\Gamma)) = \{((b, e), c)\}$. But then, φ cannot be invariant with respect to both π_1 and π_2. ■

Since the normal form perfect equilibrium solution, quasi-perfect equilibrium solution and the proper equilibrium solution are undominated, it follows that these solutions cannot be invariant with respect to all player splittings. Consider the game Γ in Figure 6.20. Let π be the player splitting which assigns the two information sets controlled by player 1 to different agents. Then, (b, d, e) is a subgame perfect equilibrium and hence a Nash equilibrium in $\pi(\Gamma)$, but $((b, d), e)$ is not a Nash equibrium and hence not a subgame perfect equilibrium in Γ. It thus follows that the Nash equilibrium solution and the subgame perfect equilibrium solution are not invariant with respect to π. These observations raise the questions whether we can characterize the class of player splittings with respect to which the above mentioned solutions *are* invariant.

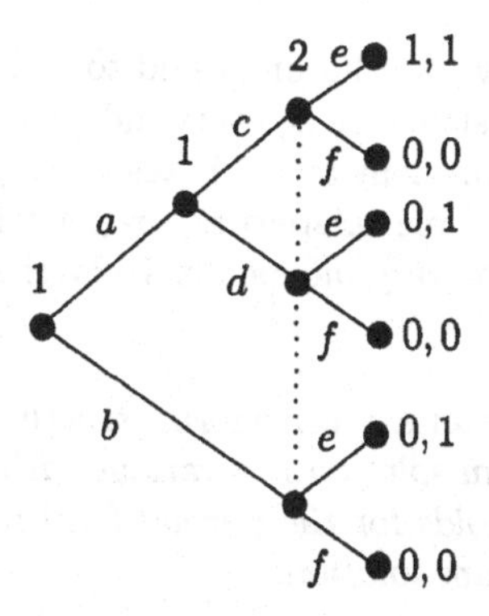

Figure 6.20

It turns out that the Nash equilibrium solution, normal form perfect equilibrium solution and quasi-perfect equilibrium solution are invariant with respect to every *independent* player splitting. Moreover, these solutions fail to be invariant with respect to every player splitting which is not independent, and thus the class of independent player splittings can be characterized as the class of *all* player splittings which leave these solutions invariant.

Definition 6.7.5 *Let S be an extensive form structure and $\pi = ((L_i)_{i \in I}, (l_i)_{i \in I})$ a player splitting on S. We say that π is independent if for every player i and every $h, h' \in H_i$ such that h' follows h, it holds that $l_i(h) = l_i(h')$.*

Hence, a player splitting is independent if player i information sets on the same path are assigned to the same agent. Or, in other words, only player i information sets which do not precede or follow one another can be assigned to different agents. In Mertens (1989) it is argued that independent player splittings do not change the "strategic features" of a game, and in accordance with this argument invariance with respect to independent player splittings should thus be viewed a desirable property. An argument that could be used to defend this point of view is the following. Suppose that an independent player splitting assigns two information sets h, h', previously controlled by player i, to different agents. If h is reached, the agent at h knows that h' cannot be reached anymore since h and h' are in different braches of the game tree. The same holds for h'. Hence, the decisions at h and h' do not affect one another, and it should therefore not matter whether these decisions are taken by the same player, or by different agents having the same preferences as this player. We shall see, however, that the proper equilibrium solution is not invariant against some independent player splittings.

Conform Mertens (1989), a solution is said to satisfy the *player splitting property* if for every extensive form structure S, every independent player splitting π on S and every game Γ' which is equivalent to $\pi(\Gamma)$ (see above) the following holds: $\sigma \in \varphi(\Gamma)$ if and only if $\sigma \in \varphi(\Gamma')$. The player splitting property is thus somewhat stronger than "invariance with respect to every independent player splitting".

Theorem 6.7.6 *Let S be an extensive form structure and π a player splitting on S. Then, the Nash equilibrium solution is invariant with respect to π if and only if π is independent. The same holds for the normal form perfect equilibrium solution and the quasi-perfect equilibrium solution.*

Before providing the proof of this result, we present a technical lemma which will be used repeatedly in the proof. Let Γ be a game, $\pi = ((L_i)_{i\in I}, (l_i)_{i\in I})$ an independent player splitting on the extensive form structure of Γ, and $\pi(\Gamma)$ the game obtained from Γ by applying π. To fix notation, let the players in Γ be denoted by i and the players in $\pi(\Gamma)$ by l. Consider a strategy s_i for player i and some agent $l \in L_i$. Let H_l be the collection of information sets controlled by agent l in $\pi(\Gamma)$. Let $s_i|_l$ be the restriction of s_i to information sets in $H_i(s_i) \cap H_l$. Let $H_l(s_i|_l)$ be the set of agent l information sets not avoided by $s_i|_l$. Since π is independent, it may be verified that $H_l(s_i|_l) = H_i(s_i) \cap H_l$, and hence $s_i|_l$ constitutes a strategy for agent l in $\pi(\Gamma)$. We denote the utilities in Γ and $\pi(\Gamma)$ by u_i and v_l respectively.

Lemma 6.7.7 *Let Γ be an extensive form game and π an independent player splitting on the extensive form structure of Γ.*
(a) Let σ be a behavioral conjecture profile, $l \in L_i$ and t_l^1, t_l^2 two strategies for agent l with $v_l(t_l^1, \sigma_{-l}) < v_l(t_l^2, \sigma_{-l})$. Then, for all strategies s_i^1, s_i^2 for player i with $s_i^1|_l = t_l^1$, $s_i^2|_l = t_l^2$ and $s_i^1|_k = s_i^2|_k$ for all $k \in L_i\backslash\{l\}$, we have that $u_i(s_i^1, \sigma_{-i}) < u_i(s_i^2, \sigma_{-i})$.
(b) Let σ be a behavioral conjecture profile, $i \in I$ and $u_i(s_i^1, \sigma_{-i}) < u_i(s_i^2, \sigma_{-i})$. Then, there is an agent $l \in L_i$ such that $v_l(s_i^1|_l, \sigma_{-l}) < v_l(s_i^2|_l, \sigma_{-l})$.

Proof. (a) Suppose that $l \in L_i$ and $v_l(t_l^1, \sigma_{-l}) < v_l(t_l^2, \sigma_{-l})$. Let $s_i^1, s_i^2 \in S_i$ be such that $s_i^1|_l = t_l^1$, $s_i^2|_l = t_l^2$ and $s_i^1|_k = s_i^2|_k$ for all $k \in L_i\backslash\{l\}$. Let Z_l be the collection of terminal nodes preceded by some information set $h \in H_l$. Then, we have

$$\begin{aligned} u_i(s_i^1, \sigma_{-i}) &= \sum_{z\in Z} \mathbb{P}_{(s_i^1,\sigma_{-i})}(z)\, u_i(z) \\ &= \sum_{z\in Z_l} \mathbb{P}_{(s_i^1,\sigma_{-i})}(z)\, u_i(z) + \sum_{z\notin Z_l} \mathbb{P}_{(s_i^1,\sigma_{-i})}(z)\, u_i(z). \end{aligned}$$

Since the player splitting π is independent, we have for every $z \in Z_l$ that $\mathbb{P}_{(s_i^1,\sigma_{-i})}(z)$ does not depend upon agent k's behavior if $k \in L_i\backslash\{l\}$. Hence, we may conclude that $\mathbb{P}_{(s_i^1,\sigma_{-i})}(z) = \mathbb{P}_{(t_l^1,\sigma_{-l})}(z)$ for all $z \in Z_l$. Hence,

$$\begin{aligned} u_i(s_i^1,\sigma_{-i}) &= \sum_{z\in Z_l} \mathbb{P}_{(t_l^1,\sigma_{-l})}(z)\,u_i(z) + \sum_{z\notin Z_l} \mathbb{P}_{(s_i^1,\sigma_{-i})}(z)\,u_i(z) \\ &= v_l(t_l^1,\sigma_{-l}) - \sum_{z\notin Z_l} \mathbb{P}_{(t_l^1,\sigma_{-l})}(z)\,u_i(z) + \sum_{z\notin Z_l} \mathbb{P}_{(s_i^1,\sigma_{-i})}(z)\,u_i(z) \\ &< v_l(t_l^2,\sigma_{-l}) - \sum_{z\notin Z_l} \mathbb{P}_{(t_l^1,\sigma_{-l})}(z)\,u_i(z) + \sum_{z\notin Z_l} \mathbb{P}_{(s_i^1,\sigma_{-i})}(z)\,u_i(z). \end{aligned}$$

For every $z \notin Z_l$, the probability $\mathbb{P}_{(\cdot,\sigma_{-i})}(z)$ does not depend upon agent l's behavior. Since s_i^1 and s_i^2 differ only at agent l's information sets, $\mathbb{P}_{(s_i^1,\sigma_{-i})}(z) = \mathbb{P}_{(s_i^2,\sigma_{-i})}(z)$ for all $z \notin Z_l$. Moreover, since (t_l^1,σ_{-l}) and (t_l^2,σ_{-l}) differ only at agent l's information sets, $\mathbb{P}_{(t_l^1,\sigma_{-l})}(z) = \mathbb{P}_{(t_l^2,\sigma_{-l})}(z)$ for all $z \notin Z_l$. Hence,

$$\begin{aligned} u_i(s_i^1,\sigma_{-i}) &< v_l(t_l^2,\sigma_{-l}) - \sum_{z\notin Z_l} \mathbb{P}_{(t_l^2,\sigma_{-l})}(z)\,u_i(z) + \sum_{z\notin Z_l} \mathbb{P}_{(s_i^2,\sigma_{-i})}(z)\,u_i(z) \\ &= u_i(s_i^2,\sigma_{-i}), \end{aligned}$$

which was to show.

(b) Suppose that $u_i(s_i^1,\sigma_{-i}) < u_i(s_i^2,\sigma_{-i})$. Let Z_i be the set of terminal nodes preceded by some player i information set. Since π is an independent player splitting, we have that $Z_i = \cup_{l\in L_i} Z_l$, where the sets Z_l are disjoint. Hence,

$$u_i(s_i^1,\sigma_{-i}) = \sum_{l\in L_i}\sum_{z\in Z_l} \mathbb{P}_{(s_i^1,\sigma_{-i})}(z)\,u_i(z) + \sum_{z\notin Z_i} \mathbb{P}_{(s_i^1,\sigma_{-i})}(z)\,u_i(z),$$

and

$$u_i(s_i^2,\sigma_{-i}) = \sum_{l\in L_i}\sum_{z\in Z_l} \mathbb{P}_{(s_i^2,\sigma_{-i})}(z)\,u_i(z) + \sum_{z\notin Z_i} \mathbb{P}_{(s_i^2,\sigma_{-i})}(z)\,u_i(z).$$

For every $z \notin Z_i$, we have that $\mathbb{P}_{(s_i^1,\sigma_{-i})}(z) = \mathbb{P}_{(s_i^2,\sigma_{-i})}(z)$. The inequality $u_i(s_i^1,\sigma_{-i}) < u_i(s_i^2,\sigma_{-i})$ thus implies that

$$\sum_{l\in L_i}\sum_{z\in Z_l} \mathbb{P}_{(s_i^1,\sigma_{-i})}(z)\,u_i(z) < \sum_{l\in L_i}\sum_{z\in Z_l} \mathbb{P}_{(s_i^2,\sigma_{-i})}(z)\,u_i(z),$$

and hence there is some agent $l \in L_i$ with

$$\sum_{z\in Z_l} \mathbb{P}_{(s_i^1,\sigma_{-i})}(z)\,u_i(z) < \sum_{z\in Z_l} \mathbb{P}_{(s_i^2,\sigma_{-i})}(z)\,u_i(z).$$

For every $z \in Z_l$, the probability $\mathbb{P}_{(\cdot,\sigma_{-i})}(z)$ does not depend upon agent k's behavior for every $k \in L_i\backslash\{l\}$, and hence

$$\sum_{z\in Z_l} \mathbb{P}_{(s_i^1|_l,\sigma_{-l})}(z)\,u_i(z) < \sum_{z\in Z_l} \mathbb{P}_{(s_i^2|_l,\sigma_{-l})}(z)\,u_i(z).$$

However, this implies that $v_l(s_i^1|_l, \sigma_{-l}) < v_l(s_i^2|_l, \sigma_{-l})$, which was to show. ■

Proof of Theorem 6.7.6. Let π be an independent player splitting on $\mathcal{S}$. We show that each of the solutions named in the theorem are invariant with respect to π. Choose an arbitrary game Γ with extensive form structure $\mathcal{S}$, and let $\pi(\Gamma)$ be the game obtained after applying the player splitting π. We prove that the Nash equilibria, normal form perfect equilibria and quasi-perfect equilibria in Γ and $\pi(\Gamma)$ coincide.
1. Nash equilibrium. It may be easily checked by the reader that every Nash equilibrium in Γ is a Nash equilibrium in $\pi(\Gamma)$. Now, let σ be a Nash equilibrium in $\pi(\Gamma)$. Suppose that σ is not a Nash equilibrium in Γ, hence there is some player i and s_i, s_i' such that $\sigma_i(s_i) > 0$ but $u_i(s_i, \sigma_{-i}) < u_i(s_i', \sigma_{-i})$. By Lemma 6.7.7 (b) there is some agent $l \in L_i$ such that $v_l(s_i|_l, \sigma_{-l}) < v_l(s_i'|_l, \sigma_{-l})$. Let σ_l be the restriction of σ_i to information sets in H_l. Then, we have that $\sigma_l(s_i|_l) > 0$ but $v_l(s_i|_l, \sigma_{-l}) < v_l(s_i'|_l, \sigma_{-l})$. However, this contradicts the fact that σ is a Nash equilibrium in $\pi(\Gamma)$. Hence, σ is a Nash equilibrium in Γ.
2. Normal form perfect equilibrium. Let σ be a normal form perfect equilibrium in Γ. Hence, there is some sequence $(\sigma^n)_{n\in\mathbb{N}}$ of strictly positive behavioral conjecture profiles converging to σ such that $\sigma_i(s_i) > 0$ only if s_i is a best response against σ^n for every n. We show for every agent l that $\sigma_l(t_l) > 0$ only if t_l is a best response against σ^n for all n, which would imply that σ is a normal form perfect equilibrium in $\pi(\Gamma)$. Suppose that t_l is not a best response against some σ^n. Then, there is some t_l' with $v_l(t_l, \sigma^n_{-l}) < v_l(t_l', \sigma^n_{-l})$. Let the strategies s_i, s_i' be such that $s_i|_l = t_l$, $s_i'|_l = t_l'$ and $s_i|_k = s_i'|_k$ for every $k \in L_i\backslash\{l\}$. Then, by Lemma 6.7.7 (a) we have that $u_i(s_i, \sigma^n_{-i}) < u_i(s_i', \sigma^n_{-i})$, and hence $\sigma_i(s_i) = 0$. Since this holds for every s_i with $s_i|_l = t_l$, we have that $\sigma_l(t_l) = 0$. We may thus conclude that σ is a normal form perfect equilibrium in $\pi(\Gamma)$.

Now, suppose that σ is a normal form perfect equilibrium in $\pi(\Gamma)$. Hence, there is some sequence $(\sigma^n)_{n\in\mathbb{N}}$ of strictly positive behavioral conjecture profiles converging to σ such that $\sigma_l(s_l) > 0$ only if s_l is a best response against σ^n for every n. We prove that $\sigma_i(s_i) > 0$ only if s_i is a best response against σ^n for every n, which would imply that σ is a normal form perfect equilibrium in Γ. Suppose that s_i is not a best response against some σ^n, hence there is some s_i' such that $u_i(s_i, \sigma^n_{-i}) < u_i(s_i', \sigma^n_{-i})$. By Lemma 6.7.7 (b) there is some agent $l \in L_i$ such that $v_l(s_i|_l, \sigma^n_{-l}) < v_l(s_i'|_l, \sigma^n_{-l})$. Then, $\sigma_l(s_i|_l) = 0$ which implies that $\sigma_i(s_i) = 0$. Hence, we may conclude that σ is a normal form perfect equilibrium in Γ.
3. Quasi-perfect equilibrium. Let σ be a quasi-perfect equilibrium in Γ. Hence, there is a sequence $(\sigma^n)_{n\in\mathbb{N}}$ of strictly positive behavioral conjecture profiles converging to σ such that for every player i and every information set $h \in H_i$ we have that $\sigma_{i|h}(s_i) > 0$ only if s_i is a sequential best response against σ^n_{-i} at h for all n. For every agent $l \in L_i$ and $h \in H_l$, let $\sigma_{l|h}$ be the revised conjecture at h induced by σ_l. We prove that $\sigma_{l|h}(t_l) > 0$ only if t_l is a sequential best response against σ^n_{-l} at h for all n. Suppose that t_l is not a sequential best response at h against σ^n_{-l}. Let $\beta^n_{lh} = \beta^n_{ih}$ be the beliefs at h induced by σ^n. Then, there is some t_l' with $v_l((t_l, \sigma^n_{-l})|\ h, \beta^n_{ih}) < v_l((t_l', \sigma^n_{-l})|\ h, \beta^n_{ih})$. Let $s_i, s_i' \in S_i(h)$ be such that s_i coincides with t_l at h and all information sets following h, s_i' coincides with t_l' at h and all information sets following h, and s_i coincides with s_i' elsewhere. Then, we have that $u_i((s_i, \sigma^n_{-i})|\ h, \beta^n_{ih}) < u_i((s_i', \sigma^n_{-i})|$

$h, \beta_{ih}^n)$ and hence $\sigma_{i|h}(s_i) = 0$. Since this holds for every $s_i \in S_i(h)$ which coincides with t_l at h and all information sets following h, it follows that $\sigma_{l|h}(t_l) = 0$. We may conclude that σ is a quasi-perfect equilibrium in $\pi(\Gamma)$.

Now, suppose that σ is a quasi-perfect equilibrium in $\pi(\Gamma)$. Hence, there is a sequence $(\sigma^n)_{n \in \mathbb{N}}$ of strictly positive behavioral conjecture profiles such that for every i, every $l \in L_i$ and every $h \in H_l$ it holds that $\sigma_{l|h}(t_l) > 0$ only if t_l is a sequential best response against σ_{-l}^n at h for every n. We prove that $\sigma_{i|h}(s_i) > 0$ only if s_i is a sequential best response against σ_{-i}^n at h for all n. Suppose that s_i is not a sequential best response against σ_{-i}^n at h. Then, there is some $s_i' \in S_i(h)$ with $u_i((s_i, \sigma_{-i}^n)| \; h, \beta_{ih}^n) < u_i((s_i', \sigma_{-i}^n)| \; h, \beta_{ih}^n)$. Since π is indendent, all player i information sets following h are controlled by agent l as well, and therefore $v_l((s_i|_l, \sigma_{-l}^n)| \; h, \beta_{ih}^n) < v_l((s_i'|_l, \sigma_{-l}^n)| \; h, \beta_{ih}^n)$. Then, $\sigma_{l|h}(s_i|_l) = 0$, which implies that $\sigma_{i|h}(s_i) = 0$. We may conclude that σ is a quasi-perfect equilibrium in Γ.

We finally prove that every player splitting which is not independent changes the set of Nash equilibria, normal form perfect equilibria and quasi-perfect equilibria for some particular choice of utilities at the terminal nodes. Let π be player splitting on S which is not independent. Then we can find two nodes x^1 and x^2 such that x^1 precedes x^2, x^1 and x^2 were previously controlled by the same player i in S, but are controlled by two different agents in $\pi(S)$, say $l^1, l^2 \in L_i$. Let $h^1, h^2 \in H_i$ be the information sets to which x^1, x^2 belong. By perfect recall, there is a unique action $a^1 \in A(h^1)$ leading to h^2. Since by assumption h^1, h^2 contain at least two actions we can choose an action $b^1 \in A(h^1) \backslash \{a^1\}$ and two different actions $a^2, b^2 \in A(h^2)$.

Construct the utilities u^1 at the terminal nodes as follows. For every terminal node z following node x^2 and action a^2, let $u_i^1(z) = 1$. For all other terminal nodes z, let $u_i^1(z) = 0$. For all players $j \neq i$, let $u_j^1(z) = 0$ for all terminal nodes z. Let $\Gamma_1 = (S, u^1)$.

Let the behavioral conjecture profile σ^1 be as follows. Player i is believed to choose b^1 with probability one at h^1, is believed to choose b^2 with probability one h^2, and at all other player i information sets preceding h^2 is believed to choose, with probability one, the unique action leading to h^2. Every player $j \neq i$, at every information set h on the path to x^2, is believed to choose with probability one the action which leads to x^2. It may be verified easily that σ^1 is a Nash equilibrium in $\pi(\Gamma_1)$, since $u_l^1(s_l, \sigma_{-l}^1) = 0$ for every agent $l \in L_i$ and all strategies s_l. However, σ^1 is not a Nash equilibrium in Γ_1. Note that σ_i^1 assigns probability one to a strategy s_i for player i with $u_i^1(s_i, \sigma_{-i}^1) = 0$. However, player i can achieve a strictly positive utility by choosing a strategy s_i' in which he chooses all player i actions leading to h^2 (including a^1) and chooses a^2 at h^2.

Construct now the utilities u^2 as follows. For every terminal node z following node x^2 and action b^2 let $u_i^2(z) = -1$. For all other terminal nodes z let $u_i^2(z) = 0$. For all players $j \neq i$, let $u_j^2(z) = 0$ for all terminal nodes z. Let $\Gamma_2 = (S, u^2)$.

Let the behavioral conjecture profile σ^2 be as follows. Player i is believed to choose, with probability one, all actions that lead to h^2 (including a^1) and is believed to choose, with probability one, a^2 at h^2. Every player $j \neq i$ is believed to choose, at every information set h on the path to x^2, with probability one the action leading to x^2. It may be verified that σ^2 is a normal form perfect equilibrium and quasi-perfect equilibrium in Γ_2. However, σ^2 is not a normal form perfect equilibrium nor a quasi-perfect equilibrium in $\pi(\Gamma_2)$. Note that $\sigma^2_{l^1}$ assigns probability one to a strategy s_{l^1} for agent l^1 that is weakly dominated in $\pi(\Gamma_2)$ by the strategy s'_{l^1} which chooses all actions leading to h^1 and chooses b^1 at h^1. The reason is that the strategy s'_{l^1} always yields utility 0 for agent l^1, whereas the strategy s_{l^1} may give him a strictly negative utility if agent l^2 chooses b^2 at h^2. Since a normal form perfect equilibrium and a quasi-perfect equilibrium always assign probability zero to weakly dominated strategies, we must conclude that σ^2 is not a normal form perfect equilibrium nor a quasi-perfect equilibrium in $\pi(\Gamma_2)$. ■

The subgame perfect equilibrium solution is invariant with respect to some player splittings that are not independent. For instance, if the extensive form structure is one of perfect information, then the subgame perfect equilibrium solution is invariant with respect to *all* player splittings on this structure since the set of subgame perfect equilibria concides with the set of sequential equilibria for games with perfect information. However, as we have seen in the example of Figure 6.20, there are player splittings that do change the set of subgame perfect equilibria in a game. We now characterize the class of player splittings to which the subgame perfect equilibrium solution is invariant. This result is due to Kline (1997).

Definition 6.7.8 *Let S be an extensive form structure and π a player splitting on S. We say that π is subgame independent if for every player i and every $h^1, h^2 \in H_i$ the following holds: if h^1 precedes h^2 and $l_i(h^1) \neq l_i(h^2)$, then there is some subgame Γ_x such that h^2 belongs to Γ_x and h^1 not.*

Hence, if two player i information sets on the same path are assigned to different agents, then there should be a node x in the game between these information sets such that all players after x know that the game has passed through x. Note that an independent player splitting is always subgame independent, but not vice versa. Moreover, in a game with perfect information, every player splitting is subgame independent.

Theorem 6.7.9 *Let S be an extensive form structure and π a player splitting on S. Then, the subgame perfect equilibrium solution is invariant with respect to π if and only if π is subgame independent.*

Proof. Let Γ be an extensive form game and π a subgame independent player splitting on the extensive form structure of Γ. We prove that the sets of subgame perfect equilibria in Γ and $\pi(\Gamma)$ coincide. We procede by induction on the number of subgames in Γ.

If Γ contains exactly one subgame, that is, the only subgame is the whole game Γ, then π is independent, and the sets of subgame perfect equilibria in Γ and $\pi(\Gamma)$ coincide with the sets of Nash equilibria. Since by Theorem 6.7.6 the sets of Nash equilibria in Γ and $\pi(\Gamma)$ coincide, it follows that the sets of subgame perfect equilibria in Γ and $\pi(\Gamma)$ coincide.

Suppose that Γ contains n subgames, and that the claim holds for all games with at most $n-1$ subgames. It is easily verified that every subgame perfect equilibrium in Γ is a subgame perfect equilibrium in $\pi(\Gamma)$. Now, let σ be a subgame perfect equilibrium in $\pi(\Gamma)$. Consider a proper subgame Γ_x in Γ. Let Γ_{-x} be the truncated game in which the node x is replaced by a terminal node z_x with utilities equal to the expected utilities generated by σ_x in Γ_x. Let σ_{-x} be the truncated behavioral conjecture profile in Γ_{-x} obtained from σ. Let π_{-x} be the induced player splitting on Γ_{-x} and π_x the induced player splitting on Γ_x. Obviously, π_{-x} and π_x are subgame independent. Since σ is a subgame perfect equilibrium in $\pi(\Gamma)$, it follows immediately that σ_{-x} and σ_x are subgame perfect equilibria in $\pi_{-x}(\Gamma_{-x})$ and $\pi_x(\Gamma_x)$ respectively. Since Γ_{-x} and Γ_x have at most $n-1$ subgames and π_{-x}, π_x are subgame independent, we know by the induction assumption that σ_{-x} is a subgame perfect equilibrium in Γ_{-x} and that σ_x is a subgame perfect equilibrium in Γ_x. But then, the conjecture profile $\sigma = (\sigma_{-x}, \sigma_x)$ is a subgame perfect equilibrium in Γ. Hence, we may conclude that the sets of subgame perfect equilibria do not change by applying player splittings that are subgame independent.

Suppose now that the player splitting π is not subgame independent. Then, there is a player i and information sets $h^1, h^2 \in H_i$ such that h^2 follows h^1, $l_i(h^1) \neq l_i(h^2)$ and for every subgame Γ_x it holds that either h^1 and h^2 are contained in Γ_x, or h^1 and h^2 are both contained in the complement of Γ_x. Let $l^1 = l_i(h^1)$ and $l^2 = l_i(h^2)$. Let x^1, x^2 be nodes in h^1, h^2 respectively such that x^1 precedes x^2. Let a^1 be the unique action at h^1 which leads to h^2. Let b^1 be some other action at h^1 and let a^2, b^2 be two different actions at h^2.

Construct the utilities u at the terminal nodes as follows. For every terminal node z following node x^2 and action a^2, let $u_i(z) = 1$. For all other terminal nodes z, let $u_i(z) = 0$. For all players $j \neq i$, let $u_j(z) = 0$ for all terminal nodes z. Let $\Gamma = (\mathcal{S}, u)$.

Let the behavioral conjecture profile σ be as follows. Player i is believed to choose b^1 with probability one at h^1, is believed to choose b^2 with probability one at h^2, and at all other player i information sets preceding h^2 he is believed to choose, with probability one, the unique action leading to h^2. Every player $j \neq i$, at every information set h on the path to x^2, is believed to choose, with probability one, the action which leads to x^2.

We first show that σ is a subgame perfect equilibrium in $\pi(\Gamma)$. Take an arbitrary subgame Γ_x in Γ. Then, either h^1 and h^2 are both contained in Γ_x or both h^1 and h^2 are contained in the complement of Γ_x. In the former case, agents l^1 and l^2 both receive zero by choosing any arbitrary strategy against σ_x in $\pi_x(\Gamma_x)$, and hence σ_x is a Nash equilibrium in $\pi_x(\Gamma_x)$. In the latter case, the utilities for all agents are constant in the subgame $\pi_x(\Gamma_x)$, and hence σ_x is trivially a Nash equilibrium in $\pi_x(\Gamma_x)$. We may thus conclude that σ is a subgame perfect equilibrium in $\pi(\Gamma)$.

We finally prove that σ is not a Nash equilibrium, and therefore not a subgame perfect equilibrium, in Γ. Note that player i in σ is believed to choose, with probability one, a strategy s_i that yields him an expected utility zero against σ. However, player i can achieve a strictly positive utility against σ by choosing a strategy s_i' in which he chooses all player i actions leading to h^2 (including a^1) and chooses a^2 at h^2. ■

We conclude with an example which shows that the proper equilibrium solution is

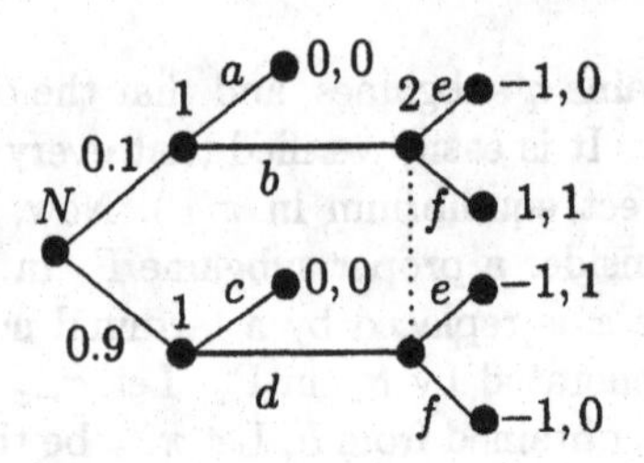

Figure 6.21

not invariant with respect to some independent player splittings. Consider the game in Figure 6.21, which is taken from Cho and Kreps (1987). The normal form of this game is depicted by the table below.

	e	f
(a,c)	$0,0$	$0,0$
(a,d)	$-0.9,0.9$	$-0.9,0$
(b,c)	$-0.1,0$	$0.1,0.1$
(b,d)	$-1,0.9$	$-0.8,0.1$

Let π be the player splitting which assigns the two player 1 information sets to different agents. It may be verified that $\sigma = (a,c,e)$ is a proper equilibrium in $\pi(\Gamma)$. However, σ is not a proper equilibrium in Γ. Take an arbitrary proper sequence $(\sigma^n)_{n\in\mathbb{N}}$ of strictly positive behavioral conjecture profiles converging to σ. Since both (a,d) and (b,d) are worse responses against σ^n than (b,c), it should hold that $\lim_{n\to\infty} \sigma_1^n(a,d)/\sigma_1^n(b,c) = 0$ and $\lim_{n\to\infty} \sigma_1^n(b,d)/\sigma_1^n(b,c) = 0$. Then, e is a worse response against σ^n than f if n is large enough, and thus $\sigma_2(e) = 0$. Hence, σ cannot be a proper equilibrium in Γ.

However, it can be shown that for every game Γ and every independent player splitting π on the extensive form structure of Γ, it holds that every proper equilibrium in Γ is a proper equilibrium in $\pi(\Gamma)$. A proof for this result can be found in Perea, Jansen and Vermeulen (2000).

Chapter 7

Rationalizability

In the concept of Nash equilibrium and its refinements, players are assumed to be informed about the opponents' subjective probability distributions about the other players' strategy choices. In this chapter we investigate the implications of dropping this assumption, while maintaining the assumption that players are informed about the opponents' utility functions at the terminal nodes. The key concept in this chapter is *rationalizability,* as each of the other concepts discussed subsequently is based upon it.

The point of departure in rationalizability is to require each player to hold some subjective probability distribution, or belief, about the opponents' strategy choices, and to impose that each player chooses a strategy which is optimal given his belief. The beliefs described here may be viewed as incomplete, however. The reason is that these beliefs do not cover all relevant events that players may be uncertain about. Namely, if a player is asked to form a belief about the opponents' behavior, he should realize that the opponents' behavior, in turn, depends upon *their* beliefs about the other players' behavior. Hence, the opponents' beliefs about the other players' behavior is a relevant source of uncertainty that a player should take into account when choosing a strategy. A player should therefore not only hold a belief about the opponents' behavior, but also about the opponents' beliefs about the other players' behavior. In contrast to the concept of Nash equilibrium and its refinements, player i's belief about the opponents' beliefs about the other players' behavior need not coincide with the opponents' "real" beliefs.

A player should now realize that his opponents will also take into account the beliefs that other players hold about the other players' behavior. The newly constructed belief space is therefore still not complete, since a player should realize that the opponents' behavior does not only depend upon their beliefs about the other players' behavior, but also about their beliefs about the other players' beliefs about the other players' behavior. Every player should thus, in addition, hold a belief about the other players' beliefs about the other players' beliefs about the other players' behavior. A recursive application of this argument leads to the requirement that for every $k \in \mathbb{N}$ players should hold beliefs about (the other players' beliefs about)^k the other players' behavior. Each player is thus supposed to hold an infinite hierarchy of beliefs.

A question which remains is whether these infinite hierarchies of beliefs exhaust all the possible uncertainty that the players may face in the game. A fundamental result due to Armbruster and Böge (1979), Böge and Eisele (1979) and Mertens and Zamir (1985) states that this is indeed the case. A formal treatment of this result may be found in Section 7.1.

An important consequence of this result is that the infinite hierarchy of beliefs held by a player may be modeled as the player's *type,* conform Harsanyi's (1967-1968) approach to games with incomplete information. The players' type spaces thus reflects all the relevant uncertainty that players may face in the game. As such, a game in which the players' basic uncertainy concerns the behavior by the other players may now be modeled as a game with incomplete information as defined by Harsanyi, which considerably facilitates the representation and analysis.

Another important assumption in the rationalizability models to be presented in this chapter is the following. A player, when forming his infinite hierarchy of beliefs, is assumed to believe that his opponents are rational, which restricts the possible beliefs he may have about the opponents' behavior. Moreover, he is assumed to believe that his opponents believe that all players are rational, which in turn restricts the players' beliefs about the other players' beliefs about the other players' behavior, and so on. By repeating this argument, one is led to the requirement that there be "common belief" about the players' rationality, that is, each player believes that each player is rational, each player believes that each player believes that each player is rational, and so on.

A concept of rationalizability then selects, for each player, those strategies that are optimal against a belief hierarchy that respects common belief of rationality. The various rationalizability concepts treated in this chapter differ, however, in their formalization of *belief hierarchies, optimal strategies* and *common belief of rationality.* For instance, an optimal strategy may be defined as a best response, but also as a sequential best response. As to belief hierarchies, we shall distinguish between *static* belief hierarchies (Section 7.1) and *conditional* belief hierarchies (Section 7.4). By the former we mean that players hold some hierarchy of beliefs at the beginning of the game, and players are not required explicitly to revise their beliefs once the game is under way. Conditional belief hierarchies, in contrast, represent the situation in which players *do* revise their beliefs during the course of the game. The latter may be important since a player, at some of his information sets, may realize that the event of reaching this information set contradicts the belief about the opponents' behavior he held until this point, and hence may be forced to revise his beliefs according to the newly acquired information. Finally, common belief of rationality may be imposed at the beginning of the game only, but may also be required at all information sets (if possible) or at some collection of information sets.

7.1 Static Belief Hierarchies

In this section, we provide a formal model of infinite hierarchies of beliefs in a static setting. By the latter, we mean that players are required to hold beliefs at the beginning of the game, but are not explicitly required to revise their beliefs during the game. In Section 7.4 we introduce belief hierarchies in which players *do* revise their beliefs. Consider an extensive form structure S in which the set of strategies for every player i is given by S_i. We assume that every player i faces uncertainty about the opponents' strategy choices, hence the basic space of uncertainty for player i is $S_{-i} = \times_{j\neq i} S_j$. Each player i is supposed to have some belief on S_{-i}, which we call player i's first-order belief. Define $X_i^1 := S_{-i}$ to be player i's *first-order state space.* Each of player i's strategies $s_i \in S_i$ may be identified with an act $f : S_{-i} \to \Delta(Z)$ that assigns to every $s_{-i} \in S_{-i}$ the probability distribution $f(s_{-i}) := \mathbb{P}_{(s_i,s_{-i})} \in \Delta(Z)$ on the set of terminal nodes. Let $\mathcal{F}(S_{-i})$ be the set of all such acts $f : S_{-i} \to \Delta(Z)$.

Player i is assumed to have a preference relation $\succeq_i^1$ on the set of acts $\mathcal{F}(S_{-i})$, which is called his *first-order preference relation.* As in Section 2.3, we make the assumption that $\succeq_i^1$ can be represented by a *utility function* $u_i : Z \to \mathbb{R}$ and a *subjective probability distribution* $p_i^1 \in \Delta(S_{-i})$. That is, for every $f, g \in \mathcal{F}(S_{-i})$, we have that $f \succeq_i^1 g$ if and only if

$$\sum_{s_{-i}\in S_{-i}} p_i^1(s_{-i}) \sum_{z\in Z} f(s_{-i})(z)\, u_i(z) \geq \sum_{s_{-i}\in S_{-i}} p_i^1(s_{-i}) \sum_{z\in Z} g(s_{-i})(z)\, u_i(z).$$

Here, $f(s_{-i})(z)$ is the objective probability that the act f assigns to terminal node z at state s_{-i}. If the act $f(s_{-i})$ corresponds to the strategy $s_i \in S_i$, then $f(s_{-i})(z)$ is simply the probability that z is reached by the strategy profile (s_i, s_{-i}). Similarly for g. We now assume that the utility functions $(u_i)_{i\in I}$ are common belief among the players. This means, each player believes that the utility functions in the opponents' first-order preferences are given by $(u_i)_{i\in I}$, each player believes that each player believes this, each player believes that each player believes that each player believes this, and so on. The subjective probability distributions, however, are *not* assumed to be common belief. Hence, it is possible that player i believes that his opponents hold subjective probability distributions $(p_j^1)_{j\neq i}$, whereas in reality the opponents hold different subjective probability distributions. If we fix the utility functions $(u_i)_{i\in I}$, player i's first-order preference relation can therefore be represented by the subjective probability distribution $p_i^1 \in \Delta(S_{-i})$, which we call player i's *first-order belief.*

Since every player j has a first-order belief $p_j \in \Delta(S_{-j})$, and player i is uncertain about both the opponents' strategy choices and the opponents' first-order beliefs, player i has uncertainty about elements in $X_i^2 := S_{-i} \times (\times_{j\neq i} \Delta(S_{-j}))$. Call X_i^2 player i's *second-order state space.* Every strategy $s_i \in S_i$ can be identified with an act $f : X_i^2 \to \Delta(Z)$, given by $f(s_{-i}, (p_j^1)_{j\neq i}) := \mathbb{P}_{(s_i,s_{-i})} \in \Delta(Z)$ for every $(s_{-i}, (p_j^1)_{j\neq i}) \in X_i^2$. Hence, $f(s_{-i}, (p_j^1)_{j\neq i})$ is simply the probability distribution on the terminal nodes induced by the strategy profile (s_i, s_{-i}). Let $\mathcal{F}(X_i^2)$ be the set of all such acts $f : X_i^2 \to \Delta(Z)$.

We assume that player i has a *second-order preference relation* on the set of acts $\mathcal{F}(X_i^2)$, representable by the utility function $u_i : Z \to \mathbb{R}$ (the same as for the first-order preferences) and some subjective probability measure $p_i^2 \in \Delta(X_i^2)$ on the compact metric space X_i^2. Note that X_i^2 is no longer a finite set, since the spaces $\Delta(S_{-j})$ are infinite. The σ-algebra on X_i^2 on which the probability measure p_i^2 is defined is as follows. Take the discrete topology τ' on S_{-i} and for every $j \neq i$ take the weak topology τ_j'' on the space of probability measures $\Delta(S_{-j})$. Let τ be the product topology on X_i^2 induced by τ' and $(\tau_j'')_{j\neq i}$. The σ-algebra on X_i^2 induced by the topology τ is the σ-algebra we use. Since there is common belief about the utility functions $(u_i)_{i\in I}$, player i's second order preferences may thus be represented by a probability measure p_i^2 on X_i^2, which we call player i's *second-order belief.* An important mathematical property is that the space of player i's second-order beliefs, endowed with the weak topology, is again a compact metric space. This follows from the fact that for a given compact metric space X, the space $\Delta(X)$ of probability measures on X, together with the weak topology on $\Delta(X)$, is a compact metric space. This property plays a fundamental role in the formal construction and analyis of infinite hierarchies of beliefs.

In the third step of the construction of the players' beliefs, we basically repeat the argument of the second step. Player i knows that each player j holds a second-order belief $p_j^2 \in \Delta(X_j^2)$, but does not know which belief. Moreover, player i is uncertain about player j's first-order belief and player j's strategy choice, which is already captured in player i's second-order state space X_i^2. Hence, player i's *third-order state space* is given by $X_i^3 = X_i^2 \times (\times_{j\neq i}\Delta(X_j^2))$. Player i's *third-order* beliefs may thus be represented by a subjective probability measure $p_i^3 \in \Delta(X_i^3)$ on the compact metric space X_i^3.

By repeating this argument recursively, each player i is assumed to hold an *infinite hierarchy of beliefs* $p_i = (p_i^1, p_i^2, ...)$ where for every $k \in \mathbb{N}$ player i's k-th order belief p_i^k is a probability measure on the k-th order state space X_i^k, and the state spaces are defined recursively by

$$\begin{aligned} X_i^1 &= S_{-i}, \\ X_i^2 &= X_i^1 \times (\times_{j\neq i}\Delta(X_j^1)), \\ &\dots \\ X_i^k &= X_i^{k-1} \times (\times_{j\neq i}\Delta(X_j^{k-1})), \\ &\dots \end{aligned}$$

We thus obtain the following definition.

Definition 7.1.1 *A belief hierarchy for player i is a vector $p_i = (p_i^k)_{k\in\mathbb{N}}$ where $p_i^k \in \Delta(X_i^k)$, and the state spaces $(X_i^k)_{k\in\mathbb{N}}$ are defined as above.*

Let B_i be the space of belief hierarchies for player i. Note that the k-th order belief p_i^k in particular induces a belief on the $k-1$-th order state space X_i^{k-1}, which is a component of the k-th order state space X_i^k. However, player i's uncertainty about X_i^{k-1} has already been captured by his $k-1$-th order belief p_i^{k-1}. In order for a belief hierarchy to be *coherent* one should thus make sure that the different layers of beliefs do not contradict one another. Formally, consider a k-th order belief $p_i^k \in \Delta(X_i^k) = \Delta(X_i^{k-1} \times (\times_{j\neq i}\Delta(X_j^{k-1}))$. Let $mrg(p_i^k|\, X_i^{k-1})$ be the marginal of the probability distribution p_i^k on the component X_i^{k-1}.

Definition 7.1.2 *The belief hierarchy $p_i = (p_i^k)_{k\in\mathbb{N}}$ is said to be coherent if for every $k \geq 2$ it holds that $mrg(p_i^k|\, X_i^{k-1}) = p_i^{k-1}$.*

Let B_i^c be the set of coherent belief hierarchies for player i. A question which arises at this point is whether the space B_i^c of coherent belief hierarchies for player i captures all possible uncertainty that player i can face within our model. One could argue, for instance, that in order to make the players' belief spaces exhaustive every player i should have a belief about the possible belief hierarchies of his opponents. Hence, player i, in addition to his belief hierarchy p_i, should hold a subjective probability measure on the space $\times_{j \neq i} B_j$ of the opponents' possible belief hierarchies. The following theorem, however, shows that every coherent belief hierarchy already induces a belief about the opponents' possible belief hierarchies, and hence the space of coherent belief hierarchies for player i may be seen as an exhaustive belief space capturing all possible uncertainty by player i. More precisely, the theorem states that the space of coherent belief hierarchies for player i is homeomorphic to the set of joint probability measures on the set of opponents' strategy choices and the set of opponents' belief hierarchies. As such, every coherent belief hierarchy can be identified with such a probability measure, which induces both a belief on the opponents' possible strategy choices, and a belief on the opponents' possible beliefs. Various versions of this theorem, which plays a fundamental role in the theory of rationalizability, can be found in Armbruster and Böge (1979), Böge and Eisele (1979), Mertens and Zamir (1985), Tan and Werlang (1988), Brandenburger and Dekel (1993) and Heifetz (1993). The papers differ in the topological restrictions put on the players' strategy spaces. The first four papers assume that the strategy spaces are compact metric spaces, Brandenburger and Dekel (1993) assume that the strategy spaces are complete, separable metric spaces whereas Heifetz (1993) assumes that the strategy spaces are Hausdorff spaces. The latter paper thus provides the more general result. Since we assume that the players' strategy spaces are finite, each of the papers above applies to our setup. The theorem is extended by Rajan (1998) to a framework in which players may deem one event infinitely more likely than some other event, while deeming both events possible. His approach requires the use of nonstandard analysis. Epstein and Wang (1996) introduce infinite hierarchies of beliefs for the case where players are not necessarily subjective expected utility maximizers, as is assumed here, and prove a theorem similar to the one below for this more general situation.

Theorem 7.1.3 *For every player i, there is a homeomorphism f_i from B_i^c to $\Delta(S_{-i} \times B_{-i})$.*

Here, B_{-i} denotes the space $\times_{j \neq i} B_j$ of the opponents' possible belief hierarchies. In order to prove this result, we use the following version of Kolmogorov's Existence Theorem, which can be found in Dellacherie and Meyer (1978). The proof for this theorem is omitted.

Theorem 7.1.4 *For every $k \in \mathbb{N}$ let Y^k be a compact metric space. Let P be the set of hierarchies of probability measures $(p^k)_{k \in \mathbb{N}}$ where (1) $p^k \in \Delta(Y^1 \times ... \times Y^k)$ for every k, and (2) for every $k \geq 2$ it holds that $mrg(p^k | \, Y^1 \times ... \times Y^{k-1}) = p^{k-1}$. Then, there is a unique probability measure $p^* \in \Delta(\times_{k \in \mathbb{N}} Y^k)$ such that $mrg(p^* | \, Y^1 \times ... \times Y^k) = p^k$ for every $k \in \mathbb{N}$.*

Proof of Theorem 7.1.3. Fix a player i. Define the sets $Y^1, Y^2, ...$ by

$$Y^1 = X_i^1, \quad Y^k = \times_{j\neq i}\Delta(X_j^{k-1}) \text{ if } k \geq 2.$$

Then, by construction, $X_i^k = Y^1 \times ... \times Y^k$ for every k, and every coherent belief hierarchy $p_i \in B_i^c$ is a hierarchy of probability measures $(p_i^k)_{k\in\mathbb{N}}$ with $p_i^k \in \Delta(Y^1 \times ... \times Y^k)$ and $mrg(p_i^k | Y^1 \times ... \times Y^{k-1}) = p_i^{k-1}$. Hence, B_i^c coincides with the set P in Theorem 7.1.4. Since all the sets Y^k constructed are compact metric spaces, Theorem 7.1.4 assures that for every coherent belief hierarchy $p_i \in B_i^c$ there is a unique probability measure $p_i^* \in \Delta(\times_{k\in\mathbb{N}} Y^k)$ such that $mrg(p_i^* | Y^1 \times ... \times Y^k) = p_i^k$ for every k. By construction,

$$\times_{k\in\mathbb{N}} Y^k = S_{-i} \times (\times_{k\in\mathbb{N}}(\times_{j\neq i}\Delta(X_j^k)))$$

which is homeomorphic to

$$S_{-i} \times (\times_{j\neq i}(\times_{k\in\mathbb{N}}\Delta(X_j^k))) = S_{-i} \times B_{-i}.$$

Moreover, $Y^1 \times ... \times Y^k = X_i^k$ for every k. Hence, for every $p_i \in B_i^c$ there is a unique $p_i^* \in \Delta(S_{-i} \times B_{-i})$ such that $mrg(p_i^* | X_i^k) = p_i^k$ for every k.

Let $f_i : B_i^c \to \Delta(S_{-i} \times B_{-i})$ be the function which assigns to every coherent belief hierarchy $p_i \in B_i^c$ the unique $p_i^* \in \Delta(S_{-i} \times B_{-i})$ with the properties above. It remains to show that f_i is a homeomorphism, that is, f_i is bijective, f_i is continuous and so is the inverse mapping f_i^{-1}.

Suppose that $f_i(p_i) = p_i^* = f_i(p_i')$, where $p_i = (p_i^k)_{k\in\mathbb{N}}$ and $p_i' = (p_i'^k)_{k\in\mathbb{N}}$. Then, $p_i^k = mrg(p_i^* | Y^1 \times ... \times Y^k) = p_i'^k$ for all k, and hence $p_i = p_i'$. Hence, f_i is one-to-one. In order to show that f_i is onto, take any $p_i^* \in \Delta(S_{-i} \times B_{-i}) = \Delta(\times_{k\in\mathbb{N}} Y^k)$. Define $p_i = (p_i^k)_{k\in\mathbb{N}}$ by $p_i^k = mrg(p_i^* | Y^1 \times ... \times Y^k)$. Then, by construction, $f_i(p_i) = p_i^*$ and hence f is onto. By the previous observation, we know that $f_i^{-1}(p_i^*) = (p_i^k)_{k\in\mathbb{N}}$ with $p_i^k = mrg(p_i^* | Y^1 \times ... \times Y^k)$. The reader may finally verify that both f_i and f_i^{-1} are continuous, which implies that f_i is a homeomorphism from B_i^c to $\Delta(S_{-i} \times B_{-i})$. This completes the proof. ■

We next impose that there be common belief of coherency of belief hierarchies. Hence, each player should believe that each player's belief hierachy is coherent, each player should believe that each player believes that each player's belief hierarchy is coherent, and so on. In order to formalize this, choose for every player i the homeomorphism f_i from B_i^c to $\Delta(S_{-i} \times B_{-i})$ constructed in the proof of Theorem 7.1.3. We recursively construct subsets $B_i^{c,1}, B_i^{c,2}, ...$ of B_i^c by

$$\begin{aligned} B_i^{c,1} &= B_i^c, \\ B_i^{c,k} &= \{p_i \in B_i^c |\ f_i(p_i)(S_{-i} \times B_{-i}^{c,k-1}) = 1\} \end{aligned}$$

for $k \geq 2$, where $B_{-i}^{c,k-1} = \times_{j\neq i} B_j^{c,k-1}$. Hence, a belief hierarchy in $B_i^{c,k}$ should assign probability one to the event that all opponents j have belief hierarchies in $B_j^{c,k-1}$, where a belief hierarchy p_i is identified, through f_i, with a probability measure on

$S_{-i} \times B_{-i}$. Now, let $B_i^{c,\infty} = \cap_{k \in \mathbb{N}} B_i^{c,k}$. Hence, a belief hierarchy in $B_i^{c,\infty}$ believes that each player has a coherent belief hierarchy, believes that each player believes that each player has a coherent belief hierarchy, and so on. Belief hierarchies in $B_i^{c,\infty}$ are said to *respect common belief of coherency.* We now prove that every belief hierarchy in $B_i^{c,\infty}$ can be identified with a probability measure on $S_{-i} \times B_{-i}^{c,\infty}$, where $B_{-i}^{c,\infty} = \times_{j \neq i} B_j^{c,\infty}$. Consequently, every belief hierarchy which respects common belief of coherency induces a belief on the possible opponents' belief hierarchies that respect common belief of coherency.

Theorem 7.1.5 *For every player i there is a homeomorphism g_i from $B_i^{c,\infty}$ to $\Delta(S_{-i} \times B_{-i}^{c,\infty})$.*

Proof. Let f_i be the homeomorphism from B_i^c to $\Delta(S_{-i} \times B_{-i})$ constructed in the proof of Theorem 7.1.3. It may be verified by the reader that $\{p_i \in B_i^c | \; f_i(p_i)(S_{-i} \times B_{-i}^{c,\infty}) = 1\} = B_i^{c,\infty}$. Since f_i is onto, it follows that $f_i(B_i^{c,\infty}) = \{p_i^* \in \Delta(S_{-i} \times B_{-i}) | \; p_i^*(S_{-i} \times B_{-i}^{c,\infty}) = 1\}$ which is homeomorphic to $\Delta(S_{-i} \times B_{-i}^{c,\infty})$. Hence, $B_i^{c,\infty}$ is homeomorphic to $\Delta(S_{-i} \times B_{-i}^{c,\infty})$. ■

An important consequence of Theorem 7.1.5 is that we may now define for each player a Bayesian decision problem in the classical sense. Formally, a *Bayesian decision problem* for player i consists of (1) a state space X_i which captures all relevant events that player i is uncertain about, (2) a set D_i of possible decisions for player i, (3) a utility function $v_i : D_i \times X_i \to \mathbb{R}$ and (4) a subjective probability distribution q_i on X_i. The utility function v_i and the subjective probability distribution q_i together define a subjective expected utility function V_i on $D_i \times \Delta(X_i)$. Player i is said to be *Bayesian rational* if he chooses a decision $d_i \in D_i$ which maximizes his subjective expected utility, given v_i and q_i.

Consider now an extensive form game in which each player i is a subjective expected utility maximizer with utility function u_i, each player holds a belief hierarchy p_i and in which there is common belief of coherency of belief hierarchies. Conform Theorem 7.1.5, player i's decision problem in the game may then be represented by a Bayesian decision problem (X_i, D_i, v_i, q_i) where the state space X_i is equal to $S_{-i} \times B_{-i}^{c,\infty}$, the set of decisions D_i is the set of strategies S_i, the utility function v_i is given by u_i, and the subjective probability distribution q_i on the state space is equal to $g_i(p_i) \in \Delta(S_{-i} \times B_{-i}^{c,\infty})$, where g_i is the homeomorphism in Theorem 7.1.5.

In the sequel, we often refer to player i's belief hierarchy p_i as player i's *type*, in accordance with Harsanyi's (1967-1968) model of games with incomplete information. In Harsanyi's framework, the players in a game may face uncertainty about some characteristics of the game, such as the players' utility functions, the players' strategy spaces, or the physical outcome of the game induced by certain profiles of strategies. Each player is assumed to form some belief, that is, a subjective probability distribution, on these uncertain events. In turn, each player is uncertain about the subjective probability distributions of the others, and hence every player is required to form a joint probability distribution on both the primary uncertain events and the subjective probability distributions of the others. All the information and uncertainty held by a player i is represented by a so-called *type* t_i, and each type holds a subjective probability distribution on the possible types of the opponents. An important advantage of the introduction of types is that it facilitates both the representation and analysis of such games with incomplete information.

Although we assume that the players' utility functions and the characteristics of the game are known the players, the notion of types is applicable to our setup. Note that a belief hierarchy for player i captures all the relevant uncertainty faced by this player in the game, and hence such a belief hierarchy may be identified with a type t_i. Moreover, we have seen that each such belief hierarchy induces a belief about the opponents' possible belief hierarchies (if there is common belief of coherency) and thus each type induces a belief about the opponents' possible types, as in Harsanyi's model. If we define, for each player i, the type space $T_i := B_i^{c,\infty}$, then, by Theorem 7.1.5, there is a homeomorphism g_i from T_i to $\Delta(S_{-i} \times T_{-i})$. Hence, each type $t_i \in T_i$ may be identified with a subjective probability distribution on the possible opponents' strategy-type pairs $(s_j, t_j)_{j \neq i}$.

7.2 Rationalizability

As mentioned at the beginning of this chapter, the idea of rationalizability concepts is, first, to model each player as a decision maker under uncertainty, whose primary uncertainty concerns the strategy choices by the opponents, and, secondly, to impose common belief or rationality within this model. The previous section presented a model for the first task. As to the second task, we first have to clarify what it means for a player to be "rational" within such a model. Roughly speaking, player i, characterized by a belief hierarchy and a strategy choice, is said to be rational if his strategy choice is "optimal" against his belief. But then, we should be explicit about the notion of "optimal strategies" in the setting of an extensive form game. Does it mean that the strategy should be a best response *at the beginning of the game,* that is, maximize the ex-ante expected utility given his initial belief about the opponents' strategy choices? Or should we use the more restrictive notion of sequential best reponse, stating that the strategy should be optimal at each of his information sets that are not avoided by the strategy itself?

Once we have decided upon the version of optimal strategy, and thus the version of rationality, we should formally define "common belief of rationality". Again, there are several possibilities here. Should we assume common belief of rationality only at the beginning of the game, or should we use a more restrictive version in which we impose common belief of rationality at every information set, or at some collection of information sets? As this chapter will demonstrate, the different versions of "optimal strategy" and "common belief of rationality" lead to substantially different rationalizability concepts.

In this section, the weakest versions of "optimal strategy" and "common belief of rationality" are chosen. Hence, within this model, players are "rational" if they choose strategies which are a best response against their belief at the beginning of the game, and common belief of rationality is imposed at the beginning of the game only. With respect to the players' beliefs, we distinguish two different cases. In the first case, players are allowed to hold *correlated* beliefs, that is, player i's subjective probability distribution on player j's strategy choices and player i's subjective probability distribution on player k's strategy choices may be correlated, if there are three or more players. In the second case, players are restricted to *uncorrelated* beliefs. The resulting solution concept in the first case is called *correlated rationalizability,* whereas the concept obtained in the second case is called *uncorrelated rationalizability,* or simply *rationalizability,* due to Bernheim (1984) and Pearce (1984).

In their seminal papers, Bernheim (1984) and Pearce (1984) defined the concept of rationalizability through an iterative procedure, in which the set of "rational" strategies for every player is reduced further and further at every round. The strategies surviving this reduction procedure are called the *rationalizable* strategies. Later, Tan and Werlang (1988) have characterized these rationalizable strategies within a model of infinite belief hierarchies as developed in the previous section. In this section, we proceed in the opposite direction by first defining rationalizable strategies in a model with infinite belief hierarchies, following Tan and Werlang (1988), and then showing that these strategies coincide with the strategies surviving the reduction procedure proposed in Bernheim (1984) and Pearce (1984). In the first part of this section we define correlated rationalizability, and prove that the correlated-rationalizable strategies are exactly the strategies surviving iterative elimination of strictly dominated strategies. The second part of the section deals with the uncorrelated case.

7.2.1 Correlated Rationalizability

Conform Section 7.1, we model the players' decision problems in the extensive form game by the space $((S_i)_{i\in I}, (T_i)_{i\in I})$, where $T_i = B_i^{c,\infty}$ for every player i. We know that for every player i there is a homeomorphism g_i from T_i to $\Delta(S_{-i} \times T_{-i})$. Hence, every type t_i can be identified with the probability measure $g_i(t_i)$ on $S_{-i} \times T_{-i}$. Let $\tilde{g}_i(t_i) := mrg(g_i(t_i)|\, S_{-i})$ be the subjective probability distribution on the opponents' strategy choices induced by type t_i. For every strategy $s_i \in S_i$, let

$$u_i(s_i, t_i) = \sum_{s_{-i}\in S_{-i}} u_i(s_i, s_{-i})\, \tilde{g}_i(t_i)(s_{-i})$$

be the expected utility for type t_i when choosing the strategy s_i. Here, $u_i(s_i, s_{-i})$ is the expected utility for player i induced by the strategy profile (s_i, s_{-i}). Note that t_i in $u_i(s_i, t_i)$ is a parameter rather than a choice variable, in contrast to s_i.

Definition 7.2.1 *A strategy-type pair* $(s_i, t_i) \in S_i \times T_i$ *is called rational if* $u_i(s_i, t_i) = \max_{s_i'} u_i(s_i', t_i)$.

In order to define common belief of rationality, we first formally define what we mean by the phrase "player i believes an event" . Consider an event $E \subseteq S_{-i} \times T_{-i}$ representing a collection of states that player i is uncertain about.

Definition 7.2.2 *We say that type* t_i *believes the event* $E \subseteq S_{-i} \times T_{-i}$ *if* $g_i(t_i)(E) = 1$.

Hence, type t_i believes an event if t_i assigns probability zero to the complement of this event. The notion of belief applied here is substantially different from the notion of *knowledge* used in Aumann (1976). First of all, both approaches differ in the way uncertainty is modeled. Within our setup, a player's uncertainty is modeled in a probabilistic way, by assuming that each player holds a subjective probability distribution over states in $S_{-i} \times T_{-i}$. In Aumann (1976), the uncertainty of each player is represented in a non-probabilistic way by means of partitions of the set of states. Here, a state is to be interpreted as a tuple $\omega = (s_i, t_i, s_{-i}, t_{-i})$ reflecting the strategy choices and beliefs of each player. In Aumann's approach, each player i is assigned a partition P_i of the set of states. The interpretation is that if the true state of the world is ω, and $P_i(\omega)$ is the partition element for player i that contains ω, then player i knows that some state in $P_i(\omega)$ occurs, without knowing which one. Another difference between both approaches is that, within our setup, a player may believe an event, and discover later that this event is not true. In Aumann's model this is not possible: if a player knows an event, then this event must be true.

Given the definition of belief above, common belief of rationality can now be defined as follows. Let

$$R_i^0 = \{(s_i, t_i) \in S_i \times T_i | \ (s_i, t_i) \text{ rational}\}.$$

Let $R_{-i}^0 = \times_{j \neq i} R_j^0$ be the event that player i's opponents are all rational. Let

$$R_i^1 = \{t_i \in T_i | \ t_i \text{ believes } R_{-i}^0\}$$

be the set of player i types which believe that all opponents are rational. Let $R_{-i}^1 = \times_{j \neq i} R_j^1$. By

$$R_i^2 = \{t_i \in R_i^1 | \ t_i \text{ believes } S_{-i} \times R_{-i}^1\}$$

we denote the set of player i types in R_i^1 which believe that all opponents have types in R_{-i}^1. Hence, by construction, each type in R_i^2 believes that all players are rational and believes that all players believe that all players are rational. Define the sets of types $R_i^3, R_i^4, \ldots$ recursively by

$$R_i^k = \{t_i \in R_i^{k-1} | \ t_i \text{ believes } S_{-i} \times R_{-i}^{k-1}\},$$

for $k \geq 3$. Let $R_i^\infty = \cap_{k \in \mathbb{N}} R_i^k$.

Definition 7.2.3 *A type t_i is said to respect common belief of rationality if $t_i \in R_i^\infty$.*

A strategy is now said to be *correlated-rationalizable* if it may be chosen by a rational player that respects common belief of rationality.

Definition 7.2.4 *A strategy $s_i \in S_i$ is called correlated-rationalizable if there is a type $t_i \in R_i^\infty$ such that (s_i, t_i) is rational.*

We next present an algorithm that can be used to compute the set of correlated-rationalizable strategies for each player. It is shown, namely, that a strategy is correlated-rationalizable if and only if it survives *iterative elimination of strictly dominated strategies.* In order to do so, we first formally define this elimination procedure, and then prove two properties of this procedure which are of interest in their own right as well. Formally, iterative elimination of strictly dominated strategies is defined as follows. Define recursively the sets of strategies $D_i^0, D_i^1, D_i^2, \ldots$ by

$$\begin{aligned} D_i^0 &= S_i, \\ D_i^k &= \{s_i \in D_i^{k-1} | \; s_i \text{ is not strictly dominated on } D_i^{k-1} \times D_{-i}^{k-1}\} \end{aligned}$$

for $k \geq 1$. Here, we say that s_i is *strictly dominated* on $D_i^{k-1} \times D_{-i}^{k-1}$ if there is some $\mu_i \in \Delta(D_i^{k-1})$ such that $u_i(\mu_i, s_{-i}) > u_i(s_i, s_{-i})$ for all $s_{-i} \in D_{-i}^{k-1} = \times_{j \neq i} D_j^{k-1}$. Define $D_i^\infty = \cap_{k \in \mathbb{N}} D_i^k$. Since every set S_i is finite, it should be clear that there exists some $K \in \mathbb{N}$ such that $D_i^k = D_i^K$ for all $k \geq K$ and all i. Moreover, D_i^K is nonempty, and hence $D_i^\infty = D_i^K \neq \emptyset$ for all players i.

Definition 7.2.5 *A strategy s_i is said to satisfy iterative elimination of strictly dominated strategies if $s_i \in D_i^\infty$.*

The following lemma states that strictly dominated strategies are exactly those strategies that are not a best response against any probability distribution on the opponents' strategy profiles. The proof below is an extension of Pearce's (1984) proof to the case of more than two players.

Lemma 7.2.6 *Let $D_j \subseteq S_j$ for all players j, and let $D_{-i} = \times_{j \neq i} D_j$. Then, $s_i \in D_i$ is not strictly dominated on $D_i \times D_{-i}$ if and only if there is some $p_i \in \Delta(D_{-i})$ such that $u_i(s_i, p_i) = \max_{s_i' \in D_i} u_i(s_i', p_i)$.*

Proof. Suppose that $s_i^* \in D_i$ is not strictly dominated. We prove that there is some $p_i \in \Delta(D_{-i})$ such that $u_i(s_i^*, p_i) = \max_{s_i \in D_i} u_i(s_i, p_i)$. Suppose not. Then, for every $p_i \in \Delta(D_{-i})$ there is some $s_i(p_i) \in D_i$ such that $u_i(s_i(p_i), p_i) > u_i(s_i^*, p_i)$. We now define a two player zero-sum game Γ^* with player set $I^* = \{i, j\}$, strategy sets $S_i^* = D_i$, $S_j^* = D_{-i}$ and utility functions u_i^*, u_j^* where $u_i^*(s_i, s_{-i}) = u_i(s_i, s_{-i}) - u_i(s_i^*, s_{-i})$ for all $(s_i, s_{-i}) \in S_i^* \times S_j^*$, and $u_j^*(s_i, s_{-i}) = -u_i^*(s_i, s_{-i})$ for all $(s_i, s_{-i}) \in S_i^* \times S_j^*$. Let $(\mu_i^*, \mu_{-i}^*) \in \Delta(S_i^*) \times \Delta(S_j^*)$ be a Nash equilibrium in the game Γ^*. Hence, by definition, $\mu_i^* \in \Delta(D_i)$ and $\mu_{-i}^* \in \Delta(D_{-i})$. Let $\mu_{-i} \in \Delta(S_j^*)$ be arbitrary. Then, since (μ_i^*, μ_{-i}^*) is a Nash equilibrium in Γ^*, we have that $u_j^*(\mu_i^*, \mu_{-i}) \leq u_j(\mu_i^*, \mu_{-i}^*)$ and hence $u_i^*(\mu_i^*, \mu_{-i}) \geq u_i(\mu_i^*, \mu_{-i}^*)$. On the other hand, we have that $u_i^*(\mu_i^*, \mu_{-i}^*) \geq u_i^*(s_i(\mu_{-i}^*), \mu_{-i}^*) > u_i^*(s_i^*, \mu_{-i}^*) = 0$, where $s_i(\mu_{-i}^*)$ is as defined above with $p_i = \mu_{-i}^*$. We may thus conclude that $u_i^*(\mu_i^*, \mu_{-i}) > 0$ for all $\mu_{-i} \in \Delta(S_j^*)$, which implies that $u_i(\mu_i^*, \mu_{-i}) > u_i(s_i^*, \mu_{-i})$ for all $\mu_{-i} \in \Delta(D_{-i})$. Hence, s_i^* is strictly dominated on $D_i \times D_{-i}$ by $\mu_i^* \in \Delta(D_i)$, which contradicts the assumption that s_i^* is not strictly dominated. We may conclude, therefore, that there is some $p_i \in \Delta(D_{-i})$ such that $u_i(s_i^*, p_i) = \max_{s_i \in D_i} u_i(s_i, p_i)$.

Suppose now that s_i^* is strictly dominated. Then, there is some $\mu_i \in \Delta(D_i)$ such that $u_i(\mu_i, s_{-i}) > u_i(s_i^*, s_{-i})$ for all $s_{-i} \in D_{-i}$, and hence $u_i(\mu_i, p_i) > u_i(s_i^*, p_i)$ for all $p_i \in \Delta(D_{-i})$. Therefore, there is no $p_i \in \Delta(D_{-i})$ with $u_i(s_i^*, p_i) = \max_{s_i \in D_i} u_i(s_i, p_i)$. This completes the proof of this lemma. ■

The following lemma states that strategies surviving iterative elimination of strictly dominated strategies can be characterized by sets of strategies satisfying the so-called *correlated-best response property*. Consider a nonempty set of strategies $X_i \subseteq S_i$ for every player i. We say that the profile $(X_i)_{i\in I}$ has the *correlated-best response property* if for every player i and every strategy $s_i \in X_i$ there exists some $\mu_i \in \Delta(X_{-i})$ such that $u_i(s_i, \mu_i) = \max_{s_i' \in S_i} u_i(s_i', \mu_i)$. Hence, every strategy in X_i can be justified by some correlated first-order belief on the opponents' strategies in X_{-i}.

Lemma 7.2.7 *A strategy s_i is in D_i^∞ if and only if for there is some profile $(X_j)_{j\in I}$ of nonempty strategy sets having the correlated-best response property such that $s_i \in X_i$.*

Proof. Let $s_i \in D_i^\infty$. By construction we have that for every player j, no strategy $s_j \in D_j^\infty$ is strictly dominated on $D_j^\infty \times D_{-j}^\infty$. Hence, by Lemma 7.2.6, for every strategy $s_j \in D_j^\infty$ there is some $p_j \in \Delta(D_{-j}^\infty)$ such that s_j is a best response against p_j. But then, the profile of strategy sets $(D_j^\infty)_{j\in I}$ has the correlated-best response property. We have thus shown the implication from left to right.

Now, suppose that $(X_j)_{j\in I}$ is a profile of nonempty strategy sets with the correlated-best response property, and that $s_i \in X_i$. We prove that $s_i \in D_i^\infty$. To this purpose we prove, by induction on k, that for every player j, every strategy $s_j \in X_j$ belongs to D_j^k. Take some $s_j \in X_j$. Since s_j is a best response, within S_j, against some $\mu_j \in \Delta(X_{-j})$, we have, by Lemma 7.2.6, that s_j is not strictly dominated on $S_j \times S_{-j}$ and hence $s_j \in D_j^1$. Now, suppose that the property holds for $k \geq 2$. Let $s_j \in X_j$. Then, s_j is a best response, within S_j, against some $\mu_j \in \Delta(X_{-j})$. Since, by assumption, $X_l \subseteq D_l^k$ for all players l, it follows that $\mu_j \in \Delta(D_{-j}^k)$, and hence, by Lemma 7.2.6, s_j is not strictly dominated on $D_j^k \times D_{-j}^k$. But then, by definition, $s_j \in D_j^{k+1}$, which completes the induction step. Hence, we may conclude that $s_i \in \cap_{k\in\mathbb{N}} D_i^k = D_i^\infty$. This completes the proof. ■

We are now in a position to prove the following characterization of correlated-rationalizable strategies, which is due to Tan and Werlang (1988).

Theorem 7.2.8 *A strategy $s_i \in S_i$ is correlated-rationalizable if and only if it survives iterative elimination of strictly dominated strategies.*

Proof. Let the strategy s_i be correlated-rationalizable. We prove that $s_i \in D_i^\infty$. We procede in three steps.

Step 1. Let $k \geq 1$ and $t_i \in R_i^k$. Let $\tilde{g}_i(t_i) \in \Delta(S_{-i})$ be the probability distribution on S_{-i} induced by type t_i. Then, $\tilde{g}_i(t_i)(D_{-i}^k) = 1$.

Proof of step 1. By induction on k. Suppose that $t_i \in R_i^1$. Then, by definition, $g_i(t_i)(R_{-i}^0) = 1$. Now, let $s_{-i} \in S_{-i} \backslash D_{-i}^1$. We show that $\tilde{g}_i(t_i)(s_{-i}) = 0$. Let $s_{-i} = (s_j)_{j\neq i}$. Since $s_{-i} \notin D_{-i}^1$, there is at least one $j \neq i$ such that $s_j \notin D_j^1$. Then, by definition, s_j is strictly dominated on $S_j \times S_{-j}$. By Lemma 7.2.6, s_j is not a best response against any $\mu_j \in \Delta(S_{-j})$. Consequently, for every type t_j the strategy-type pair (s_j, t_j) is not rational, and hence $(\{s_j\} \times T_j) \cap R_j^0 = \emptyset$. Since $g_i(t_i)(R_{-i}^0) = 1$, it follows that $\tilde{g}_i(t_i)(s_{-i}) = 0$.

Now, suppose that $k \geq 2$ and assume that the claim holds for $k-1$. Let $t_i \in R_i^k$ and $s_{-i} \in S_{-i} \backslash D_{-i}^k$. We show that $\tilde{g}_i(t_i)(s_{-i}) = 0$. Let $s_{-i} = (s_j)_{j\neq i}$. Then, there is some j with $s_j \notin D_j^k$. Hence, by definition, s_j is strictly dominated on $D_j^{k-1} \times D_{-j}^{k-1}$. By Lemma 7.2.6, there is no $\mu_j \in D_{-j}^{k-1}$ such that s_j is a best response, within D_j^{k-1}, against μ_j. Since, by the induction hypothesis, every type $t_j \in R_j^{k-1}$ satisfies $\tilde{g}_j(t_j)(D_{-j}^{k-1}) = 1$, it follows that there is no type $t_j \in R_j^{k-1}$ such that (s_j, t_j) is rational. Hence, $(\{s_j\} \times R_j^{k-1}) \cap R_j^0 = \emptyset$. Since $t_i \in R_i^k$ we have, by definition, that $\tilde{g}_i(t_i)(S_{-i} \times R_{-i}^{k-1}) = 1$. Since $R_i^k \subseteq R_i^1$ we have, moreover, that $t_i \in R_i^1$ and hence $\tilde{g}_i(t_i)(R_{-i}^0) = 1$. It thus follows that $\tilde{g}_i(t_i)((S_{-i} \times R_{-i}^{k-1}) \cap R_{-i}^0) = 1$. However, we have seen that $(\{s_j\} \times R_j^{k-1}) \cap R_j^0 = \emptyset$ and hence $\tilde{g}_i(t_i)(s_{-i}) = 0$. This completes the proof of step 1.

Step 2. Let (s_i, t_i) be rational and $t_i \in R_i^k$. Then, $s_i \in D_i^{k+1}$.

Proof of step 2. Since $t_i \in R_i^k$, we know by step 1 that $\tilde{g}_i(t_i) \in \Delta(D_{-i}^k)$. Since (s_i, t_i) is rational, we know that s_i is best reponse against $\tilde{g}_i(t_i)$ on S_i. By Lemma 7.2.6, this implies that $s_i \in D_i^{k+1}$, which completes the proof of step 2.

Step 3. Let s_i be correlated-rationalizable. Then, $s_i \in D_i^\infty$.

Proof of step 3. If s_i is correlated-rationalizable then, by definition, there is some type $t_i \in R_i^\infty$ such that (s_i, t_i) is rational. Since $R_i^\infty = \cap_k R_i^k$, it follows by step 2 that $s_i \in \cap_k D_i^{k+1} = D_i^\infty$, which completes the proof of step 3. Hence, we have shown that every correlated-rationalizable strategy s_i belongs to D_i^∞.

It remains to prove that every strategy in D_i^∞ is correlated-rationalizable. Let $s_i^* \in D_i^\infty$. Then, by Lemma 7.2.7, there is a profile of nonempty strategy sets $(Y_j)_{j \in I}$ satisfying the correlated-best response property such that $s_i^* \in Y_i$. Hence, for every player j and every $s_j \in Y_j$ there is some $\mu_j^{s_j} \in \Delta(Y_{-j})$ such that s_j is a best response against $\mu_j^{s_j}$. We now define for every player j and every strategy $s_j \in Y_j$ a coherent belief hierarchy $p_j^{s_j} = (p_j^{s_j,k})_{k \in \mathbb{N}}$ where $p_j^{s_j,k} \in \Delta(X_j^k)$ for every $k \in \mathbb{N}$ and the sets X_j^k are defined as in the previous section. To this purpose we define, by induction on k, functions $g_j^k : Y_{-j} \to X_j^k$ for all players j, and probability distributions $p_j^{s_j,k} \in \Delta(X_j^k)$ for all $s_j \in Y_j$. For $k = 1$, we have that $X_j^1 = S_{-j}$. Let $g_j^1 : Y_{-j} \to X_j^1$ be given by $g_j^1(s_{-j}) = s_{-j}$. For every $s_j \in Y_j$ define $p_j^{s_j,1} \in \Delta(X_j^1)$ by

$$p_j^{s_j,1} = \mu_j^{s_j}.$$

Now, let $k \geq 2$ and suppose that the functions $g_j^{k-1} : Y_{-j} \to X_j^{k-1}$ and the probability distributions $p_j^{s_j,k-1} \in \Delta(X_j^{k-1})$ have been defined for all players j and all $s_j \in Y_j$. By definition $X_j^k = X_j^{k-1} \times (\times_{l \neq j} \Delta(X_l^{k-1}))$. Let $g_j^k : Y_{-j} \to X_j^k$ be given by

$$g_j^k(s_{-j}) = (g_j^{k-1}(s_{-j}), (p_l^{s_l,k-1})_{l \neq j}) \in X_j^{k-1} \times (\times_{l \neq j} \Delta(X_l^{k-1}))$$

for every $s_{-j} = (s_l)_{l \neq j} \in Y_{-j}$. For $s_j \in Y_j$, let $p_j^{s_j,k} \in \Delta(X_j^k)$ be defined by

$$p_j^{s_j,k}(E_j^k) = \mu_j^{s_j}((g_j^k)^{-1}(E_j^k))$$

for all measurable subsets $E_j^k \subseteq X_j^k$.

It may be verified that the belief hierarchy $p_j^{s_j} = (p_j^{s_j,k})_{k\in\mathbf{N}}$ is coherent for every j and every $s_j \in Y_j$. Write $t_j^{s_j}$ instead of $p_j^{s_j}$. Note that for every $s_j \in Y_j$, the strategy-type pair $(s_j, t_j^{s_j})$ is rational since the induced marginal probability distribution on S_{-j} equals $\mu_j^{s_j}$ and s_j is a best response against $\mu_j^{s_j}$. By induction on k it may be verified that $t_j^{s_j} \in R_j^k$ for all k, and hence $t_j^{s_j}$ respects common belief of rationality for all $s_j \in Y_j$. Since the strategy-type pair $(s_i^*, t_i^{s_i^*})$ is rational, it follows that s_i^* is correlated-rationalizable, which completes the proof. ■

7.2.2 Uncorrelated Rationalizability

In the previous subsection we assumed that the players' beliefs may be correlated, that is, player i's belief about player j's strategy choice and player i's belief about player k's strategy choice may not be independent. Let us now assume that player i's belief about player j's strategy choice should be independent from his belief about player k's strategy choice. Belief hierarchies satisfying this requirement are called *uncorrelated*.

Definition 7.2.9 *A type $t_i \in T_i$ is called uncorrelated if $mrg(g_i(t_i)|\ S_j \times S_k)(s_j, s_k) = mrg(g_i(t_i)|\ S_j)(s_j) \cdot mrg(g_i(t_i)|\ S_k)(s_k)$ for all $j, k \neq i$ and all $(s_j, s_k) \in S_j \times S_k$.*

Hence, the probability measure induced by t_i on S_{-i} can be written as the product of the probability measures induced on S_j for $j \neq i$. Or, stated formally, $mrg(g_i(t)|\ S_{-i}) \in \times_{j\neq i}\Delta(S_j)$. Let T_i^u be the set of uncorrelated types for player i. We assume that players do not only hold correlated types, but in addition believe that every player holds an uncorrelated type, believe that every player believes that every player holds an uncorrelated type, and so on. Hence, we assume *common belief of uncorrelated types*. This restriction is formalized as follows. For every player i, define the sets of types $T_i^{u,1}, T_i^{u,2}, \ldots$ by

$$\begin{aligned} T_i^{u,1} &= T_i^u, \\ T_i^{u,k} &= \{t_i \in T_i^{u,k-1} \mid t_i \text{ believes } S_{-i} \times T_{-i}^{u,k-1}\} \end{aligned}$$

for $k \geq 2$. Let $T_i^{u,\infty} = \cap_{k\in\mathbf{N}} T_i^k$.

Definition 7.2.10 *A type t_i is said to respect common belief of uncorrelated types if $t_i \in T_i^{u,\infty}$.*

A strategy is now called *rationalizable* if it may be chosen by a rational player which respects both common belief of rationality and common belief of uncorrelated types.

Definition 7.2.11 *A strategy $s_i \in S_i$ is called rationalizable if there is some type $t_i \in R_i^\infty \cap T_i^{u,\infty}$ such that (s_i, t_i) is rational.*

Bernheim (1984) and Pearce (1984) originally defined rationalizable strategies through an iterative elimination procedure similar to iterative elimination of strictly dominated strategies. For every player i define the sets of strategies $S_i^1, S_i^2, \ldots$ by

$$\begin{aligned} S_i^1 &= S_i, \\ S_i^k &= \{s_i \in S_i^{k-1} | \; \exists \mu_i \in \times_{j \neq i} \Delta(S_j^{k-1}) \text{ with } u_i(s_i, \mu_i) = \max_{s_i' \in S_i^{k-1}} u_i(s_i', \mu_i)\} \end{aligned}$$

for $k \geq 2$. For every player i, let $S_i^\infty = \cap_{k \in \mathbb{N}} S_i^k$. Hence, at every round only those strategies in S_i^{k-1} survive that are best responses against some *uncorrelated* belief over strategies in S_{-i}^{k-1}. If one would allow for correlated beliefs on S_{-i}^{k-1}, then Lemma 7.2.6 implies that the reduction procedure above would coincide with iterative elimination of strictly dominated strategies. Since for two-player games all beliefs are uncorrelated, both reduction procedures coincide for such games. The following theorem, which is due to Tan and Werlang (1988), states that the reduction procedure above yields exactly the set of rationalizable strategies for each player.

Theorem 7.2.12 *A strategy s_i is rationalizable if and only if $s_i \in S_i^\infty$.*

The proof for this result is similar to the proof of Theorem 7.2.8 and is therefore omitted. Since the sets S_i^∞ are, by construction, nonempty, it follows that every player has at least one rationalizable strategy in every game. For the proof of Theorem 7.2.12, one can use the following characterization of strategies in S_i^∞ by means of a best response property, similar to the one used for correlated beliefs. A profile $(X_i)_{i \in I}$ of sets of strategies is said to have the *(uncorrelated) best response property* if for every player i and every strategy $s_i \in X_i$ there exists some $\mu_i \in \times_{j \neq i} \Delta(X_j)$ such that $u_i(s_i, \mu_i) = \max_{s_i' \in S_i} u_i(s_i', \mu_i)$. Hence, every strategy in X_i can be justified by some uncorrelated first-order belief on the opponents' strategies in X_{-i}.

Lemma 7.2.13 *A strategy s_i is in S_i^∞ if and only if there is some profile $(X_j)_{j \in I}$ of nonempty strategy sets having the (uncorrelated) best response property such that $s_i \in X_i$.*

The proof of this lemma is similar to the one for Lemma 7.2.7 and is therefore omitted. We conclude this section by illustrating a relationship between the concept of Nash equilibrium and the concept of rationalizability. It turns out, namely, that every Nash equilibrium strategy is rationalizable. Recall, from Section 3.2, that a strategy is called a Nash equilibrium strategy if it is a best response against some Nash equilibrium σ. Since every rationalizable strategy is obviously correlated-rationalizable, the same result holds for correlated-rationalizable strategies as well. This result may thus be used as an alternative proof for the nonemptiness of the set of correlated-rationalizable and rationalizable strategies.

Lemma 7.2.14 *Every Nash equilibrium strategy is rationalizable.*

Proof. Let s_i^* be a Nash equilibrium strategy. Then, there is some Nash equilibrium $\sigma = (\sigma_j)_{j \in I}$ such that s_i^* is a best response against σ. For every player j, let $X_j = \{s_j \in S_j | \; s_j \text{ best response against } \sigma\}$ and let $\mu_j \in \times_{k \neq j} \Delta(X_k)$ be given by $\mu_j((s_k)_{k \neq j}) = \prod_{k \neq j} \sigma_k(s_k)$ for all $(s_k)_{k \neq j} \in \times_{k \neq j} X_k$. Note that, by the definition of Nash equilibrium, σ_k assigns positive probability only to strategies in X_k. Then,

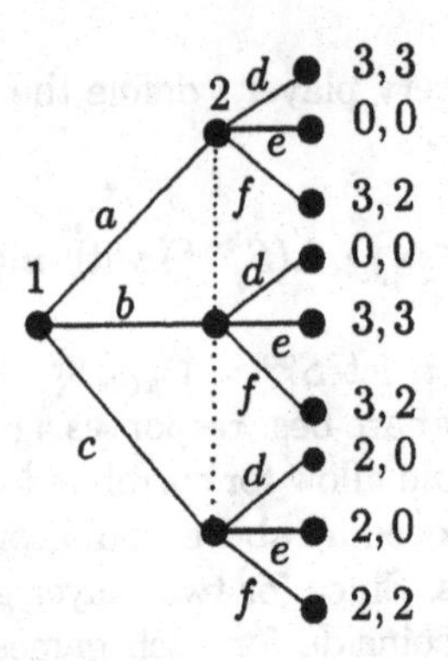

Player 1	(s_1, t_1)	belief
1	(a, t_1^1)	$(1, 0, 0)$
2	(b, t_1^2)	$(0, 1, 0)$
3	(c, t_1^3)	$(0.5, 0.5, 0)$

Player 2	(s_2, t_2)	belief
1	(d, t_2^1)	$(1, 0, 0)$
2	(e, t_2^2)	$(0, 1, 0)$
3	(f, t_2^3)	$(0, 0, 1)$

Figure 7.1

by construction, every $s_j \in X_j$ is a best response against μ_j, and hence $(X_j)_{j \in I}$ has the best response property. Since $s_i^* \in X_i$ it follows by Lemma 7.2.13 that s_i^* is rationalizable, which completes the proof. ■

The converse of Lemma 7.2.14 is not true in general. Consider the game in Figure 7.1. In this game, every strategy is rationalizable. In order to see this, consider the tables in Figure 7.1 which indicate for both players three strategy-type pairs. The tables are to be read as follows. The belief $(0.5, 0.5, 0)$ after the strategy-type pair (c, t_1^3) indicates that t_1^3 assigns probability 0.5 to the strategy-type pair (d, t_2^1), probability 0.5 to (e, t_2^2) and probability zero to all other strategy-type pairs for player 2. The other beliefs are to be read in the same way. It may be checked that each of the strategy-type pairs in the tables is rational, and therefore every strategy is rationalizable. We show, however, that c is not a Nash equilibrium strategy. Suppose the opposite. Then, there would be a Nash equilibrium σ such that c is a best response against σ. This is only possible if $\sigma_2(d) > 0$ and $\sigma_2(e) > 0$. Hence, both d and e should be best responses against σ, which is impossible.

A key difference between Nash equilibrium and rationalizability is the fact that in a Nash equilibrium, the players' beliefs about the conjectures held by the opponents about the other players' strategy choices coincide with the opponents' "real" conjectures, whereas this need not be true in the concept of rationalizability. We refer to this event as "mutual belief of true conjectures". In the example above, for instance, we could consider a state of the world in which the players' true types are t_1^3 and t_2^1, respectively. We know that both types respect common belief of rationality and uncorrelated types. Since t_2^1 believes that player 1 has type t_1^1 with probability one, player 2 believes that player 1 believes that player 2 chooses d. However, player 1's "real" conjecture about player 2's behavior is $0.5d + 0.5e$, and thus there is no mutual belief of true conjectures in this state.

For two-player games, it can be shown that mutual belief of true conjectures is exactly the condition which separates Nash equilibrium from rationalizability. Aumann and Brandenburger (1995) show that for two-player games, mutual belief of rationality and mutual belief of true conjectures imply that the players' conjectures about the opponent's strategy choice constitute a Nash equilibrium. For games with more than two players, these conditions are no longer sufficient to yield a Nash equilibrium. The reason is that for more than two players, one must require that player i's and player j's conjecture about player k's behavior coincide. Aumann and Brandenburger (1995) provide sufficient conditions for Nash equilibrium in games with three or more players as well. Alternative sufficient epistemic conditions for Nash equilibrium can be found in Armbruster and Böge (1979) and Tan and Werlang (1988).

7.3 Permissibility

The concept of rationalizability discussed in the previous section may allow for strategies which are intuitively unreasonable in a given extensive form game. Similarly to the concept of Nash equilibrium, it may allow for strategies which are weakly dominated, as well as strategies which prescribe unreasonable behavior at certain information sets of the game. In this section we present a refinement of rationalizability which does not allow for weakly dominated strategies. The idea is similar to the one adopted in the concept of normal form perfect equilibrium. The motivation behind normal form perfect equilibrium is that players should take into account each of the opponents' strategies, and hence players are believed to choose strategies which are optimal against conjectures that attach positive probability to all opponents' strategies. A player which attaches positive probability to each of the opponents' strategies is called *cautious*. In this section we implement a similar idea within the framework of rationalizability. More precisely, we require that there be common belief about the event that (1) each player is cautious, and (2) each player believes that the opponents are rational "with high probability". By letting the "high probability" tend to one, we obtain the concept of *permissibility*.

These conditions are very similar to, but slightly different from, the ones imposed in Börgers (1994). Indeed, we shall prove in this section that the set of permissible strategies defined here coincides with the set of strategies satisfying Börgers' criterion. A key role in both the present and Börgers' criterion is the fact that players are assumed to believe that opponents are rational with high probability. In this case we say that there is *approximate certainty* about the rationality of the players. By iterating the argument of approximate certainty we would obtain the concept of *approximate common certainty* of an event, reflecting the fact that players believe the event with high probability (which is to be read as "with probability converging to one") , players believe with high probability that the players believe the event with high probability, and so on. This concept of approximate common certainty first appeared in Monderer and Samet (1989) and Stinchcombe (1988). Börgers' criterion may be viewed as an expression of approximate common certainty since it requires players to believe, with high probability, that players are cautious and rational, it requires players to believe,

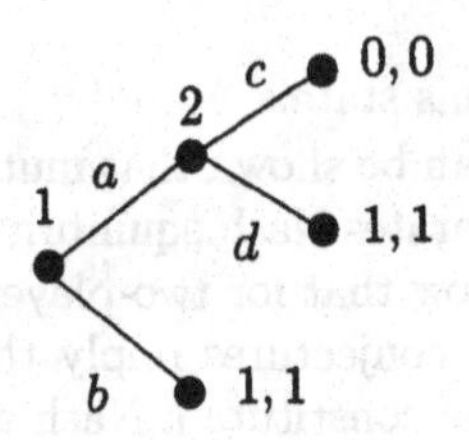

Figure 7.2

with high probability, that players believe with high probability that everyone is cautious and rational, and so on. In contrast, the criterion used in this section is *not* an expression of approximate common certainty since we require players to believe with probability one that players are cautious, we require players to believe that the opponents are rational with high probability, but then this event is assumed to be *commonly believed* with probability one. A similar approach can be found in Asheim and Dufwenberg (2000) in which "fully permissible" sets of strategies are defined by means of a "certain belief" operator. In their paper, a player is said to *certainly believe* an event if he deems the complement of this event subjectively impossible, or, in formal terms, if he deems the complement *Savage-null*.

At this stage, the reader might wonder why we do not use *full* common belief of the event that players are cautious and rational (that is, players believe with probability one that players are cautious and rational, and so on) instead of imposing the weaker condition that players should believe that the opponents are rational with high probability. The reason is that full common belief of cautiousness and rationality is in general impossible. Consider for instance the game in Figure 7.2. In this game, a cautious and rational player 1 will never choose a, since a is not a best response if player 1 attaches positive probability to both c and d. Hence, if player 2 believes with probability one that player 1 is cautious and rational, player 2 should believe with probability one that player 1 chooses b. Hence, if player 1 believes with probability one that player 2 believes with probability one that player 1 is cautious and rational, player 1 believes with probability one that player 2 believes with probability one that player 1 chooses b. But then, player 1 believes with probability one that player 2 is not cautious, since player 2 does not attach positive probability to a. This is clearly a contradiction.

Formally, the concept of permissibility is defined as follows. Conform Section 7.1, let $T_i = B_i^{c,\infty}$ be the set of static coherent belief hierarchies, or types, for player i, and let g_i be the homeomorphism from T_i to $\Delta(S_{-i} \times T_{-i})$. In this and all subsequent sections, we concentrate solely on the case of correlated beliefs. All of the rationalizability concepts developed in the sequel can be adapted easily to the situation in which belief hierarchies are assumed to be uncorrelated. Let C_i^0 be the set of types which assign positive probability to each of the opponents' strategy profiles, hence,

$$C_i^0 = \{t_i \in T_i | \; \tilde{g}_i(t_i)(s_{-i}) > 0 \text{ for all } s_{-i} \in S_{-i}\},$$

where $\tilde{g}_i(t_i)$ is the probability distribution on S_{-i} induced by $g_i(t_i)$. A type $t_i \in C_i^0$ is called *cautious*. Let CR_i^0 be the set of strategy-type pairs (s_i, t_i) which are rational and in which t_i is cautious. Now, fix a probability $p \in (0,1)$. Let $C_i^1(p)$ be the set of cautious types which believe, with probability at least p, that each opponent is cautious and rational. Hence,

$$C_i^1(p) = \{t_i \in C_i^0 | \ g_i(t_i)(CR_{-i}^0) \geq p\},$$

where $CR_{-i}^0 = \times_{j \neq i} CR_j^0$. For every $k \in \mathbb{N}$, the set $C_i^k(p)$ of types is inductively defined by

$$C_i^k(p) = \{t_i \in C_i^{k-1}(p) | \ t_i \text{ believes } S_{-i} \times C_{-i}^{k-1}(p)\}.$$

Intuitively, $C_i^k(p)$ is the set of cautious types which believe with probability one that each player believes with probability one that ... that each player believes with probability p that players are cautious and rational. Define $C_i^\infty(p) = \cap_{k \in \mathbb{N}} C_i^k(p)$. Hence, $C_i^\infty(p)$ contains those types that respect common belief about the event that the players' types are in $C_j^1(p)$. Clearly, the set $C_i^\infty(p)$ is decreasing in p since the restrictions on the types at the first step of the procedure become more stringent if p becomes larger. We are interested in those strategies that are best responses for types in $C_i^\infty(p)$ if p tends to one. For every $p \in (0,1)$, let

$$P_i(p) = \{s_i | \ \exists t_i \in C_i^\infty(p) \text{ such that } (s_i, t_i) \text{ rational}\}.$$

Clearly, the set $P_i(p)$ is decreasing in p. Let $P_i = \cap_{p \in (0,1)} P_i(p)$.

Definition 7.3.1 *A strategy s_i is said to be permissible if $s_i \in P_i$.*

We shall now prove that the set of permissible strategies coincides with the set of strategies satisfying the so-called Dekel-Fudenberg procedure (Dekel and Fudenberg, 1990). Since the set of strategies surviving this procedure is always nonempty, this characterization implies in particular that permissible strategies always exist.

The Dekel-Fudenberg procedure consists of first eliminating all strategies that are weakly dominated, and then iteratively eliminating strategies that are strictly dominated. It is thus a refinement of iterative elimination of strictly dominated strategies, used to characterize correlated-rationalizable strategies in Section 7.2. It follows therefore that a permissible strategy is always correlated-rationalizable. Formally, let

$$\begin{aligned} DF_i^1 &= \{s_i \in S_i | \ s_i \text{ not weakly dominated on } S_i \times S_{-i}\}, \\ DF_i^k &= \{s_i \in DF_i^{k-1} | \ s_i \text{ not strictly dominated on } DF_i^{k-1} \times DF_{-i}^{k-1}\} \end{aligned}$$

for $k \geq 2$. Here, we say that s_i is weakly dominated on $S_i \times S_{-i}$ if there is some $\mu_i \in \Delta(S_i)$ such that $u_i(s_i, s_{-i}) \leq u_i(\mu_i, s_{-i})$ for all $s_{-i} \in S_{-i}$ and $u_i(s_i, \tilde{s}_{-i}) < u_i(\mu_i, \tilde{s}_{-i})$ for some $\tilde{s}_{-i} \in S_{-i}$. Let $DF_i^\infty = \cap_{k \in \mathbb{N}} DF_i^k$. A strategy $s_i \in DF_i^\infty$ is said to *survive the Dekel-Fudenberg procedure*. It should be clear that the sets DF_i^∞ are all nonempty.

Theorem 7.3.2 *A strategy is permissible if and only if it survives the Dekel-Fudenberg procedure.*

In order to prove this result, we need the following lemma.

Lemma 7.3.3 *A strategy s_i is not weakly dominated on $S_i \times S_{-i}$ if and only if there exists a strictly positive probability measure $p_i \in \Delta(S_{-i})$ such that $u_i(s_i, p_i) = \max_{s'_i \in S_i} u_i(s'_i, p_i)$.*

Proof. Suppose that $s_i^* \in S_i$ is not weakly dominated. We prove that there is some strictly positive probability measure $p_i \in \Delta(S_{-i})$ such that $u_i(s_i^*, p_i) = \max_{s_i \in S_i} u_i(s_i, p_i)$. Assume that this were not the case. Similar to the proof of Lemma 7.2.6, we construct a two player zero-sum game Γ^* with player set $I^* = \{i, j\}$, strategy sets $S_i^* = S_i$, $S_j^* = S_{-i}$ and utilityf functions u_i^*, u_j^* where $u_i^*(s_i, s_{-i}) = u_i(s_i, s_{-i}) - u_i(s_i^*, s_{-i})$ for all $(s_i, s_{-i}) \in S_i^* \times S_j^*$, and $u_j^*(s_i, s_{-i}) = -u_i^*(s_i, s_{-i})$ for all $(s_i, s_{-i}) \in S_i^* \times S_j^*$. Let $(\mu_i^*, \mu_{-i}^*) \in \Delta(S_i) \times \Delta(S_{-i})$ be a normal form perfect equilibrium for Γ^*. We prove that μ_i^* weakly dominates s_i^*. Let $\mu_{-i} \in \Delta(S_{-i})$ be arbitrary. Since (μ_i^*, μ_{-i}^*) is a Nash equilibrium, we have that $u_j^*(\mu_i^*, \mu_{-i}) \leq u_j^*(\mu_i^*, \mu_{-i}^*)$ and hence $u_i^*(\mu_i^*, \mu_{-i}) \geq u_i(\mu_i^*, \mu_{-i}^*)$. On the other hand, we have that $u_i^*(\mu_i^*, \mu_{-i}^*) \geq u_i^*(s_i^*, \mu_{-i}^*) = 0$. We may thus conclude that $u_i^*(\mu_i^*, \mu_{-i}) \geq 0$ for all $\mu_{-i} \in \Delta(S_{-i})$, which implies that $u_i(\mu_i^*, \mu_{-i}) \geq u_i(s_i^*, \mu_{-i})$ for all $\mu_{-i} \in \Delta(S_{-i})$.

Since (μ_i^*, μ_{-i}^*) is a normal form perfect equilibrium in Γ^*, there exists some sequence $(\mu_{-i}^n)_{n \in \mathbf{N}}$ of strictly positive probability measures converging to μ_{-i}^* such that $u_i^*(\mu_i^*, \mu_{-i}^n) = \max_{s_i \in S_i} u_i^*(s_i, \mu_{-i}^n)$ for every n. Hence, we have that $u_i(\mu_i^*, \mu_{-i}^n) = \max_{s_i \in S_i} u_i(s_i, \mu_{-i}^n)$ for every n. Take some arbitrary μ_{-i}^n from this sequence. Since μ_{-i}^n is strictly positive, and s_i^* is, by assumption, never a best reply against a strictly positive probability measure, we have that $u_i(\mu_i^*, \mu_{-i}^n) > u_i(s_i^*, \mu_{-i}^n)$. We thus have that $u_i(\mu_i^*, \mu_{-i}) \geq u_i(s_i^*, \mu_{-i})$ for all $\mu_{-i} \in \Delta(S_{-i})$ and $u_i(\mu_i^*, \mu_{-i}) > u_i(s_i^*, \mu_{-i})$ for at least one μ_{-i}. This implies that μ_i^* weakly dominates s_i^*, which contradicts the assumption that s_i^* is not weakly dominated.

Suppose now that s_i^* is weakly dominated. Then, there is some $\mu_i \in \Delta(S_i)$ such that $u_i(\mu_i, s_{-i}) \geq u_i(s_i^*, s_{-i})$ for all $s_{-i} \in S_{-i}$, with strict inequality for some $s_{-i} \in S_{-i}$. Hence, $u_i(\mu_i, \mu_{-i}) > u_i(s_i^*, \mu_{-i})$ for all strictly positive probability measures $\mu_{-i} \in \Delta(S_{-i})$. This implies that s_i^* can never be a best response against a strictly positive probability measure $\mu_{-i} \in \Delta(S_{-i})$, which completes the proof of this lemma. ■

Proof of Theorem 7.3.2. For a given $p \in (0, 1)$, let

$$S_i(p) = \{s_i | \ \exists t_i \in C_i^1(p) \text{ such that } (s_i, t_i) \text{ rational}\}$$

be the set of strategies that are best responses for types in $C_i^1(p)$. Here, $C_i^1(p)$ is the set of types surviving the first round in the procedure leading to the definition of permissibility. On the other hand, let DF_i^1 and DF_i^2 be the sets of strategies surviving the first and the second round of the Dekel-Fudenberg procedure, respectively. Hence, DF_i^1 are the strategies which are not weakly dominated on $S_i \times S_{-i}$ and DF_i^2 are those strategies in DF_i^1 that are not strictly dominated on $DF_i^1 \times DF_{-i}^1$.

Claim 1. There is a number $\bar{p} \in (0, 1)$ such that for every $p \geq \bar{p}$ we have $S_i(p) = DF_i^2$.

Proof of claim 1. Since the set $C_i^1(p)$ is nonempty for all p and decreasing in p, it follows that $S_i(p)$ is nonempty for all p and decreasing in p. Since $S_i(p)$ is a subset of the finite set S_i, there must be some $\bar{p}$ such that for all players i and all $p \geq \bar{p}$ we have that $S_i(p) = S_i(\bar{p})$. We show that $S_i(\bar{p}) = DF_i^2$, which implies that $S_i(p) = DF_i^2$ for all $p \geq \bar{p}$.

Let $s_i \in S_i(\bar{p})$. Then, there is some $t_i \in C_i^1(\bar{p})$ such that s_i is a best reponse for t_i. Since t_i is cautious, it follows that there is some strictly positive probability measure $\mu_i \in \Delta(S_{-i})$ such that s_i is a best reponse against μ_i. By Lemma 7.3.3 it follows that s_i is not weakly dominated on $S_i \times S_{-i}$ and hence $s_i \in DF_i^1$. Suppose that $s_i \notin DF_i^2$. Then, s_i is strictly dominated on $DF_i^1 \times DF_{-i}^1$ and hence there is some $\mu_i \in \Delta(DF_i^1)$ with $u_i(\mu_i, s_{-i}) > u_i(s_i, s_{-i})$ for all $s_{-i} \in DF_{-i}^1$. Let $\Delta^q(S_{-i}) = \{\mu_{-i} \in \Delta(S_{-i})| \ \mu_{-i}$ strictly positive and $\mu_{-i}(DF_{-i}^1) \geq q\}$. We can find some $\bar{q} \in (0,1)$ such that that $u_i(\mu_i, \mu_{-i}) > u_i(s_i, \mu_{-i})$ for all $\mu_{-i} \in \Delta^{\bar{q}}(S_{-i})$. It may be verified that for $p \geq \bar{p}$ large enough, every $t_i \in C_i^1(p)$ induces a probability distribution $\tilde{g}_i(t_i) \in \Delta^{\bar{q}}(S_{-i})$. Choose such a p. Then, since $S_i(p) = S_i(\bar{p})$, we have that s_i is a best response for some type $t_i \in C_i^1(p)$, and hence s_i is a best response against some $\mu_{-i} \in \Delta^{\bar{q}}(S_{-i})$. However, this contradicts the fact that $u_i(\mu_i, \mu_{-i}) > u_i(s_i, \mu_{-i})$ for all $\mu_{-i} \in \Delta^{\bar{q}}(S_{-i})$. We may thus conclude that $s_i \in DF_i^2$.

Let $s_i \in DF_i^2$. Since $s_i \in DF_i^1$ we have, by definition, that s_i is not weakly dominated on $S_i \times S_{-i}$ and hence, by Lemma 7.3.3, s_i is a best response within S_i against some strictly positive probability distribution $\mu_{-i}^1 \in \Delta(S_{-i})$. Moreover, since $s_i \in DF_i^2$, we have that s_i is not strictly dominated on $DF_i^1 \times DF_{-i}^1$ and hence, by Lemma 7.2.6, s_i is a best response within DF_i^1 against some $\mu_{-i}^2 \in \Delta(DF_{-i}^1)$. It may be verified easily that for every $q \in (0,1)$ the strategy s_i is a best response within DF_i^1 against $\mu_{-i}^3(q) = (1-q)\mu_{-i}^1 + q\mu_{-i}^2$. We show that s_i is a best response within S_i against $\mu_{-i}^3(q) = (1-q)\mu_{-i}^1 + q\mu_{-i}^2$. Suppose not. Then, there is some $s_i' \in S_i$ with $u_i(s_i', \mu_{-i}^3(q)) > u_i(s_i', \mu_{-i}^3(q))$ and s_i' best response against $\mu_{-i}^3(q)$. Since $\mu_{-i}^3(q)$ is strictly positive, this implies that s_i' is not weakly dominated on $S_i \times S_{-i}$, and hence $s_i' \in DF_i^1$. However, this contradicts the fact that s_i is a best response within DF_i^1 against $\mu_{-i}^3(q)$. Hence, we may conclude that s_i is a best response within S_i against $\mu_{-i}^3(q)$ for every $q \in (0,1)$.

If one chooses q large enough, the probability distribution $\mu_{-i}^3(q)$ puts large weight on the set DF_{-i}^1 of strategy profiles that consist of best responses of cautious types. Hence, for large enough q there will exist some type $t_i \in C_i^1(\bar{p})$ such that $\tilde{g}_i(t_i) = \mu_{-i}^3(q)$. Since s_i is a best response against $\mu_{-i}^3(q)$, it follows by definition that $s_i \in S_i(\bar{p})$. We may thus conclude that $S_i(\bar{p}) = DF_i^2$.

Claim 2. For every $p \geq \bar{p}$, we have $P_i(p) = DF_i^\infty$.

Proof of claim 2. For every p the set $P_i(p)$ consists of those strategies that are best responses in a world in which there is common belief about the event that types t_j are in $C_j^1(p)$. Since $S_j(p)$ consists of those strategies that are best responses in a world in which all types t_j are in $C_j^1(p)$, common belief of rationality and common belief about types belonging to $C_j^1(p)$ imply common belief about the event that chosen strategies belong to $S_j(p)$. On the other hand, common belief about the event that chosen strategies belong to $S_j(p)$, together with common belief of rationality, is equivalent to iterative elimination of strictly dominated strategies starting from the sets $S_j(p)$. The latter follows immediately from Theorem 7.2.8. For $p \geq \bar{p}$, we have seen that $S_j(p) = DF_j^2$. Consequently, for $p \geq \bar{p}$ common belief about the event that chosen strategies belong to $S_j(p)$ and common belief of rationality is equivalent to iterative elimination of strictly dominated strategies, starting with the sets DF_j^2, which, by construction, yields DF_j^∞. It thus follows that for $p \geq \bar{p}$, every $s_i \in P_i(p)$ belongs to DF_i^∞.

On the other hand, let $s_i \in DF_i^\infty$. Then, $s_i \in DF_i^2$ and claim 2 thus implies that $s_i \in S_i(p)$ for all $p \geq \bar{p}$. Choose an arbitrary $p \geq \bar{p}$. Since $s_i \in S_i(p)$, there is some type $t_i \in C_i^1(p)$ such that (s_i, t_i) is rational. Since $s_i \in DF_i^\infty$, the strategy s_i survives iterative elimination of strictly dominated strategies starting from the sets DF_j^2. Hence, s_i survives iterative elimination of strictly dominated strategies starting from the sets $S_j(p)$. By basically mimicking the proof of Theorem 7.2.8, it can then be shown that there is some type $t_i \in C_i^\infty(p)$ such that (s_i, t_i) is rational. But then, $s_i \in P_i(p)$.

Claim 3. $P_i = DF_i^\infty$.

Proof of claim 3. Since $P_i(p) = DF_i^\infty$ for all $p \geq \bar{p}$ we have that $P_i = \bigcap_{p \in (0,1)} P_i(p) = DF_i^\infty$. ■

According to Theorem 7.3.2, the Dekel-Fudenberg procedure may thus be viewed as an algorithm which yields exactly those strategies that are compatible with the following two events: (1) every player is cautious and rational and (2) there is common belief about the event that each player believes that each of the opponents is cautious and rational with "high probability". In the original paper by Dekel and Fudenberg (1990), the motivation for this procedure is quite different. Their paper focusses on games in which players face a slight uncertainty about the opponents' utilities. For such incomplete information games, *iterative maximal elimination of weakly dominated strategies* is used as a rationality concept. By the latter, we mean the procedure in which first all weakly dominated strategies are eliminated simultaneously, then all weakly dominated strategies within the reduced game are eliminated, and so on, until no strategies are weakly dominated within the reduced game. Since at each stage all weakly dominated strategies are eliminated simultaneously, the outcome of this procedure does not depend on any specific order of elimination, in contrast to the concept of iterated weak dominance discussed in Section 5.2. It is shown that, if the degree of uncertainty about the utilities tends to zero, the set of strategies surviving iterative maximal elimination of weakly dominated strategies within the incomplete information game coincides with the set of strategies surviving the Dekel-Fudenberg procedure applied to the original game (in which there is no uncertainty about utilities).

Papers that provide alternative foundations for strategies surviving the Dekel-Fudenberg procedure are, for instance, Brandenburger (1992), Börgers (1994), Gul (1996), Ben-Porath (1997), Rajan (1998) and Herings and Vannetelbosch (2000). In Brandenburger (1992) it is assumed that players hold "lexicographic" belief hierarchies, and it is shown that common first-order knowledge of rationality yields exactly those strategies surviving the Dekel-Fudenberg procedure. Börgers (1994) shows that the Dekel-Fudenberg procedure selects exactly those strategies compatible with approximate common knowledge of cautiousness and rationality. In Gul (1996) it is shown that the Dekel-Fudenberg procedure can be expressed as a so-called τ-theory. Ben-Porath (1997) demonstrates that for a game of perfect information in generic po-

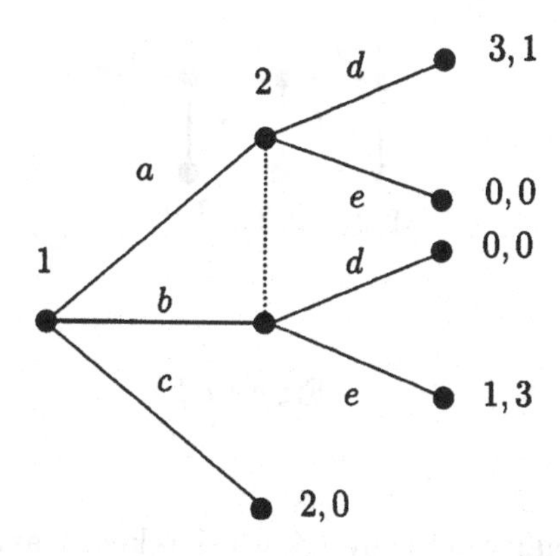

Figure 7.3

sition, imposing common belief of sequential rationality at the beginning of the game is equivalent to the Dekel-Fudenberg procedure. Rajan (1998) assumes that players hold non-standard belief hierarchies in which "infinitesemal" probabilities may be assigned to events, and shows that almost common knowledge of near Bayesian rationality yields the Dekel-Fudenberg procedure. Herings and Vannetelbosch (1999) introduce the concept of weak perfect rationalizability, and prove in Herings and Vannetelbosch (2000) that it is equivalent to the Dekel-Fudenberg procedure.

At a first glance, it may seem somewhat surprising that permissible strategies are not characterized by iterative maximal elimination of weakly dominated strategies, but instead by the weaker Dekel-Fudenberg procedure. Let us illustrate the difference between permissible strategies and iterative maximal elimination of weakly dominated strategies by means of two examples. Consider first the Battle-of-the-sexes game with outside option in Figure 7.3. In this example, the Dekel-Fudenberg procedure solely eliminates the strategy b for player 1, and hence a and c are permissible strategies for player 1, whereas d and e are permissible for player 2. In contrast, iterative maximal elimination of weakly dominated strategies singles out the forward induction strategies a and d. The reason why, for instance, strategy e is permissible is that player 2 may believe that strategy a is less likely than the strictly dominated strategy b, which in turn is perceived much less likely than c. Such beliefs are compatible with permissibility and given these beliefs, player 2's optimal strategy is e. In the concept of iterative maximal elimination of weakly dominated strategies, once b is eliminated in the first round, it is assumed that in the next round player 2 should deem both a and c (the surviving strategies) infinitely more likely than b (the deleted strategy). Given such beliefs, and given that player 2 is cautious, e cannot be a best response, and hence is eliminated in the second round. The key difference is thus that iterative maximal elimination of weakly dominated strategies requires players to deem a strategy infinitely less likely than some other strategy if the former is eliminated earlier than the latter in the elimination process. Permissibility, in contrast, does not put such restrictions.

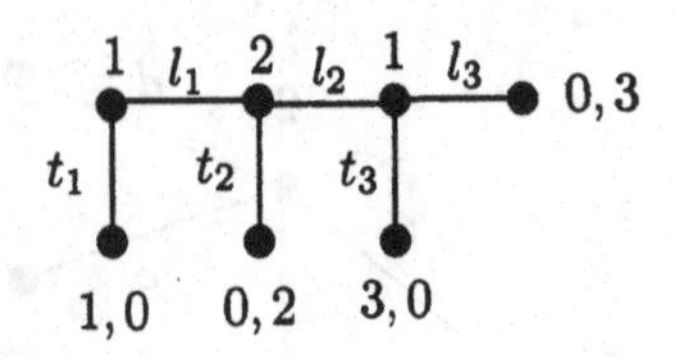

Figure 7.4

Consider next the game in Figure 7.4, which is known as a Take-it-or-leave-it game, and can be found, for instance, in Reny (1992a). In this example, all strategies except (l_1, l_3) are permissible, whereas iterative maximal elimination of weakly dominated strategies uniquely selects the backward induction strategies t_1 and t_2. The reason why, for instance, l_2 is permissible is that player 2 may believe that strategy (l_1, t_3) is much less likely than (l_1, l_3), which in turn is believed to be much less likely than t_1. Such beliefs are compatible with permissibility, and given these beliefs l_2 is optimal. In the procedure of iterative maximal elimination of weakly dominated strategies, (l_1, l_3) is eliminated in the first round. Hence, in the second round player 2 is required to believe that both t_1 and (l_1, t_3) are infinitely more likely than the deleted strategy (l_1, l_3). Consequently, l_2 can no longer be optimal and is thus deleted in the second round. Again, the difference lies in the restrictions that iterative maximal elimination of weakly dominated strategies puts on the relative likelihoods attached to the opponents' strategies.

Stahl (1995) and Brandenburger and Keisler (2000) characterize the set of strategies surviving iterative maximal elimination of weakly dominated strategies by means of restrictions on the players' belief hierarchies. Notions of rationalizability that are related to permissibility and iterated maximal elimination of weakly dominated strategies are, for instance, perfect rationalizability (Bernheim, 1984), cautious rationalizability (Pearce, 1984), weakly perfect rationalizability (Herings and Vannetelbosch, 1999) and fully permissible sets of strategies (Asheim and Dufwenberg, 2000). The concept of proper rationalizability (Schuhmacher, 1999) is a refinement of permissibility which is based on the requirement that a player should deem one strategy of the opponent infinitely more likely than another strategy if the opponent strictly prefers the former over the latter, given the opponent's beliefs. A characterization of properly rationalizable strategies in terms of restrictions on lexicographic belief hierarchies is given in Asheim (2000b).

7.4 Conditional Belief Hierarchies

The static belief hierarchies constructed in Section 7.1 do not require a player to revise his beliefs about the opponents' strategies and types if past play has contradicted these beliefs. The absence of belief revision may, however, lead to unreasonable behavior at certain information sets in the game. Consider, for instance, the game in Figure 7.5.

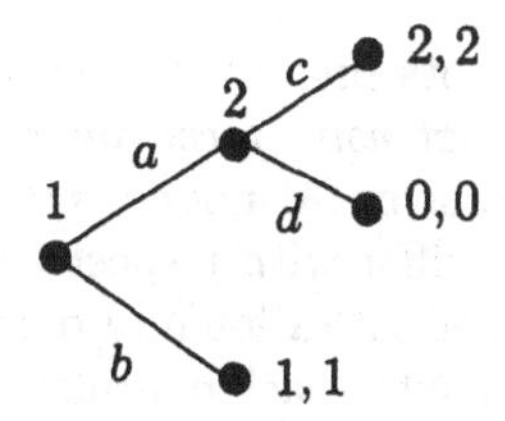

Player 1	(s_1, t_1)	belief
1	(a, t_1^1)	(1,0)
2	(b, t_1^2)	(0,1)

Player 2	(s_2, t_2)	belief
1	(c, t_2^1)	(1,0)
2	(d, t_2^2)	(0,1)

Figure 7.5

In this game, strategy d for player 2 is rationalizable. In order to see this, consider the two tables in Figure 7.5 which represent for each player two strategy-type pairs. The tables should be read as in Section 7.2. Since all strategy-type pairs in the tables are rational, it follows that all types above respect common belief of rationality. Since (d, t_2^2) is rational, it follows that d is rationalizable. The reason is thus that d is a best response against the "rationalizable" static belief that player 1 chooses b. However, if player 2's information set is reached, it is evident that player 1 has chosen a and thus the belief held above by player 2 is contradicted. If player 2 were required to revise his belief at this instance, player 2's revised belief should attach probability one to the event that player 1 has chosen a, and therefore the unique optimal strategy for player 2 would be c. Player 1, anticipating on this reasoning, should thus conclude that his unique optimal strategy is a.

In the context of equilibrium refinements, the requirement that players should revise their beliefs during the game if necessary, and should act optimally against such beliefs at each information set, has been formalized by the notion of sequential best response. By imposing the requirement that players should believe that opponents choose sequential best responses we obtained the concepts of sequential rationality and weak sequential rationality. In the present and the following two sections we translate the requirement of weak sequential rationality to the context of rationalizability. To this purpose, we first construct belief hierachies in which players are explicitly asked to revise their beliefs about the opponents' strategies and types whenever their previous belief has been contradicted by the course of the game. Such belief hierarchies will be called *conditional belief hierarchies,* and their construction is based on Battigalli and Siniscalchi (1997 and 1999). Formally, in such a conditional belief hierarchy a player holds, at each of his information sets, a subjective probability distribution on the opponents' strategies and types. The basic rationality requirement is then that players be *sequentially rational,* that is, a player should choose a strategy which is optimal, given his conditional belief hierarchy, at each information set that is not avoided by the strategy itself.

The rationalizability concepts presented in Sections 7.5 and 7.6, called *weak sequential rationalizability* and *extensive form rationalizability,* respectively, may both be viewed as rationality criteria based upon "common belief of sequential rationality". The concepts, however, differ with respect to the following two questions: (1) should common belief of sequential rationality only be imposed at the beginning of the game, or should it be imposed at each information set, or, more generally, at some collection of information sets? and (2) how should a player revise his beliefs if his previous beliefs have been contradicted?

Before turning to these questions in Sections 7.5 and 7.6, let us first formally introduce conditional belief hierarchies. Its definition is based on the notion of *conditional probability systems* due to Rênyi (1956). Consider a compact metric space X, and let $\mathcal{B}$ be the Borel σ-algebra on X. Choose a finite collection $\mathcal{A}$ of Borel subsets in $\mathcal{B}$.

Definition 7.4.1 *A conditional probability system on* $(X, \mathcal{A})$ *is a function* $\mu(\cdot|\cdot) : \mathcal{B} \times \mathcal{A} \to [0,1]$ *with the following properties:*
(1) for every $A \in \mathcal{A}$, $\mu(\cdot|\ A)$ *is a probability measure on* X *with* $\mu(A|A) = 1$,
(2) for all $A \in \mathcal{B}$, $B, C \in \mathcal{A}$ *with* $A \subseteq B \subseteq C$ *we have that* $\mu(A|B)\mu(B|C) = \mu(A|C)$.

Let the set of conditional probability systems on $(X, \mathcal{A})$ be denoted by $\Delta(X|\mathcal{A})$. With the application we have in mind, the space X may be interpreted as the space of uncertainty faced by player i, that is, the set of opponents' strategy-type pairs. The collection $\mathcal{A}$ of subsets may be viewed as the collection of observable events for player i upon which he can condition his beliefs. In the context of extensive form games, this coincides with the collection of player i information sets, or, to be more precise, for each player i information set it coincides with the set of opponents' strategy-type pairs that are compatible with the event of reaching this information set. Condition (1) states that for every information set $h \in H_i$, the beliefs held by player i at h should attach full probability to those strategy-type pairs that are compatible with the event of reaching h. It thus reflects the requirement that a player should always update, or revise his beliefs in accordance with the newly acquired information, namely that the information set h has been reached. Condition (2) simply states that player i should update his beliefs according to Bayes' rule, whenever possible. Consider namely two information sets h_1 and h_2 for player i, where h_2 comes after h_1. Let C be the event that h_1 is reached, and let B be the event that h_2 is reached. Then, $B \subseteq C$. Assume that player i's beliefs at h_1 deem possible the event that h_2 will be reached, that is, $\mu(B|C) > 0$. In this case, condition (2) states that for every event A compatible with reaching h_2 we have that $\mu(A|B) = \mu(A|C)/\mu(B|C)$. Hence, player i's beliefs at h_2 should be derived from his beliefs at h_1 by Bayesian updating.

Similar to Section 7.1, we now construct for each player i an infinite hierarchy of state spaces, consisting of a first-order state space, a second-order state space, and so on, and require the player to form a belief on each of these state spaces. In contrast to Section 7.1, beliefs are now represented by conditional probability systems instead of ordinary probability distributions. We assume in the sequel that we are dealing with a game without chance moves. The basic source of uncertainty for player i is still the profile of strategies chosen by the opponents. Hence, $X_i^1 = S_{-i}$ is player i's first-order state space. We next require each player i to form a belief about the opponents' strategies *before the game* and *at each of his information sets.* The first-order belief can thus be represented by a conditional probability system $p_i^1 \in \Delta(X_i^1|\mathcal{A}_i^1)$, where

$$\mathcal{A}_i^1 = \{A \subseteq X_i^1 | \ A = S_{-i} \text{ or } A = S_{-i}(h) \text{ for some } h \in H_i\}.$$

Hence, $\mathcal{A}_i^1$ is the collection of sets of opponents' strategies that are compatible with the event of reaching a player i information set. Recall that $S_{-i}(h)$ is the set of opponents' strategy profiles which lead to h. The set $A = S_{-i}$ reflects the event where player i has to form a belief without any additional information, that is, before the game starts. Within the second-order state space, player i faces uncertainty about the opponents' strategies and the opponents' first-order beliefs. Player i's second-order state space is thus given by

$$X_i^2 = X_i^1 \times (\times_{j \neq i} \Delta(X_j^1 | \mathcal{A}_j^1)).$$

The space X_i^2 can thus be written as $X_i^2 = S_{-i} \times Y_i^2$ for some Y_i^2. Player i's second-order belief is represented by a conditional probability system $p_i^2 \in \Delta(X_i^2|\mathcal{A}_i^2)$, where

$$\mathcal{A}_i^2 = \{A \subseteq X_i^2 | \ A = X_i^2 \text{ or } A = S_{-i}(h) \times Y_i^2 \text{ for some } h \in H_i\}.$$

Here, $S_{-i}(h) \times Y_i^2$ represents the event in X_i^2 that corresponds to the observation that h has been reached, whereas $A = X_i^2$ represents the beginning of the game. The state spaces $X_i^1, X_i^2, ...$ and the collections of events $\mathcal{A}_i^1 \subseteq X_i^1, \mathcal{A}_i^2 \subseteq X_i^2, ...$ on which beliefs are to be conditioned, can thus be defined as follows:

$$\begin{aligned} X_i^1 &= S_{-i}, \\ \mathcal{A}_i^1 &= \{A \subseteq X_i^1 | \ A = X_i^1 \text{ or } A = S_{-i}(h) \text{ for some } h \in H_i\}, \\ &\cdots \\ X_i^k &= X_i^{k-1} \times (\times_{j \neq i} \Delta(X_j^{k-1} | \ \mathcal{A}_j^{k-1})), \\ \mathcal{A}_i^k &= \{A \subseteq X_i^k | \ A = X_i^k \text{ or } A = S_{-i}(h) \times Y_i^k \text{ for some } h \in H_i\}, \end{aligned}$$

where Y_i^k is such that $X_i^k = S_{-i} \times Y_i^k$.

Definition 7.4.2 *A conditional belief hierarchy for player i is a vector $p_i = (p_i^k)_{k \in \mathbb{N}}$ of conditional probability systems, where $p_i^k \in \Delta(X_i^k|\mathcal{A}_i^k)$ for all $k \in \mathbb{N}$.*

Note that each set $\mathcal{A}_i^k$ contains a finite collection of events. More precisely, each event in $\mathcal{A}_i^k$ corresponds to either the observation that some information set $h \in H_i$ has been reached, or to the "empty" observation (the situation before the game starts). Let B_i be the set of conditional belief hierarchies for player i. As in Section 7.1, we require each conditional belief hierarchy to be coherent, that is, the k-th order belief should coincide with the $(k-1)$-th order belief on the $(k-1)$-th order state space. In order to formalize this, we need the following notation. For every $h \in H_i \cup \{\emptyset\}$, let $\mathcal{A}_i^k(h)$ be the unique event in $\mathcal{A}_i^k$ which corresponds to observing information set h. If $h = \emptyset$, then $\mathcal{A}_i^k(h)$ corresponds to the "empty" observation, that is, $\mathcal{A}_i^k = X_i^k$.

Definition 7.4.3 *A conditional belief hierarchy* $p_i = (p_i^k)_{k\in\mathbb{N}}$ *is called coherent if for every* $k \geq 2$, *every* $h \in H_i \cup \{\emptyset\}$ *it holds that* $mrg(p_i^k(\cdot|\ A_i^k(h))|\ X_i^{k-1}) = p_i^{k-1}(\cdot|\ A_i^{k-1}(h))$.

Hence, upon reaching information set h, the k-th order belief and the $(k-1)$-th order belief coincide on events in X_i^{k-1}. Let B_i^c be the set of coherent conditional belief hierarchies for player i. Similar to Section 7.1, it may be shown that the set of coherent conditional belief hierarchies for player i is homeomorphic to the set of conditional probability systems on $S_{-i} \times B_{-i}$, where $B_{-i} = \times_{j\neq i} B_j$ is the set of opponents' conditional belief hierarchies. The collection of events upon which such conditional probability systems on $S_{-i} \times B_{-i}$ are to be conditioned is given by $\mathcal{A}_i^\infty = \{A \subseteq S_{-i} \times B_{-i}|\ A = S_{-i} \times B_{-i} \text{ or } A = S_{-i}(h) \times B_{-i} \text{ for some } h \in H_i\}$.

Theorem 7.4.4 *For every player* i *there is a homeomorphism* f_i *from* B_i^c *to* $\Delta(S_{-i} \times B_{-i}|\ \mathcal{A}_i^\infty)$.

A proof of this result for the class of multi-stage games with observed actions can be found in Battigalli and Siniscalchi (1999). We next impose the requirement that there be common belief of coherency. Let the sets of conditional belief hierarchies $B_i^{c,1}, B_i^{c,2}, \ldots$ be given by

$$
\begin{aligned}
B_i^{c,1} &= B_i^c, \\
B_i^{c,k} &= \{p_i \in B_i^{c,k-1}|\ f_i(p_i)(S_{-i} \times B_{-i}^{c,k-1}|\ A) = 1 \text{ for all } A \in \mathcal{A}_i^\infty\}
\end{aligned}
$$

for all $k \geq 2$. Here, $B_{-i}^{c,k-1} = \times_{j\neq i} B_j^{c,k-1}$. Let $B_i^{c,\infty} = \cap_{k\in\mathbb{N}} B_i^{c,k}$ be the set of conditional belief hierarchies which respect common belief of coherency. Denote by

$$\mathcal{A}_i = \{A \subseteq S_{-i} \times B_{-i}^{c,\infty}|\ A = S_{-i} \times B_{-i}^{c,\infty} \text{ or } A = S_{-i}(h) \times B_{-i}^{c,\infty} \text{ for some } h \in H_i\}$$

the collection of observable events on which player i can condition his beliefs, given that there is common belief about the event that all players have coherent conditional belief hierarchies. Similar to Section 7.1, we obtain the following result.

Lemma 7.4.5 *For every player* i *there is a homeomorphism* g_i *from* $B_i^{c,\infty}$ *to* $\Delta(S_{-i} \times B_{-i}^{c,\infty}|\ \mathcal{A}_i)$.

From now on, let $T_i = B_i^{c,\infty}$ denote the set of possible *types* for player i, where each type can be identified, through the homeomorphism g_i, with the conditional probability system $g_i(t_i) \in \Delta(S_{-i} \times T_{-i}|\ \mathcal{A}_i)$. In words, each type holds, before the game starts, and at each of his information sets, a belief about the possible strategies and types of the opponents. Ben-Porath (1997) proposes an epistemic model for games with perfect information, in which each type is required to hold, at each of his decision nodes, a belief about the opponents' strategy-type pairs. His model may thus be seen as a special case of the model developed here. In contrast to the approach in this section, however, Ben-Porath (1997) does not *explicitly* construct the conditional belief hierarchies corresponding to such types. Instead, Ben-Porath (1997) *implicitly* defines types by means of the above correspondence between types and beliefs about strategy-type pairs. Subsequently, conditional belief hierarchies can be *derived* from this definition of types.

7.5 Weak Sequential Rationalizability

Given the requirement that a player updates and revises his beliefs by means of a conditional belief hierarchy, it is natural to impose that such a player chooses a strategy which is optimal, given his beliefs, at each of his information sets that are not avoided by this strategy itself. In this case, we say that the player is *sequentially rational.* In order to formalize this definition, consider a strategy-type pair (s_i, t_i), where the type t_i corresponds to a conditional probability system $g_i(t_i) \in \Delta(S_{-i} \times T_{-i} | \mathcal{A}_i)$. For every $h \in H_i$, let $g_i(t_i)(\cdot | h) := g_i(t_i)(\cdot | S_{-i}(h) \times T_{-i})$ be the probability measure on $S_{-i}(h) \times T_{-i}$ induced at information set h. Let $g_i(t_i)(\cdot | \emptyset) := g_i(t_i)(\cdot | S_{-i} \times T_{-i})$ be the initial probability measure at the beginning of the game. For every $h \in H_i \cup \{\emptyset\}$, let $\tilde{g}_i(t_i)(\cdot | h) \in \Delta(S_{-i}(h))$ be the marginal probability distribution on $S_{-i}(h)$ induced by $g_i(t_i)(\cdot | h)$. Let $H_i(s_i)$ be the set of player i strategies that are not avoided by s_i. For a given $h \in H_i(s_i) \cup \{\emptyset\}$, let

$$u_i(s_i, t_i | h) = \sum_{s_{-i} \in S_{-i}(h)} u_i(s_i, s_{-i}) \, \tilde{g}_i(t_i)(s_{-i} | h)$$

be the expected utility for type t_i at information set h by choosing strategy s_i. If $h = \emptyset$, then $u_i(s_i, t_i | h)$ is simply $u_i(s_i, t_i)$; the expected utility at the beginning of the game.

Definition 7.5.1 *A strategy-type pair (s_i, t_i) is said to be sequentially rational if at every $h \in H_i(s_i) \cup \{\emptyset\}$ we have that $u_i(s_i, t_i|h) = \max_{s_i' \in S_i(h)} u_i(s_i', t_i|h)$.*

If $u_i(s_i, t_i | h) = \max_{s_i' \in S_i(h)} u_i(s_i', t_i | h)$, then we say that (s_i, t_i) is sequentially rational at h. If $h = \emptyset$, the strategy-type pair (s_i, t_i) is sequentially rational at h if and only if it is rational. It may be verified that if (s_i, t_i) is sequentially rational at every $h \in H_i(s_i)$, then (s_i, t_i) is rational, and hence is sequentially rational at $h = \emptyset$. Therefore, we could as well write "at every $h \in H_i(s_i)$" in the definition of sequential rationality.

From Theorem 3.1 in Perea (2001) it easily follows that a strategy-type pair is sequentially rational if and only if it is *locally* sequentially rational. By the latter, we mean that at each information set $h \in H_i(s_i)$, player i cannot improve his expected utility by changing his behavior *only at h*. For every type t_i one can construct easily a strategy s_i such that (s_i, t_i) is locally sequentially rational. To this purpose, one starts at the final information sets for player i, that is, those information sets not followed by any other player i information set. At these information sets, choose actions that maximize player i's utility, given his beliefs about the opponents' behavior there. Afterwards, turn to player i's penultimate information sets that are followed by only one (final) player i information set. Given that actions at player i's final information sets have already been chosen, one can choose at each penultimate information set an action that maximizes player i's utility, given his belief about the opponents' behavior there and given his action chosen at the player i information set that follows. By working backwards in this way, one constructs a strategy s_i such that (s_i, t_i) is locally sequentially rational. As a consequence, for every type t_i there is a strategy s_i such that (s_i, t_i) is sequentially rational.

In this section, we are interested in the implications of *common belief of sequential rationality*. That is, we not only impose that players be sequentially rational, but also players should believe that all opponents are sequentially rational, all players should believe that all players should believe that all opponents are sequentially rational, and so on. More precisely, we wish to investigate the following question: given an arbitrary collection of information sets H^* (which may include the "empty" information set, reflecting the situation at the beginning of the game), which strategies are compatible with imposing common belief of sequential rationality at information sets in H^*? Such strategies are called *weakly sequentially rationalizable* at H^*. We shall provide an algorithm, similar to the algorithms for rationalizability and permissibility, which yields these strategies. Moreover, as two polar cases, we will see that weak sequential rationalizability only at the beginning of the game (i.e. $H^* = \{\emptyset\}$) is always possible, whereas weak sequential rationalizability at all information sets is in general impossible.

Formally, common belief of sequential rationality at a collection of information sets is defined as follows (see Battigalli and Siniscalchi, 1999). Choose some collection of information sets $H^* \subseteq H \cup \{\emptyset\}$, a type t_i for player i, and some event $E \subseteq S_{-i} \times T_{-i}$. We say that t_i *believes* E *at* H^* if at all $h \in H_i^*$ it holds that $g_i(t_i)(E|\ h) = 1$, where H_i^* is the set of player i information sets in H^*. Here, we use the convention that the empty information set is owned by every player, hence H_i^* contains the empty information set whenever H^* contains it. Hence, t_i believes the event E at H^* if it assigns probability one to this event at each of his information sets in H^*.

For every player i, let R_i^0 be the set of strategy-type pairs that are sequentially rational. For a given collection of information sets $H^* \subseteq H \cup \{\emptyset\}$ we recursively define the sets $R_i^1(H^*), R_i^2(H^*), \ldots$ of types as follows:

$$\begin{aligned} R_i^1(H^*) &= \{t_i \in T_i|\ t_i \text{ believes } R_{-i}^0 \text{ at } H^*\}, \\ R_i^k(H^*) &= \{t_i \in R_i^{k-1}(H^*)|\ t_i \text{ believes } S_{-i} \times R_{-i}^{k-1}(H^*) \text{ at } H^*\}, \end{aligned}$$

for $k \geq 2$. Hence, $R_i^k(H^*)$ contains those types which believe at each information set in H^* that each player believes at each information set in H^* that ... that each player is sequentially rational. Let $R_i^\infty(H^*) = \cap_{k\in\mathbb{N}} R_i^k(H^*)$.

Definition 7.5.2 *A type t_i is said to respect common belief of sequential rationality at H^* if $t_i \in R_i^\infty(H^*)$. A strategy s_i is called weakly sequentially rationalizable at H^* if there is some $t_i \in R_i^\infty(H^*)$ such that (s_i, t_i) is sequentially rational.*

Similar to rationalizability and permissibility, we now provide an algorithm that computes the set of weakly sequentially rationalizable strategies at H^*, without any reference to types. Let $\mathcal{A}_i^1 = \{A \subseteq S_{-i}|\ A = S_{-i} \text{ or } A = S_{-i}(h) \text{ for some } h \in H_i\}$ be the collection of sets of opponents' strategy profiles representing either the beginning of the game or the event of reaching a player i information set. Let $\Delta(S_{-i}|\ \mathcal{A}_i^1)$ be the set of conditional probability systems on $(S_{-i}, \mathcal{A}_i^1)$. For a given set of opponents' strategy profiles $X_{-i} \subseteq S_{-i}$, let $\Delta^*(X_{-i}, H^*)$ be the set of conditional probability systems $\mu_i \in \Delta(S_{-i}|\ \mathcal{A}_i^1)$ such that for every $h \in H_i^*$

$$\mu_i(X_{-i}|\ S_{-i}(h)) = 1.$$

Hence, a conditional probability system μ_i in $\Delta^*(X_{-i}, H^*)$ assigns full probability to strategy profiles in X_{-i} at every $h \in H_i^*$. Note that $\Delta^*(X_{-i}, H^*)$ may be empty if $S_{-i}(h) \cap X_{-i}$ is empty for some $h \in H_i^*$. A pair (s_i, μ_i) consisting of a strategy s_i and a conditional probability system $\mu_i \in \Delta(S_{-i}|\ \mathcal{A}_i^1)$ is called *sequentially rational* if at every $h \in H_i(s_i)$, the strategy s_i is a best response against $\mu_i(\cdot|\ S_{-i}(h))$. Let the strategy subsets $S_i^0(H^*), S_i^1(H^*), \ldots$ be defined recursively by

$$\begin{aligned} S_i^0(H^*) &= S_i, \\ S_i^k(H^*) &= \{s_i \in S_i^{k-1}(H^*) |\ \exists \mu_i \in \Delta^*(S_{-i}^{k-1}(H^*), H^*) \text{ such that} \\ &\qquad (s_i, \mu_i) \text{ sequentially rational}\}, \end{aligned}$$

for $k \geq 2$. Let $S_i^\infty(H^*) = \cap_{k \in \mathbb{N}} S_i^k(H^*)$. Note that $S_i^\infty(H^*)$ may be empty, since $\Delta^*(S_{-i}^{k-1}(H^*), H^*)$ may be empty for reasons mentioned above. By construction, $S_i^1(H^*)$ contains those strategies that are a sequential best response against *some* conditional probability system. For $k \geq 2$, $S_i^k(H^*)$ contains those strategies that have survived so far, and that are in addition a sequential best response against some conditional probability system that at each information set in H_i^* assigns full probability to the opponents' strategies that have survived so far. The following theorem is due to Battigalli and Siniscalchi (1999).

Theorem 7.5.3 *Let $H^* \subseteq H \cup \{\emptyset\}$. Then, a strategy s_i is weakly sequentially rationalizable at H^* if and only if $s_i \in S_i^\infty(H^*)$.*

The proof is omitted for the sake of brevity. The interested reader is referred to the proof in Battigalli and Siniscalchi (1999). We shall now investigate some special choices of the collection H^* of information sets at which common belief of sequential rationality is imposed. First, consider the case where $H^* = \{\emptyset\}$. Hence, common belief of sequential rationality is required at the beginning of the game only, but may be violated at certain information sets once the game is under way. Ben-Porath (1997) investigates this case for the class of games with perfect information, and calls it *common certainty of rationality at the beginning of the game*. A first observation is that for every game, there is at least one strategy for every player that is weakly sequentially rationalizable at the beginning of the game. This easily follows from Theorem 7.5.3. Note that $S_i^1(\{\emptyset\})$ contains those strategies that are a sequential best response against *some* conditional updating system, and hence $S_i^1(\{\emptyset\})$ is nonempty, as we have seen above. At stage 2, $S_i^2(\{\emptyset\})$ selects among $S_i^1(\{\emptyset\})$ those strategies that are a best response, at the beginning of the game, against a conjecture that assigns full weight to strategy profiles in $S_{-i}^1(\{\emptyset\})$, and hence $S_i^2(\{\emptyset\})$ is nonempty. By repeating this argument, it follows that $S_i^k(\{\emptyset\})$ is nonempty for all k, and hence $S_i^\infty(\{\emptyset\})$ is nonempty for all players i.

In fact, by Lemma 7.2.6 and the argument above, it follows that the sets $S_i^k(\{\emptyset\})$ for $k \geq 2$ are obtained by iterative elimination of strictly dominated strategies, starting from the sets $S_i^1(\{\emptyset\})$. Hence, strategies that are weakly sequentially rationalizable at the beginning of the game are obtained by first eliminating all strategies that are never a sequential best response, and afterwards iteratively eliminating strictly dominated strategies. Ben-Porath (1997) shows that in a game with perfect information

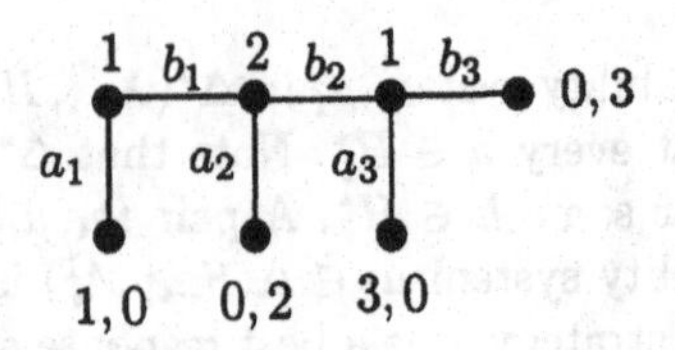

Player 1	(s_1, t_1)	belief at beginning	belief at h_3
1	(a_1, t_1^1)	(1,0)	(0,1)
2	$((b_1, a_3), t_1^2)$	(0,1)	(0,1)
3	$((b_1, b_3), t_1^3)$	(0,1)	(0,1)

Player 2	(s_2, t_2)	belief at beginning	belief at h_2
1	(a_2, t_2^1)	(0,1,0)	(0,1,0)
2	(b_2, t_2^2)	(1,0,0)	(0,0,1)

Figure 7.6

in generic position (see Section 3.1), a strategy is never a sequential best response if and only if the strategy is weakly dominated. Consequently, for such games weak sequential rationalizability at the beginning of the game is equivalent to first eliminating all weakly dominated strategies, followed by iterative elimination of strictly dominated strategies, that is, is equivalent to the Dekel-Fudenberg procedure. Since, by Theorem 7.3.2, a strategy survives the Dekel-Fudenberg procedure if and only if it is permissible, we obtain the following characterization.

Lemma 7.5.4 *Consider a game with perfect information in generic position. Then, a strategy is weakly sequentially rationalizable at the beginning of the game if and only if it is permissible.*

It is important to note that weak sequential rationalizability at the beginning of the game allows players to drop the conjecture that opponents are sequentially rational, once their initial belief about the opponents' strategies has been contradicted by the course of the game. In order to illustrate the consequences of this fact, consider the Take-it-or-leave-it game in Figure 7.6. The unique backward induction strategies in this game are a_1 for player 1 and a_2 for player 2. We prove, however, that the strategies (b_1, a_3) and b_2 are weakly sequentially rationalizable at the beginning of the game. Consider the two tables in Figure 7.6, which represent three strategy-type pairs for player 1 and two strategy-type pairs for player 2. By h_1, h_2, h_3 we indicate the first, second and third information set, respectively. The tables are to be read as follows. For instance, the row corresponding to strategy-type pair (a_1, t_1^1) indicates that at the beginning of the game (hence, at h_1), t_1^1 believes that player 2 has type t_2^1 and chooses a_2, whereas at h_3 type t_1^1 believes that player 2 has type t_2^2 and has chosen b_2. The reason why t_1^1 is allowed to revise his belief at h_3 is because t_1^1's initial belief is contradicted if h_3 is reached. It may be checked that all the above strategy-type pairs, except $((b_1, b_3), t_1^3)$, are sequentially rational. Since at the beginning of the game, both player 2 types above assign probability zero to $((b_1, b_3), t_1^3)$, it follows that

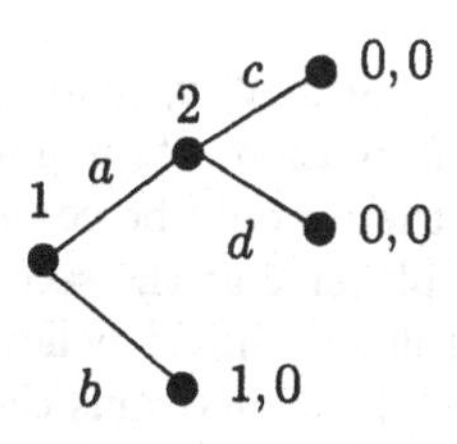

Figure 7.7

t_1^1, t_1^2, t_2^1 and t_2^2 respect common belief of sequential rationality at the beginning of the game. Consequently, all strategies but (b_1, b_3) are weakly sequentially rationalizable at the beginning of the game. In particular, (b_1, a_3) and b_2 are weakly sequentially rationalizable at the beginning of the game. Note that b_2 is weakly sequentially rationalizable since type t_2^2, for which b_2 is a sequential best response, does no longer believe at h_2 that player 1 is sequentially rational. Player 2 is allowed to revise his belief about player 1 at h_2 since t_2^2's initial belief, namely that player 1 chooses a_1, is contradicted at h_2. If t_2^2 were restricted to believe at h_2 that player 1 is still sequentially rational, t_2^2 should at h_2 believe that player 1 will choose a_3 at his final node, and hence b_2 would no longer be optimal. The latter requirement, that a player should maintain the belief that his opponents are sequentially rational, whenever possible, is formalized in the concept of *extensive form rationalizability,* to be discussed in the next section.

Weak sequential rationalizability at the beginning of the game may thus be viewed as a rather weak concept, since common belief of sequential rationality is only required at the beginning of the game, but may be violated during the course of the game. The other extreme case would be to require common belief of sequential rationality *at each information set*, that is, to choose $H^* = H \cup \{\emptyset\}$. However, it is easily seen that this concept may lead to impossibilities, as the game in Figure 7.7 demonstrates. In this game, common belief of sequential rationality at each information set is clearly impossible since at player 2's information set it is evident that player 1 has chosen a, and hence player 2 should believe at his information set that player 1 is not sequentially rational. However, this example is a bit trivial since player 2's decision node is simply incompatible with player 1 being sequentially rational, and hence common belief of sequential rationality at this node is, without any doubt, impossible. In order to judge whether weak sequential rationalizability is possible at a given collection H^* of information sets, we may thus, as a first criterion, check whether H^* contains information sets that are incompatible with sequential rationality of one of the players. If this is so, then weak sequential rationalizability is clearly impossible. Formally, we say that an information set h is *compatible with sequential rationality* if there is a profile $(s_i, t_i)_{i \in I}$ of sequentially rational strategy-type pairs such that $(s_i)_{i \in I}$ reaches h. Let H^c be the collection of information sets compatible with sequential rationality. The question which remains is thus whether common belief of sequential rationality at H^c is always possible. Again, the answer is no. Consider, to this purpose, again the Take-it-or-leave-it game in Figure 7.6. It may be checked that all information sets are compatible with sequential rationality, however we show that common belief of sequential rationality at all information sets is impossible (see also Reny, 1992a and 1993). Suppose, namely, that there would be common belief of sequential rationality at all information sets. Then, player 2 at the second node should believe that player 1 is sequentially rational, and hence should believe that player 1 chooses (b_1, a_3). If player 2 is sequentially rational, he should thus choose a_2. But then, there cannot be common belief of sequential rationality at the last node, since reaching the last node would contradict the fact that player 2 is sequentially rational and believes, at the second node, that player 1 is sequentially rational.

The argument in this example can even be strengthened. One can prove, namely, that common belief of sequential rationality *at the first and second node only* is already impossible (see, again, Reny, 1992a and 1993). Suppose, namely, that there would be common belief of sequential rationality at the first and second node. Then, at the second node, player 2 should believe that player 1 chooses (b_1, a_3) and hence player 2 should choose a_2 at the second node. Player 1, at the first node, believes that player 2 is sequentially rational, and believes that player 2, at his second node, believes that player 1 is sequentially rational. Hence, player 1, at the first node, should believe that player 2 chooses a_2. As such, player 1 should choose a_1 at the first node. But then, reaching the second node would contradict common belief of sequential rationality at the first and the second node.

Reny (1992a) considers take-it-or-leave-it games of arbitrary length and shows, by a similar argument as above, that for such games common belief of sequential rationality at all nodes is always impossible. An even more general point is made in Reny (1993). In this paper attention is restricted to games with perfect information with a unique backward induction outcome. In such a game, a node is called *relevant* if it is compatible with sequential rationality, and if the player at this node does not have a strictly dominant action. The set of relevant nodes may be seen as the collection of nodes for which requiring common belief of sequential rationality could be possible and "interesting". Note that, if a player has a strictly dominant action at one of his nodes, then he would always choose this action, irrespective of his belief about future behavior. Reny (1993) then characterizes those games with perfect information in which common belief of sequential rationality *at all relevant nodes* is possible: these are exactly the games in which all nodes off the backward induction path are not relevant. In the example of Figure 7.6, for instance, the second node lies off the equilibrium path, but is relevant. Consequently, common belief of sequential rationality at all relevant nodes (which in this case are the first and second node) is not possible in this game.

Suppose now, that H^{1*} and H^{2*} are collections of information sets such that common belief of sequential rationality is possible at H^{1*}, as well as at H^{2*}. Is it then true that common belief of sequential rationality is possible at the union of both collections of information sets? The example in Figure 7.8 shows that this is not true. This example is taken from Reny (1993) and Battigalli and Siniscalchi (1999). Note that x_1 and x_2 are relevant nodes in the sense of Reny (1993). We show that common belief of sequential rationality is possible at $\{\emptyset, x_1\}$ and at $\{\emptyset, x_2\}$, but not at $\{\emptyset, x_1, x_2\}$. Consider the two tables in Figure 7.8, that represent for each player four strategy-type pairs. The types t_1^1 for player 1 and t_2^2 for player 2 respect common

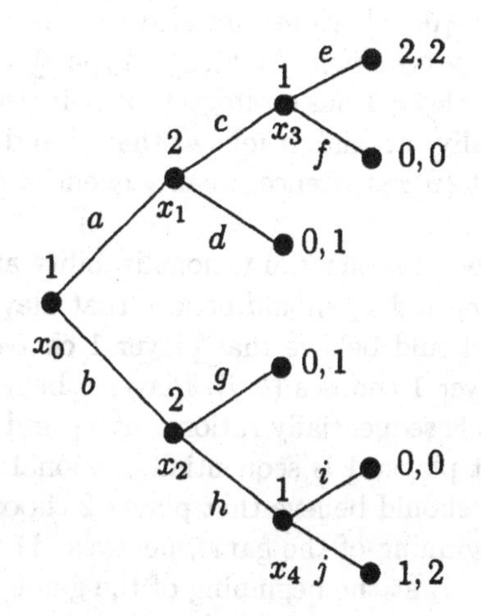

Player 1	(s_1, t_1)	belief at beginning	belief at x_3	belief at x_4
1	$((a, e), t_1^1)$	$(0, 1, 0, 0)$	$(0, 1, 0, 0)$	$(0, 1, 0, 0)$
2	$((a, f), t_1^2)$	$(0, 1, 0, 0)$	$(0, 1, 0, 0)$	$(0, 1, 0, 0)$
3	$((b, i,), t_1^3)$	$(0, 1, 0, 0)$	$(0, 1, 0, 0)$	$(0, 1, 0, 0)$
4	$((b, j), t_1^4)$	$(0, 0, 0, 1)$	$(0, 1, 0, 0)$	$(0, 0, 0, 1)$

Player 2	(s_2, t_2)	belief at beginning	belief at x_1	belief at x_2
1	$((c, g,), t_2^1)$	$(1, 0, 0, 0)$	$(1, 0, 0, 0)$	$(0, 0, 1, 0)$
2	$((c, h), t_2^2)$	$(1, 0, 0, 0)$	$(1, 0, 0, 0)$	$(0, 0, 0, 1)$
3	$((d, g), t_2^3)$	$(1, 0, 0, 0)$	$(1, 0, 0, 0)$	$(0, 0, 0, 1)$
4	$((d, h), t_2^4)$	$(0, 0, 0, 1)$	$(0, 1, 0, 0)$	$(0, 0, 0, 1)$

Figure 7.8

belief of sequential rationality at $\{\emptyset, x_1\}$. In order to see this, note that t_1^1, at the beginning of the game, believes that player 2 is of strategy-type pair $((c,h), t_2^2)$. On the other hand, t_2^2 at the beginning of the game and at x_1, believes that player 1 has strategy-type pair $((a,e), t_1^1)$. Since $((a,e), t_1^1)$ and $((c,h), t_2^2)$ are sequentially rational, it follows that both t_1^1 and t_2^2 respect common belief of sequential rationality at $\{\emptyset, x_1\}$. Hence, common belief of weak sequential rationality is possible at $\{\emptyset, x_1\}$.

Moreover, types t_1^4 for player 1 and t_2^4 for player 2 respect common belief of sequential rationality at $\{\emptyset, x_2\}$. Note that t_1^4, at the beginning of the game, believes that player 2 has strategy-type pair $((d,h), t_2^4)$. Type t_2^4, at the beginning of the game and at x_2, believes that player 1 has strategy-type pair $((b,j), t_1^4)$. Since $((b,j), t_1^4)$ and $((d,h), t_2^4)$ are sequentially rational, it follows that t_1^4 and t_2^4 respect common belief of sequential rationality at $\{\emptyset, x_2\}$. Hence, weak sequential rationalizability at $\{\emptyset, x_2\}$ is possible.

Now, suppose that weak sequential rationalizability at $\{\emptyset, x_1, x_2\}$ would be possible. Then, player 2 at x_1 and x_2 should believe that player 1 is sequentially rational. Hence, player 2 at x_1 should believe that player 1 chooses (a,e) and player 2 at x_2 should believe that player 1 chooses (b,j). Player 1 believes, at the beginning of the game, (1) that player 2 is sequentially rational at x_1 and x_2, and (2) that player 2 at x_1 and x_2 believes that player 1 is sequentially rational. Therefore, player 1, at the beginning of the game, should believe that player 2 chooses c at x_1 and chooses h at x_2. Player 2, at the beginning of the game, believes (1) that player 1 is sequentially rational, (2) that player 1, at the beginning of the game, believes that player 2 is sequentially rational at x_1 and x_2, and (3) that player 1, at the beginning of the game, believes that player 2, at x_1 and x_2, believes that player 1 is sequentially rational. Consequently, player 2, at the beginning of the game, should believe that player 1 chooses (a,e). But then, common belief of sequential rationality is not possible at x_2. Hence, we may conclude that common belief of sequential rationality is not possible at $\{\emptyset, x_1, x_2\}$.

The examples above show that imposing common belief of sequential rationality at all information sets, or even at some subcollection of information sets, often leads to impossibilities. As a consequence, it is often necessary for players to believe that opponents are acting irrationally, or to believe that opponents believe that *their* opponents are acting irrationally, at some instances of the game. This observation raises the following question: when should players conclude that their opponents are being irrational? There does not exist a clear answer to this question, since different rationality concepts answer this question differently. Weak sequential rationalizability at the beginning of the game, for instance, allows players to believe that opponents are acting irrationally once their initial conjecture about the opponents' behavior has been contradicted. In the following section we shall present an alternative rationalizability concept, called extensive form rationalizability, in which players cannot immediately drop the belief that opponents are being rational after observing that their initial belief is falsified. Basu (1990) provides an axiomatic approach to this problem for games with perfect information. More precisely, he defines, for a given solution, an irrationality map which assigns to every node in the game-tree the set of players which is to be viewed irrational once this node is reached. He then imposes some axioms on the solution, making use of this irrationality map, and shows that there is no solution which satisfies all these axioms.

7.6 Extensive Form Rationalizability

In the previous section, we have investigated two extreme cases of weak sequential rationalizability: weak sequential rationalizability at the beginning of the game, and weak sequential rationalizability at all information sets. We have seen that the first requirement is always "possible" in every game, that is, for every player there is always at least one strategy that is weakly sequentially rationalizable at the beginning of the game. However, this concept puts no restriction on the belief revision process of a player if the initial belief of this player has been contradicted. Recall, namely, that within this concept players are only required to believe that opponents are sequentially rational at the beginning of te game. However, if the initial belief of a player has been contradicted by the history of play, then the player may from now on believe that his opponents are not sequentially rational.

On the other hand, weak sequential rationalizability at all information sets may be viewed "too strong" a requirement since in many games common belief of sequential rationality at all information sets is simply impossible. In this section we introduce a concept, called *extensive form rationalizability* (Pearce, 1984), which is more restrictive than weak sequential rationalizability at the beginning of the game, but less restrictive than weak sequential rationalizability at all information sets. Important is that for every player there is always at least one strategy which is extensive form rationalizable, hence extensive form rationalizability is always "possible" in every game.

The idea behind extensive form rationalizability is as follows. At the beginning of the game, we impose common belief of sequential rationality, that is, each player believes that each player is sequentially rational, each player believes that each player believes that each player is sequentially rational, and so on. A player's type is then assumed to respect common belief of sequential rationality as long as the history of play does not contradict this. However, if the event of reaching some information set contradicts common belief of sequential rationality, then the player at this information set is required to seek the "highest possible degree of common belief of sequential rationality" that is compatible with the event of reaching this information set, and should then believe in this "degree of common belief of sequential rationality" until it will be contradicted by the history of play. In case the new belief is contradicted by the event of reaching some information set, the player again seeks the "highest possible degree of common belief of sequential rationality" that is compatible with the event of reaching this information set, and so on. In comparison to weak sequential rationalizability at the beginning of the game, we thus put extra conditions on the "new" beliefs of players, once their previous beliefs have been contradicted. Roughly speaking, this extra condition states that, if the previous beliefs have been contradicted by the event of reaching some information set, the player is required to find a "best possible" explanation for the event of reaching this information set. This condition on the players' revision of beliefs is called the *best rationalization principle* (Battigalli, 1996a). Strategies that are sequential best responses for types that revise

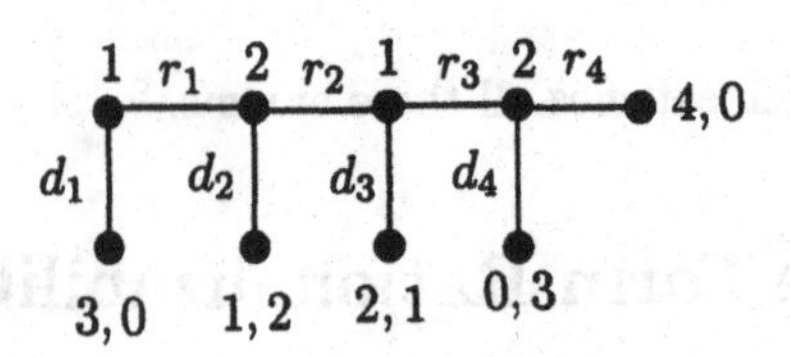

Figure 7.9

their beliefs according to the best rationalization principle are called extensive form rationalizable.

In order to illustrate what we mean above by the "highest possible degree of common belief of sequential rationality", consider the game in Figure 7.9, which is taken from Reny (1992b). Let the nodes in this game be denoted by x_0, x_1, x_2 and x_3, where x_0 is the initial node and x_3 is the last node. It may be verified that common belief of sequential rationality is not possible at $\{\emptyset, x_1\}$. Namely, assume that there would be common belief of sequential rationality at $\{\emptyset, x_1\}$. Then, player 2, at x_1, should believe (1) that player 1 is sequentially rational and (2) that player 1, at he beginning of the game, believes that player 2 is sequentially rational. Hence, player 2, at x_1, should believe (1) that player 1 is sequentially rational, and (2') that player 1, at the beginning of the game, believes that player 2 chooses d_4 at the last node. Therefore, player 2 should believe that player 1 chooses d_1. Hence, the second node is incompatible with common belief of sequential rationality at $\{\emptyset, x_1\}$.

Assume, now, that player 2 observes that his node x_1 has been reached. Then, player 2 should conclude that one of the following two events is not true: (1) player 1 is sequentially rational, (2) player 1 is sequentially rational, and player 1, at the beginning of the game, believes that player 2 is sequentially rational. Here, the second event has a higher degree of common belief of sequential rationality then the first event. The best rationalization principle then states that player 2 should seek for the highest degree of common belief of sequential rationality that is compatible with x_1 being reached. Clearly, x_1 is compatible with (1) but is not compatible with (2). Hence, the best rationalization principle requires player 2 to believe, at x_1, that (1) is true. In other words, player 2 at x_1, should believe that player 1 is sequentially rational. Since (r_1, r_3) is the only strategy reaching x_1 that is sequentially rational for player 1, player 2, at x_1, should believe that player 1 chooses (r_1, r_3). But then, player 2 should choose the strategy (r_2, d_4). Hence, the unique strategy for player 2 that is extensive form rationalizable is (r_2, d_4). Note that the unique backward induction strategies for the players are d_1 and d_2, and hence extensive form rationalizability prescribes a unique strategy for player 2 which is different from his backward induction strategy. Weak sequential rationalizability at the beginning of the game, however, allows the strategy d_2 for player 2. The reason is that according to the latter concept, player 2 may believe at the beginning of the game that player 1 chooses d_1. Upon observing x_1, player 2 may then as well believe that player 1 is not sequentially rational, and may therefore believe that he chooses (r_1, d_3). Since d_2 is a sequential best response against these beliefs, d_2 is weakly sequentially rationalizable at the beginning of the game.

We will now formally define the concept of extensive form rationalizability. Pearce (1984) introduces the concept of extensive form rationalizability as an elimination procedure within the space of strategies, similar to the elimination procedure yielding rationalizable strategies. Later, Battigalli (1997) provides an alternative procedure which puts progressively stronger conditions on the players' revision of beliefs, and which yields as a final result the same strategies as Pearce's procedure. However, both procedures do not work with an epistemic model of conditional belief hierarchies. In Battigalli and Siniscalchi (1997 and 2000) the strategies surviving the procedures by Pearce and Battigalli have been characterized within a model of conditional belief hierarchies. In this section, we reverse this chronological order by first providing the characterization of Battigalli and Siniscalchi and using this as the definition of extensive form rationalizability. Subsequently, we discuss the procedures by Pearce and Battigalli and state that they yield exactly the extensive form rationalizable strategies *à la* Battigalli and Siniscalchi.

A key role in the definition of extensive form rationalizability is played by the notion of *strong belief* of an event. Intuitively, a player strongly believes an event if he believes this event until the game reaches an information set that contradicts this event. Formally, consider a type t_i for player i and an event $E \subseteq S_{-i} \times T_{-i}$. Then, we say that t_i *strongly believes* the event E if for every information set $h \in H_i \cup \{\emptyset\}$ with $E \cap (S_{-i}(h) \times T_{-i}) \neq \emptyset$, it holds that $g_i(t_i)(E|\, h) = 1$. It is important to note that "t_i strongly believes E" and "$E \subseteq F$" does *not* imply "t_i strongly believes F". We say that the strong belief operator is *not monotonic*. This property thus distinguishes the strong belief operator from the belief operator "t_i believes E at a given collection of information sets" used in the previous section, which *is* monotonic. In order to see why the strong belief operator is not monotonic, let us return to the example in Figure 7.9. Suppose that some type t_2 for player 2 believes, at the beginning of the game, that player 1 chooses d_1, and believes at x_1 that player 1 chooses (r_1, d_3). Let E be the event "player 1 chooses d_1", and let F be the event "player 1 is sequentially rational". Then, $E \subseteq F$. We then have that t_2 strongly believes E, since t_2 believes E at $h = \emptyset$, and $h = \emptyset$ is the only instance in $H_2 \cup \{\emptyset\}$ which is compatible with E. However, t_2 does not strongly believe F since x_1 is compatible with F, but t_2 does not believe F at x_1. Let

$$R_i^0 = \{(s_i, t_i) | (s_i, t_i) \text{ sequentially rational}\}.$$

By

$$R_i^1 = \{t_i \in T_i |\, t_i \text{ strongly believes } R_{-i}^0\}$$

we denote those player i types which believe that all opponents are sequentially rational, until the event of reaching some player i information set contradicts it. Define

$$R_i^2 = \{t_i \in R_i^1 |\, t_i \text{ strongly believes } R_{-i}^0 \cap (S_{-i} \times R_{-i}^1)\}.$$

Hence, R_i^2 contains those types which strongly believe that (1) all players are sequentially rational and, in addition, strongly believe the more restrictive event (2) which states that (1) holds and all players strongly believe that all players are sequentially rational. In other words, a type in R_i^2 believes (2) as long as possible, and if (2) has been contradicted by the course of the game, he believes (1) as long as possible. Define the sets of types $R_i^3, R_i^4, \ldots$ recursively by

$$R_i^k = \{t_i \in R_i^{k-1} | \ t_i \text{ strongly believes } R_{-i}^0 \cap (S_{-i} \times R_{-i}^1) \cap \ldots \cap (S_{-i} \times R_{-i}^{k-1})\},$$

for $k \geq 3$. In order to provide an intuition for R_i^k, consider the events (0), (1),...$(k-1)$ where (0) is the event that all players are sequentially rational, (1) is the event that all players strongly believe event (0), (2) is the event that all players strongly believe event (0)&(1),...., and $(k-1)$ is the event that all players strongly believe event (0)&(1)&...&$(k-2)$. Then, R_i^k contains those types that believe event (0)&(1)...&$(k-1)$ as long as possible. If this event is contradicted, then they believe the less restrictive event (0)&(1)...&$(k-2)$ as long as possible, and so on. Hence, a type in R_i^k always seeks for the highest possible degree of common strong belief of sequential rationality that is compatible with the information set being reached. The definition of these sets R_i^k thus formalizes the *best rationalization principle* mentioned earlier on. Let $R_i^\infty = \cap_{k \in \mathbb{N}} R_i^k$.

Definition 7.6.1 *A type t_i is said to respect common strong belief of sequential rationality if $t_i \in R_i^\infty$.*

A strategy is called *extensive form rationalizable* if it is a sequential best response for a type that respects common strong belief of sequential rationality.

Definition 7.6.2 *A strategy s_i is called extensive form rationalizable if there is some type $t_i \in R_i^\infty$ such that (s_i, t_i) is sequentially rational.*

The reader may wonder at this stage why we did not define the sets of types R_i^k in the following, simpler way:

$$\begin{aligned} \tilde{R}_i^1 &= \{t_i \in T_i | \ t_i \text{ strongly believes } R_{-i}^0\} \text{ and} \\ \tilde{R}_i^k &= \{t_i \in \tilde{R}_i^{k-1} | \ t_i \text{ strongly believes } \tilde{R}_{-i}^{k-1}\} \end{aligned}$$

for $k \geq 2$. The reason is that, with this definition, some sets $\tilde{R}_i^k$ may become empty, and hence $\tilde{R}_i^\infty$ may be empty. Consider again the game in Figure 7.9. We show that $\tilde{R}_2^2$ is empty. Suppose that $t_2 \in \tilde{R}_2^2$. Then, $t_2 \in \tilde{R}_2^1$, and hence t_2 strongly believes that player 1 is sequentially rational. Since x_1 is compatible with player 1 being sequentially rational, t_2 should believe at x_1 that player 1 is sequentially rational. On the other hand, t_2 strongly believes $\tilde{R}_1^1$, hence t_2 strongly believes that player 1 strongly believes that player 2 is sequentially rational. Since x_1 is compatible with the event that player 1 strongly believes that player 2 is sequentially rational, t_2 should believe at x_1 that player 1 strongly believes that player 2 is sequentially rational. Consequently, t_2 at x_1 believes (1) that player 1 is sequentially rational and (2) that player 1 believes that player 2 is sequentially rational. However, (1) and (2) imply that player 1 chooses d_1, and hence t_2 cannot believe (1) and (2) at x_1, which is a contradiction. We may thus conclude that $\tilde{R}_2^2$ is empty. The problem here is that x_1 is compatible with player 1 being sequentially rational, and is compatible with player 1 strongly believing that player 2 is sequentially rational, but x_1 is not compatible with the intersection of both events.

In order to illustrate the definition of extensive form rationalizability, let us explicitly compute the sets R_i^k in the example of Figure 7.9. Let the decision nodes be denoted by x_0, x_1, x_2, x_3. Clearly, R_1^0 contains those strategy-type pairs for player 1 in which he chooses d_1 or (r_1, r_3), whereas R_2^0 contains those strategy-type pairs in which player 2 chooses d_2 or (r_2, d_4). Types in R_1^1 believe, at x_2, that player 2 chooses (r_2, d_4), whereas types in R_2^1 believe, at x_1, that player 1 chooses (r_1, r_3). Types in R_1^2 should thus believe, at the beginning of the game, that player 2 chooses (r_2, d_4), whereas types in R_2^2 should believe, at the beginning of the game, that player 1 chooses d_1. Hence, the unique sequential best responses for such types are d_1 for player 1, and (r_2, d_4) for player 2. These strategies are then, by construction, the unique extensive form rationalizable strategies in this game. In this example, the unique extensive form rationalizable strategy (r_2, d_4) for player 2 thus differs from his unique backward induction strategy d_2. However, the *outcome* induced by the unique pair of extensive form rationalizable strategies coincides with the backward induction outcome. Battigalli (1997) shows that this is always the case for games with perfect information in generic position. More precisely, he shows that for such games the path induced by a profile of extensive form rationalizable strategies always coincides with the unique backward induction path.

The best rationalization principle in the concept of extensive form rationalizability has a strong forward induction flavour, since it requires a player to revise his beliefs in such a way as to always find a best possible explanation for "unexpected" behavior. As to illustrate its relation to forward induction, consider again the Battle-of-the-sexes game with outside option in Figure 7.3. In this game, strategy-type pairs in R_1^0 yield strategies a and c, whereas strategy-type pairs in R_2^0 contain both strategies for player 2. A type in R_2^1 should strongly believe the event R_1^0. Hence, upon reaching player 2's information set, a type in R_2^1 should believe that player 1 chooses a. It follows that every type in R_1^2 should believe that player 2 chooses d. The unique sequential best responses to such beliefs are a and d, and hence a and d are the unique extensive form rationalizable strategies. In this example, extensive form rationalizability thus singles out the forward induction strategies for players 1 and 2.

We now discuss the above mentioned iterative procedure in Battigalli (1997). Let $\Delta(S_{-i}|\ \mathcal{A}_i^1)$ be the set of all conditional probability systems c_i, assigning to every information set $h \in H_i \cup \{\emptyset\}$ some probability distribution $c_i(\cdot|\ h) \in \Delta(S_{-i}(h))$ on the opponents' strategy profiles. For $k = 0, 1, 2, ...$ define the sets $\Delta^k(S_{-i}|\ \mathcal{A}_i^1)$ of conditional probability systems, and the sets S_i^k of strategies as follows:

$$\begin{aligned}
\Delta^0(S_{-i}|\, \mathcal{A}_i^1) &= \Delta(S_{-i}|\, \mathcal{A}_i^1), \\
S_i^0 &= S_i, \\
\Delta^k(S_{-i}|\, \mathcal{A}_i^1) &= \{c_i \in \Delta^{k-1}(S_{-i}|\, \mathcal{A}_i^1)|\; c_i(S_{-i}^{k-1}|\, h) = 1 \text{ if } S_{-i}(h) \cap S_{-i}^{k-1} \neq \emptyset\}, \\
S_i^k &= \{s_i \in S_i^{k-1}|\; \exists c_i \in \Delta^k(S_{-i}|\, \mathcal{A}_i^1) \text{ for which} \\
&\qquad s_i \text{ is sequential best response}\},
\end{aligned}$$

for $k \geq 2$. Hence, for every k the set $\Delta^k(S_{-i}|\, \mathcal{A}_i^1)$ contains those conditional probability systems that at every information set attach positive probability only to those strategies not yet deleted, if possible. On the other hand, S_i^k contains those strategies that are sequential best responses for some conditional probability system not yet deleted. In this procedure, one thus puts stronger and stronger conditions on the conditional probability systems. Let $S_i^\infty = \cap_{k \in \mathbb{N}} S_i^k$. By construction, the sets $\Delta^k(S_{-i}|\, \mathcal{A}_i^1)$ are always nonempty, and hence S_i^∞ is nonempty. The following result is due to Battigalli and Siniscalchi (1997 and 2000).

Theorem 7.6.3 *A strategy s_i is extensive form rationalizable if and only if $s_i \in S_i^\infty$.*

The proof is omitted here, for the sake of brevity. Pearce's original procedure (Pearce, 1984) is similar to Battigalli's. The crucial difference between both procedures lies in the collection of information sets at which optimality of a given strategy s_i against a conditional probability system c_i is required. In Battigalli's procedure, a strategy s_i, in order to survive the next round, should be a sequential best response against a conditional probability system c_i not deleted yet. Hence, by definition, s_i should be a best response against c_i at every information set $h \in H_i$ *not avoided by s_i itself.* In Pearce's procedure, a strategy s_i, in order to survive the next round, should be a best response against such a conditional probability system c_i *only at those information sets $h \in H_i$ that can be reached by s_i and some profile of opponents' strategies not deleted yet.* Hence, the collection of information sets at which optimality of s_i is required is smaller than in Battigalli's procedure. Battigalli (1997) has shown, however, that both procedures lead to the same set of strategies for each player. Shimoji and Watson (1998) provide an alternative procedure which yields the set of extensive form rationalizable strategies for each player. They introduce the concept of *conditional dominance* of strategies, and propose an elimination procedure which iteratively eliminates strategies that are conditionally dominated. It is shown that the set of strategies surviving this elimination procedure is equal to the set of extensive form rationalizable strategies.

We conclude this section by discussing in more detail the difference between extensive form rationalizability on the one hand, and backward induction on the other hand. The example if Figure 7.9 has shown that extensive form rationalizability may *uniquely* select a strategy for a player that is not a backward induction strategy. The crucial difference with backward induction lies in the fact that in extensive form rationalizability, at each node of the game the corresponding player "interprets" the history that has led to this node, and bases his conjecture about future play upon this interpretation. Backward induction, in contrast, requires this player to "ignore" the history that has led to his node, and face the remainder of the game as if the game would actually start at this node. In order to illustrate this difference, let us return to the game in Figure 7.9. Let the nodes in this game be denoted by x_0, x_1, x_2 and x_3. Consider the following three events: (1) player 1, at x_2, believes that player 2 is sequentially rational at x_3, (2) player 2, at x_1, believes (1) and believes that player 1 is sequentially rational at x_2, (3) player 1, at the beginning of the game, believes (2) and believes that player 2 is sequentially rational. Now, let (4) be the event that (1), (2) and (3) hold, and that players 1 and 2 are sequentially rational. Then, event (4) is a sufficient condition for backward induction in this game. Namely, (1) implies that player 1, at x_2, believes that player 2 will choose d_4. Event (2) implies that player 2, at x_1, believes that player 1 will choose d_3. Event (3) implies that player 1, at the beginning of the game, believes that player 2 chooses d_2. Event (4) then implies that player 1 chooses d_1 and player 2 chooses d_2.

Suppose now that player 2 observes that the node x_1 has been reached. Then, player 2 must conclude that either player 1 is not sequentially rational, or event (3) is violated. Backward induction requires that this observation should have no consequences for what player 2 believes in the remainder of the game. Hence, according to the backward induction criterion, player 2 should still believe, at x_1, that (1) holds and that player 1 is sequentially rational at x_2. Consequently, player 2 should believe, at x_1, that player 1 chooses (r_1, d_3).

Extensive form rationalizability requires player 2 to seek for the "highest possible degree of common strong belief of sequential rationality" on the part of player 1 that is compatible with x_1 being reached, and to revise his beliefs according to this criterion. Since x_1 is compatible with player 1 being sequentially rational, but not with event (3) plus the event of player 1 being sequentially rational, player 2 is required to believe, at x_1, that player 1 is sequentially rational but that event (3) is false. Consequently, player 2 should believe, at x_1, that player 1 chooses (r_1, r_3).

Papers that provide sufficient epistemic conditions for backward induction in games of perfect information are, for instance, Aumann (1995), Samet (1996), Balkenborg and Winter (1997), Stalnaker (1998) and Asheim (1999). The concept of proper rationalizability (Schuhmacher, 1999 and Asheim, 2000b) induces backward induction when applied to games with perfect information, and Asheim's epistemic foundation for proper rationalizability thus establishes a sufficient condition for backward induction as well. Asheim and Perea (2000) present two rationalizability concepts for general two-player extensive form games, called *sequential rationalizability* and *quasi-perfect rationalizability*, which both induce backward induction in games with perfect information. The two concepts may be viewed as "non-equilibrium analogues" to sequential equilibrium and quasi-perfect equilibrium, respectively. By the latter we mean that, if we would add to the definition of sequential rationalizability the requirement that players hold correct beliefs about the opponent's real type, then we would obtain exactly the concept of sequential equilibrium. Similarly for quasi-perfect rationalizability. The epistemic definitions of sequential and quasi-perfect rationalizability, when applied to games with perfect information, establish alternative sufficient conditions for backward induction.

the history that has led to his node, and face the remainder of the game as if the game would actually start at this node. In order to illustrate this difference, let us return to the game in Figure 6.9. Let the nodes in this game be denoted by x_0, x_1, x_2 and x_3. Consider the following three events: (1) player 1 at x_2 believes that player 2 is sequentially rational at x_3, (2) player 2 at x_1 believes (1) and believes that player 1 is sequentially rational at x_2, (3) player 1 at the beginning of the game believes (2) and believes that player 2 is sequentially rational. Now, let (4) be the event that (1), (2) and (3) hold, and that players 1 and 2 are sequentially rational. Then, event (4) is a sufficient condition for backward induction in this game. Namely, (1) implies that player 1 at x_2 believes that player 2 will choose c_2. Event (2) implies that player 2, at x_1, believes that player 1 will choose d_1. Event (3) implies that player 1, at the beginning of the game, believes that player 2 chooses d_2. Event (4) then implies that player 1 chooses b_1 and player 2 chooses d_2.

Suppose now that player 2 observes that the node x_1 has been reached. Then, player 2 must conclude that either player 1 is not sequentially rational, or event (3) is violated. Backward induction requires that this observation should have no consequences for what player 2 believes in the remainder of the game. Hence, according to the backward induction criterion, player 2 should still believe, at x_1, that (1) holds and that player 1 is sequentially rational. Consequently, player 2 should believe, at x_1, that player 1 chooses (c_1, b_1).

Extensive form rationalizability requires player 2 to search for the highest possible degree of common strong belief of sequential rationality on the part of player 1 that is compatible with x_1 being reached, and to revise his beliefs accordingly. To this criterion, the event x_1 is compatible with player 1 being sequentially rational, but not with event (3). Given the event that player 1 behaves sequentially rational, player 2 is required to believe at x_1 that player 1 is sequentially rational, and that event (3) is false. Consequently, player 2 should believe, at x_1, that player 1 chooses (c_1, a_1).

Papers that provide sufficient epistemic conditions for backward induction in games with perfect information are, for instance, Aumann (1995), Samet (1996), Balkenborg and Winter (1997), Stalnaker (1998) and Asheim (2002). The concept of proper rationalizability (Schuhmacher, 1999 and Asheim, 2001) induces backward induction when applied to games with perfect information, and Asheim's epistemic foundation for proper rationalizability thus establishes a sufficient condition for backward induction as well. Asheim and Perea (2004) present two rationalizability concepts for general extensive form games, called sequential rationalizability and quasi-perfect rationalizability, which both induce backward induction in games with perfect information. The two concepts may be viewed as non-equilibrium analogues to sequential equilibrium and quasi-perfect equilibrium, respectively. By the latter we mean that, if we would add to the definition of sequential rationalizability the requirement that players hold correct beliefs about the opponent's real types, then we would obtain exactly the concept of sequential equilibrium. Similarly for quasi-perfect rationalizability. The epistemic definitions of sequential and quasi-perfect rationalizability, when applied to games with perfect information, establish alternative sufficient conditions for backward induction.

Bibliography

ARMBRUSTER, W. AND W. BÖGE (1979), Bayesian game theory, in: *Game Theory and Related Topics* (O. Moeschlin and D. Pallaschke, Eds.), North-Holland, Amsterdam.

ASHEIM, G.B. (1999), On the epistemic foundation for backward induction, Memorandum No. 25, Department of Economics, University of Oslo.

ASHEIM, G.B. (2000a), On the epistemic foundation for backward induction, Memorandum No. 30, Department of Economics, University of Oslo. Revised.

ASHEIM, G.B. (2000b), Proper consistency, Memorandum No. 31, Department of Economics, University of Oslo.

ASHEIM, G.B. AND M. DUFWENBERG (2000), Asmissibility and common belief, Memorandum No. 7, Department of Economics, University of Oslo. Revised.

ASHEIM, G.B. AND A. PEREA (2000), Lexicographic probabilities and rationalizability in extensive form games, Memorandum No. 38, Department of Economics, University of Oslo, and Research Memorandum RM/00/033, Maastricht University.

AUMANN, R. (1976), Agreeing to disagree, *Annals of statistics* **4,** 1236-1239.

AUMANN, R. (1995), Backward induction and common knowledge of rationality, *Games and Economic Behavior* **8,** 6-19.

AUMANN, R. AND A. BRANDENBURGER (1995), Epistemic conditions for Nash equilibrium, *Econometrica* **63,** 1161-1180.

BALKENBORG, D. AND E. WINTER (1997), A necessary and sufficient epistemic condition for playing backward induction, *Journal of Mathematical Economics* **27,** 325-345.

BANKS, J.S. AND J. SOBEL (1987), Equilibrium selection in signaling games, *Econometrica* **55,** 647-661.

BASU, K. (1990), On the non-existence of a rationality definition for extensive games, *International Journal of Game Theory* **19,** 33-44.

BATTIGALLI, P. (1996a), Strategic rationality orderings and the best rationalization principle, *Games and Economic Behavior* **13,** 178-200.

BATTIGALLI, P. (1996b), Strategic independence and perfect Bayesian equilibria, *Journal of Economic Theory* **70,** 201-234.

BATTIGALLI, P. (1997), On rationalizability in extensive games, *Journal of Economic Theory* **74,** 40-61.

BATTIGALLI, P. AND M. SINISCALCHI (1997), An epistemic characterization of extensive form rationalizability, Social Science Working Paper 1009, California Institute of Technology.

BATTIGALLI, P. AND M. SINISCALCHI (1999), Hierarchies of conditional beliefs and interactive epistemology in dynamic games, *Journal of Economic Theory* **88,** 188-230.

BATTIGALLI, P. AND M. SINISCALCHI (2000), Interactive beliefs and forward induction, Manuscript.

BEN-PORATH, E. (1997), Rationality, Nash equilibrium and backwards induction in perfect-information games, *Review of Economic Studies* **64.** 23-46.

BEN-PORATH, E. AND E. DEKEL (1992), Signaling future actions and the potential for sacrifice, *Journal of Economic Theory* **57,** 36-51.

BERNHEIM, B.D. (1984), Rationalizable strategic behavior, *Econometrica* **52,** 1007-1028.

BLUME, L.E., BRANDENBURGER, A. AND E. DEKEL (1991a), Lexicographic probabilities and choice under uncertainty, *Econometrica* **59,** 61-79.

BLUME, L.E., BRANDENBURGER, A. AND E. DEKEL (1991b), Lexicographic probabilities and equilibrium refinements, *Econometrica* **59,** 81-98.

BLUME, L.E., AND W.R. ZAME (1994), The algebraic geometry of perfect and sequential equilibrium, *Econometrica* **62,** 783-794.

BÖGE, W. AND T.H. EISELE (1979), On solutions of bayesian games, *International Journal of Game Theory* **8,** 193-215.

BÖRGERS, T. (1994), Weak dominance and approximate common knowledge, *Journal of Economic Theory* **64,** 265-276.

BRANDENBURGER, A. (1992), Lexicographic probabilities and iterated admissibility, In: Dasgupta, P. *et al.* (eds), *Economic Analysis of Markets and Games,* pp. 282-290, Cambridge, MA, MIT Press.

BRANDENBURGER, A. AND E. DEKEL (1993), Hierarchies of beliefs and common knowledge, *Journal of Economic Theory* **59,** 189-198.

BRANDENBURGER, A. AND H.J. KEISLER (2000), Epistemic conditions for iterated admissibility, Harvard Business School.

CHO, I.-K. (1986), *Refinement of Sequential Equilibrium: Theory and Application,* Ph.D. dissertation, Princeton University.

CHO, I.-K. (1987), A refinement of sequential equilibrium, *Econometrica* **55,** 1367-1389.

CHO, I.-K., AND D.M. KREPS (1987), Signaling games and stable equilibria, *Quarterly Journal of Economics* **102,** 179-221.

DALKEY, N. (1953), Equivalence of information patterns and essentially determinate games, In: *Contributions to the Theory of Games,* Vol. 2, pp. 217-245, Princeton University Press, Princeton.

DEKEL, E. AND D. FUDENBERG (1990), Rational behavior with payoff uncertainty, *Journal of Economic Theory* **52,** 243-267.

DELLACHERIE, C. AND P.-A. MEYER (1978), *Probabilities and Potential,* North-Holland, New York.

EILENBERG, S. AND D. MONTGOMERY (1946), Fixed-point theorems for multivalued transformations, *American Journal of Mathematics* **68,** 214-222.

ELMES, S. AND P.J. RENY (1994), On the strategic equivalence of extensive form games, *Journal of Economic Theory* **62,** 1-23.

EPSTEIN, L.G. AND T. WANG (1996), "Beliefs about beliefs" without probabilities, *Econometrica* **64,** 1343-1373.

FUDENBERG, D. AND J. TIROLE (1991), Perfect Bayesian equilibrium and sequential equilibrium, *Journal of Economic Theory* **53,** 236-260.

GLAZER, J. AND A. RUBINSTEIN (1996), An extensive game as a guide for solving a normal game, *Journal of Economic Theory* **70,** 32-42.

GROSSMAN, S.J. AND M. PERRY (1986), Perfect sequential equilibrium, *Journal of Economic Theory* **39,** 97-119.

GUL, F. (1996), Rationality and coherent theories of strategic behavior, *Journal of Economic Theory* **70,** 1-31.

HAMMOND, P.J. (1994), Elementary non-archimedean representations of probability for decision theory and games, in: Patrick Suppes: Scientific Philosopher, Vol. 1, Probability and Probabilistic Causality (P. Humphreys, Ed.), Chap. 2, pp.25-59, Kluwer Academic Publishers, Dordrecht.

HARSANYI, J.C. (1967-1968), Games with incomplete information played by "bayesian" players, I-III, *Management Science* **14,** 159-182, 320-334, 486-502.

HAUK, E. AND S. HURKENS (1999), On forward induction and evolutionary and strategic stability, Manuscript.

HEIFETZ, A. (1993), The Bayesian formulation of incomplete information - The non-compact case, *International Journal of Game Theory* **21,** 329-338.

HENDON, E., JACOBSEN, H.J. AND B. SLOTH (1996), The one-shot-deviation principle for sequential rationality, *Games and Economic Behavior* **12,** 274-282.

HERINGS, J.J. AND V.J. VANNETELBOSCH (1999), Refinements of rationalizability for normal-form games, *International Journal of Game Theory* **28,** 53-68.

HERINGS, J.J. AND V.J. VANNETELBOSCH (2000), The equivalence of the Dekel-Fudenberg iterative procedure and weakly perfect rationalizability, *Economic Theory* **15,** 677-687.

HILLAS, J. (1990), On the definition of the strategic stability of equilibria, *Econometrica* **58,** 1365-1390.

HILLAS, J. (1994), Sequential equilibria and stable sets of beliefs, *Journal of Economic Theory* **64,** 78-102.

HILLAS, J. (1996), On the relation between perfect equilibria in extensive form games and proper equilibria in normal form games, Manuscript.

HILLAS, J., JANSEN, M., POTTERS, J. AND D. VERMEULEN (2000), On the relation among some definitions of strategic stability, Working Paper.

KAKUTANI, S. (1941), A generalisation of Brouwer's fixed point theorem, *Duke Mathematical Journal* **8,** 457-458.

KLINE, J.J. (1997), Information structures and decentralizability of equilibria, *Economic Theory* **9,** 81-96.

KOHLBERG, E. (1981), Some problems with the concept of perfect equilibrium, *Rapp. Rep. NBER Conf. Theory Gen. Econ. Equilibr. K. Dunz N. Singh, Univ. Calif., Berkely.*

KOHLBERG, E. (1990), Refinement of Nash equilibrium: The main ideas, In: *Game Theory and Applications,* Economic Theory, Econometrics, and Mathematical Economics Series, Edited by T. Ichiishi, A. Neyman and Y. Taumann, Academic Press, New York, London, Sydney and Toronto, pp. 3-45.

KOHLBERG, E. AND J.-F. MERTENS (1986), On the strategic stability of equilibria, *Econometrica* **54,** 1003-1037.

KOHLBERG, E. AND P.J. RENY (1997), Independence on relative probability spaces and consistent assessments in game trees, *Journal of Economic Theory* **75,** 280-313.

KREPS, D. AND G. RAMEY (1987), Structural consistency, consistency, and sequential rationality, *Econometrica* **55,** 1331-1348.

KREPS, D. AND J. SOBEL (1994), Signalling, In: *Handbook of Game Theory*, Chapter 25, Volume 2, Edited by R.J. Aumann and S. Hart, Elsevier Science.

KREPS, D. AND R. WILSON (1982a), Sequential equilibria, *Econometrica* **50,** 863-894.

KREPS, D. AND R. WILSON (1982b), Reputation and imperfect information, *Journal of Economic Theory* **27,** 253-279.

KUHN, H.W. (1953), Extensive games and the problem of information, *Annals of Mathematics Studies* **28,** 193-216.

MAILATH, G.J., OKUNO-FUJIWARA, M. AND A. POSTLEWAITE (1993), Belief-based refinements in signalling games, *Journal of Economic Theory* **60,** 241-276.

MAILATH, G.J., SAMUELSON, L. AND J.M. SWINKELS (1993), Extensive form reasoning in normal form games, *Econometrica* **61,** 273-302.

MAILATH, G.J., SAMUELSON, L. AND J.M. SWINKELS (1997), How proper is sequential equilibrium?, *Games and Economic Behavior* **18,** 193-218.

MARX, L.M. AND J.M. SWINKELS (1997), Order independence for iterated weak dominance, *Games and Economic Behavior* **18,** 219-245.

MCLENNAN, A. (1985), Justifiable beliefs in sequential equilibria, *Econometrica* **53,** 889-904.

MCLENNAN, A. (1989a), The space of conditional systems is a ball, *International Journal of Game Theory* **18,** 125-139.

MCLENNAN, A. (1989b), Consistent conditional systems in noncooperative games, *International Journal of Game Theory* **18,** 141-174.

MCLENNAN, A. (1989c), Fixed points of contractible valued correspondences, *International Journal of Game Theory* **18,** 175-184.

MERTENS, J.-F. (1989), Stable equilibria - a reformulation, Part I: Definition and basic properties, *Mathematics of Operations Research* **14,** 575-624.

MERTENS, J.-F. (1991), Stable equilibria - a reformulation, Part II: Discussion of the definition and further results, *Mathematics of Operations Research* **16,** 694-753.

MERTENS, J.-F. AND S. ZAMIR (1985), Formulation of bayesian analysis for games with incomplete information, *International Journal of Game Theory* **14,** 1-29.

MONDERER, D. AND D. SAMET (1989), Approximating common knowledge with common beliefs, *Games and Economic Behavior* **1,** 170-190.

MYERSON, R.B. (1978), Refinements of the Nash equilibrium concept, *International Journal of Game Theory* **7,** 73-80.

NASH, J.F. (1950), Equilibrium points in n-person games, *Proc. Nat. Acad. Sci. U.S.A.* **36,** 48-49.

NASH, J.F. (1951), Non-cooperative games, *Annals of Mathematics* **54,** 286-295.

OKADA, A. (1981), On stability of perfect equilibrium points, *International Journal of Game Theory* **10,** 67-73.

PEARCE, D. (1984), Rationalizable strategic behavior and the problem of perfection, *Econometrica* **52,** 1029-1050.

PEREA, A. (2001), A note on the one-deviation property in extensive form games, Forthcoming in *Games and Economic Behavior.*

PEREA, A., JANSEN, M. AND H. PETERS (1997), Characterization of consistent assessments in extensive form games, *Games and Economic Behavior* **21,** 238-252.

PEREA, A., JANSEN, M. AND D. VERMEULEN (2000), Player splitting in extensive form games, *International Journal of Game Theory* **29,** 433-450.

RAJAN, U. (1998), Trembles in the bayesian foundations of solution concepts of games, *Journal of Economic Theory* **82,** 248-266.

RAMSEY, F.P. (1931), Truth and probability, *The Foundations of Mathematics and Other Logical Essays,* Routledge and Kegan Paul, London, 156-198.

RENY, P.J. (1992a), Rationality in extensive-form games, *Journal of Economic Perspectives* **6,** 103-118.

RENY, P.J. (1992b), Backward induction, normal form perfection and explicable equilibria, *Econometrica* **60,** 627-649.

RENY, P.J. (1993), Common belief and the theory of games with perfect information, *Journal of Economic Theory* **59,** 257-274.

RÊNYI, A. (1956), On conditional probability spaces generated by a conditionally ordered set of measures, *Theory of Probability and its Applications* **1,** 61-71.

ROCKAFELLAR, R.T. (1970), *Convex Analysis,* Princeton, Princeton University Press.

ROSENTHAL, R. (1981), Games of perfect information, predatory pricing, and the chain-store paradox, *Journal of Economic Theory* **25,** 92-100.

SAMET, D. (1996), Hypothetical knowledge and games with perfect information, *Games and Economic Behavior* **17,** 230-251.

SAVAGE, L.J. (1954), *The Foundations of Statistics,* Wiley, New York.

SAVAGE, L.J. (1972), *The Foundations of Statistics,* New York: Dover, 2nd revised edition.

SCHUHMACHER, F. (1999), Proper rationalizability and backward induction, *International Journal of Game Theory* **28,** 599-615.

SCHWALBE, U. AND P. WALKER (2001), Zermelo and the early history of game theory, *Games and Economic Behavior* **34,** 123-137.

SELTEN, R. (1965), Spieltheoretische Behandlung eines Oligopolmodells mit Nachfragezeit, *Zeitschrift für die Gesammte Staatswissenschaft* **121,** 301-324, 667-689.

SELTEN, R. (1975), Reexamination of the perfectness concept for equilibrium points in extensive games, *International Journal of Game Theory* **4,** 25-55.

SELTEN, R. (1978), The chain-store paradox, *Theory and Decision* **9,** 127-159.

SHIMOJI, M. AND J. WATSON (1998), Conditional dominance, rationalizability, and game forms, *Journal of Economic Theory* **83,** 161-195.

STAHL, D. (1995), Lexicographic rationality, common knowledge, and iterated admissibility, *Economics Letters* **47,** 155-159.

STALNAKER, R. (1998), Belief revision in games: forward and backward induction, *Mathematical Social Sciences* **36,** 31-56.

STINCHCOMBE, M.B. (1988), Approximate common knowledge, mimeo, University of California at San Diego.

TAN, T. AND S.R.C. WERLANG (1988), The bayesian foundations of solution concepts of games, *Journal of Economic Theory* **45,** 370-391.

THOMPSON, F.B. (1952), Equivalence of games in extensive form, RM 759, The Rand Corporation.

VAN DAMME, E. (1984), A relation between perfect equilibria in extensive form games and proper equilibria in normal form games, *International Journal of Game Theory* **13,** 1-13.

VAN DAMME, E. (1989), Stable equilibria and forward induction, *Journal of Economic Theory* **48,** 476-496.

VAN DAMME, E. (1991), *Stability and Perfection of Nash Equilibria,* Springer Verlag, Berlin.

VERMEULEN, D., POTTERS, J. AND M. JANSEN (1997), On stable sets of equilibria, In: *Game Theory and Applications to Economics and Operations Research,* 133-148, edited by Parthasarathy a.o., Kluwer Academic Publishers, The Netherlands.

VON NEUMANN, J. (1928), Zur Theorie der Gesellschaftsspiele, *Mathematische Annalen* **100,** 295-320.

VON NEUMANN, J., AND O. MORGENSTERN (1947), *Theory of Games and Economic Behavior,* Princeton: Princeton University Press, 2nd edition.

WERLANG, S.R.C. (1986), *Common Knowledge and Game Theory,* Ph. D. thesis, Princeton University.

ZERMELO, E. (1913), Über eine Anwendung der Mengenlehre auf die Theorie des Schachspiels, *Proceedings Fifth International Congress of Mathematicians* **2,** 501-504.

Index